More Landmarks In Biochemistry

FOUNDATIONS OF MODERN BIOCHEMISTRY
A Multi-Volume Treatise, Volume 4

Editors: **MARGERY G. ORD and LLOYD A. STOCKEN,** *Department of Biochemistry, University of Oxford, Oxford, England*

More Landmarks In Biochemistry

Edited by **MARGERY G. ORD**
LLOYD A. STOCKEN
Department of Biochemistry
University of Oxford
Oxford, England

JAI PRESS INC.

Stamford, Connecticut *London, England*

Library of Congress Cataloging-in-Publication Data

Foundations of modern biochemistry / editors, Margery G. Ord and Lloyd A. Stocken
p. cm.
Includes bibliographical references and indexes.
ISBN 1-55938-960-5 (v. 1)
1. Biochemistry–History. I. Ord, Margery G. II. Stocken, Lloyd A.
QD415.F68 1995
574.19'09—dc20 95-17048
CIP

100 Prospect Street
Stamford, Connecticut 06904

JAI PRESS LTD.
38 Tavistock Street
Covent Garden
London WC2E 7PB
England

ISBN: 0-7623-0351-4

Library of Congress Catalog Number: 97-41660

Transferred to digital printing 2006

CONTENTS

LIST OF CONTRIBUTORS

Richard Boyd	Department of Human Anatomy University of Oxford Oxford, England
Brian Cox	Department of Biosciences University of Kent Canterbury, England
Phillip S. Nutman	St Leonards' Exeter, England
Margery G. Ord	Department of Biochemistry University of Oxford Oxford, England
Daphne J. Osborne	Oxford Research Unit The Open University Oxford, England
Peter Schuster	Institut für Theoretische Chemie und Strahlenchemie der Universität Wien Wien, Austria
Lisa A. Steiner	Department of Biology Massachusetts Institute of Technology Cambridge, Massachusetts
Lloyd A. Stocken	Department of Biochemistry University of Oxford Oxford, England
Cheryll Tickle	Wellcome Trust Building University of Dundee Dundee, Scotland

ACKNOWLEDGMENTS*

We are very grateful to Professors Newell and Radda for continuing their generous provision of office space in the department. Once more we are indebted to the staff of the Radcliffe Science Library for their assistance, and also to the librarians of the Departments of Biochemistry, Physiology and Plant Sciences. Dr. Jeremy Rowntree and Mr. Martin Ackland gave us invaluable and patient help with computing problems and MS A. Morgan did a splendid job with the photographs. Our thanks are also due to Professors Armitage and Whatley and Dr.Sebastian Bonhoeffer for their help when we were considering the selection of topics for this, the final volume of the series. Mr. Piers Allen of the London branch of JAI Press has been extremely helpful in liasing with the parent company in the U.S.

We also wish to thank Professors Hodgkin, Keynes and Wolpert for their photographs and Wolfgang Filser, Studio Archiv, Gottingen, for that of Dr. Manfred Eigen. Mrs Spiegelman very kindly sent us the photo of her husband. The Director, the Rothamsted Experimental Station, U.K., gave us permission to reproduce photographs of Jean Baptiste Boussingault, Joseph Henry Gilbert, Her-

*Superscript numbers next to surnames throughout the volume refer to photographs, pages 151–158. Superscripts in **bold** refer to earlier volumes in this series where photographs may be found.

mann Hellriegel, John Benet Lawes, Justus von Liebig and Herman Wilfarth, and Professor Thomas, the Physiology department, Cambridge, U.K., permitted us to copy the photograph of Charles Darwin by Julia Margaret Cameron. Professor Osborne provided us with photos of James Bonner, Anton Lang, Kenneth Thimann and Fritz Went. That of Philip Wareing was kindly given to Professor Osborne by Professor Michael Hall and that of Eishii Kurosawa by Professor Bernard Phinney. The photo of Dimitri Neljubov appeared on page 3 of the introduction to "Ethylene in Plant Biology" (A.B. Abeles, 1973, Academic Press) where it had been reproduced from an original by courtesy of K.V.Riazanskaya in Proc. Inst. Hist. Nat. Sci. Tec., Academy of Science, USSR, Vol. 24. (Hist. Biol. Sci. 5, p. 85). We thank also Professor Ellory and Dr. A. Bangham for making the photograph of Jens Skou available. The Department of Biochemistry, Oxford, allowed us to copy their photo of Rudolph Peters.

We thank the Nobel Foundation for letting us reproduce their photographs of Melvin Calvin, Tom Cech, Edward Lewis, Thomas Hunt Morgan, Erwin Neher, Christiane Nusslein-Volhard, Bert Sakmann, Hans Spemann and Eric Weischaus. Photographs of Frederick Steward (Annu. Rev. Plant Physiol. 12, 1971) and Dan Koshland (Annu. Rev. Biochem. 65, 1996) are reproduced with permission from Annual Reviews Inc.

Margery G. Ord
Lloyd A. Stocken
Editors

Chapter 1

ANTIBODY SPECIFICITY AND DIVERSITY
PART II: THE GENES*

Lisa A. Steiner

INTRODUCTION

Part I of this chapter (Volume 3 of this series) reviewed the background of our present understanding of the basis for antibody specificity and diversity, the gradual emergence and acceptance of the clonal selection hypothesis and the fundamental structural features of the antibody molecule. By the mid-1960s, it had been

*Part I appeared in Volume 3 of this series

Foundations of Modern Biochemistry,
Volume 4, pages 1–95.

ISBN: 0-7623-0351-4

established that the distinctive characteristic of antibodies is that each of their constituent polypeptide chains consists of a segment that is constant in amino acid sequence from one molecule to another and a segment that varies extensively, accounting for the exquisite selectivity and diversity of antigen recognition. The attention of many investigators was now focussed on the genetic basis for these unique properties of antibodies. The status of the debates about the origins of antibody specificity was discussed at this time in an influential review by Lennox and Cohn (1967).

To account for constant (C) and variable (V) regions in the same polypeptide, it had been proposed that each light or heavy chain is encoded by two distinct genes, one for the V and another for the C region (Dreyer and Bennett, 1965). There were some problems with this hypothesis, particularly the inheritance of allotypic[a] (allelic) determinants of rabbit immunoglobulins (see Part I, Two Genes, One Polypeptide and the section here, Bias in Rearrangement). Moreover, the proposal flew in the face of the dogma, "one gene–one polypeptide." Nonetheless, the idea that more than a single gene must encode an antibody polypeptide chain gradually became accepted by most immunologists, if not by other biologists.

There was less agreement about the number and nature of the genes required to encode all the different V regions. The "germline position" was simple: one gene for each V region (Hood and Talmage, 1970). It followed that the total number of genes required must be large. Taking into account variable pairing of heavy and light chains, several thousand *V* genes could, at least in principle, generate well over a million combining sites. The "somatic position" was that there are relatively few *V* genes in the germline, the extensive diversity needed being generated in somatic cells during the lifetime of each individual (Cohn et al., 1974). Despite theoretical arguments that generated much heat, it was clear that light would come only from direct analysis of the genes themselves. Techniques for enumerating genes and for cloning and characterizing them were being developed at this time and were soon brought to bear on the antibody problem.

In this chapter, the steps leading to our current view of the genetic basis of the structural features of antibody molecules will be reviewed. Roughly the first half of the chapter examines the processes by which the preimmune repertoire of antibody combining regions is generated before initial encounter with antigen. Confirming the "two-gene, one polypeptide" hypothesis, it was shown that non-contiguous DNA segments rearrange in antibody-forming cells, generating genes that encode complete heavy and light chains. The repertoire generated by the rearrangement of these gene segments[b] ensures that more or less appropriate antibodies will be available to respond to any antigenic challenge.

The second half of the chapter considers restrictions on and elaborations of the process of gene rearrangement. Allelic exclusion—the expression of an allele from only one chromosome of a pair—assures that each antibody-forming cell will produce antibodies of only a single specificity, as postulated by the clonal selection hypothesis. The preimmune repertoire is modified after exposure to antigen: an elaborate process of somatic mutation generates antibody variants that may bind to the immunizing antigen with improved affinity, and cells expressing new gene rearrangements may emerge. A variety of secondary rearrangements of already-rearranged gene segments may occur, and their role in modulating an initial response is now being explored. Complex mechanisms enable a cell to express different heavy-chain C regions, while keeping the V region fixed; these permit antibodies of the same specificity to be expressed as secreted molecules and as membrane-bound receptors, and also allow expression of a variety of immunoglobulin classes of that same specificity.

Almost all of the experimental work cited in this chapter has been carried out in the last two decades. With few exceptions, the topics discussed are the subject of continuing research. Consequently, in addition to providing historical background and connections, in most cases I have attempted to indicate, at least briefly, the current status (as of 1997).

COUNTING GENES

Determination of the amino acid sequences of immunoglobulin polypeptide chains in the late 1960s led to estimates of the minimum number of germline genes required to encode the V and C regions of κ, λ, and heavy chains, and also to approximations of the total number of distinct V regions of each type (see Part I, sections Two Genes, One Polypeptide and Many Germline Genes or Few?). At about this time, the technique for estimating the multiplicity of repeated sequences in DNA by analysis of hybridization kinetics was introduced (Britten and Kohne, 1968) and was applied to determine the number of immunoglobulin genes, first by Storb (1972) and Delovitch and Baglioni (1973). In one approach, the rate of association of a highly labeled specific nucleic acid probe with an excess of unlabeled cellular DNA is measured (Melli et al., 1971). Purified specific mRNA is used either directly as a probe or as a template for the synthesis of cDNA. The rate of hybridization is a function of the number of genes complementary to the probe.

For hybridization to immunoglobulin genes, probes were generally prepared from mouse plasmacytomas (plasma cell tumors, also referred to as myelomas). Early efforts utilizing this approach encountered problems in technique and interpretation, in particular the difficulty of obtaining pure mRNA probes, as well as their intrinsic instability. The results did indicate, however, that there are few *C* genes for κ light chains, perhaps even only one (Faust et al., 1974; Honjo et al., 1974; Rabbitts, 1974; Stavnezer et al., 1974).

Regarding *V* genes there was considerable uncertainty, although estimates generally indicated that there are more *V* than *C* genes, thereby supporting the idea of separate genetic control for these regions. Initially, data from several laboratories were interpreted as consistent with the germline theory, i.e., that the number of V_κ

and V_H genes is very large. However, when more highly purified preparations of mRNA became available, the data for mouse V_κ appeared to indicate that a relatively small number of genes encodes the sequences in a V-region subgroup (Farace et al., 1976; Tonegawa, 1976). (A subgroup is a set of closely related V-region sequences; see Part I, Two Genes, One Polypeptide, for a discussion of V region subgroups.) Since there generally appeared to be more expressed sequences in a subgroup than germline genes encoding these sequences, these results supported somatic diversification, not germline, theories. Similar conclusions were reached by saturation hybridization analysis (Valbuena et al., 1978). This study also demonstrated that mRNAs encoding sequences belonging to the same subgroup cross-hybridized extensively in the reaction conditions used, so that a probe corresponding to any member could be used to estimate the number of germline *V* genes encoding all the expressed V regions in the subgroup.

Experiments to estimate the total number of genes encoding V regions of mouse κ chains were hampered by the difficulty of determining exactly how many sequences there are in each subgroup as well as the total number of subgroups. Mouse λ chains offered a more clearly defined system for these analyses. By the time the hybridization experiments were carried out in the mid-1970s, all known V_λ regions had been classified into two groups, λ1 and λ2, based on the amino acid sequence is of their associated C regions. Eighteen light chains of the λ1 type had been studied by peptide mapping and partial sequence analysis; 12 of the 18 V regions seemed identical and the remaining six had one to four amino acid replacements, all in the complementarity-determining regions (CDRs) (Appella et al., 1967; Appella and Perham, 1968; Weigert et al., 1970; Appella, 1971; Cesari & Weigert, 1973; Cohn et al., 1974). (The CDRs are the segments within the V region that are the most variable in amino acid sequence; they form the major part of the binding site for antigen.) As will be discussed in more detail in the section Somatic Hypermutation, Hypothesis and Evidence, it was proposed that all of these $V_{\lambda 1}$ are encoded by a single germline gene (Weigert et al., 1970), and that the variant sequences result from somatic mutation of that gene.

At about this time, an unusual light chain was identified in a myeloma protein produced by the plasmacytoma, MOPC-315. This protein attracted considerable interest because it was the first myeloma protein found to have high affinity for a defined hapten (Eisen et al., 1968). By the criterion of peptide fingerprinting, the light chain did not resemble the known λ chains, and it was originally classified as a κ chain. However, an affinity-labeled peptide was shown to resemble a segment of λ light chains, and amino- and carboxy-terminal analyses indicated that the MOPC-315 light chain was not a κ chain (Goetzl and Metzger, 1970). The nature of the MOPC-315 light chain was clarified when its complete amino acid sequence was determined (Schulenberg et al., 1971; Dugan et al., 1973). Its C region was found to resemble $C_{\lambda 1}$ more closely than C_κ, but differed at about 30 of 110 positions from the C region of λ1 chains. Thus, the light chain of MOPC-315 appeared to be the first identified member of a new type of mouse λ chain, designated λ2. The V region of the MOPC-315 λ chain differed in only about 10 amino acid residues from the V regions of λ1 chains and mRNA encoding the MOPC-315 light chain exhibited a high degree of cross-hybridization with $V_{\lambda 1}$ cDNA (Honjo et al., 1976).

It was not clear, initially, whether the MOPC-315 $V_{\lambda 2}$ region was encoded by a somatically mutated version of the gene thought to encode all the $V_{\lambda 1}$ or by a different germline gene. In any event, it seemed likely that the total number of genes encoding all the known mouse V_{λ} regions was less than the number of expressed V_{λ} regions and might even be as small as one or two. This hypothesis was supported by the hybridization analyses, which indicated that the number of V_{λ} genes was no more than three (Honjo et al., 1976; Tonegawa, 1976). Finding fewer mouse V_{λ} genes than expressed V_{λ} regions implied that these V_{λ} genes are diversified by a somatic process. However, see Smith (1976) for a discussion of the difficulties in estimating gene number by hybridization kinetics.

REARRANGING GENES

Two inferences regarding the genetic basis for antibody diversity were drawn from the "gene counting" experiments: 1) antibody *V* genes are diversified somatically, and 2) the two gene–one polypeptide hypothesis must be correct since there appeared to be a greater number of V_{κ} than of C_{κ} genes. Nonetheless these ideas were not yet universally accepted because the conclusions drawn from the hybridization experiments were indirect. What was needed was characterization of the genes themselves.

In addition, although it was agreed that there must be two genes, at least, to encode each antibody chain, it was not clear how the single polypeptide chain is produced. Evidence was presented in part I (Two Genes, One Polypeptide) that separate V and C protein segments are not fused. Therefore, the joining must be either at the RNA or the DNA level, the latter being considered more likely. Fortunately, tools (e.g., restriction enzymes) for carrying out the necessary analyses had become available and were now used to determine whether there are differences between the arrangement of *V* and *C* genes in the germline and in a tissue, such as a plasmacytoma, that expresses a specific immunoglobulin.

Three Gene Segments Encode a Light Chain

Hozumi and Tonegawa, working at the Basel Institute for Immunology, were the first to describe a somatic rearrangement of antibody genes. They prepared *Bam*HI digests of DNA from an embryo, representing germline DNA, and from the MOPC-321 plasmacytoma, a tumor that synthesizes a κ light chain. The digests were fractionated by gel electrophoresis and the DNA fragments were eluted from the gel and hybridized to two different probes: ^{125}I-labeled mRNA encoding the whole MOPC-321 κ chain and a probe consisting of approximately the 3′ half of this mRNA, corresponding only to C_{κ} sequences.[c] A probe for *V* alone was not available; consequently, the presence of *V* sequences was determined indirectly from the difference in hybridization of the whole mRNA and the 3′ segment.

The results of the experiment showed a clear difference in the hybridization patterns of *Bam*HI digests of embryo and tumor DNA. In embryo DNA, two restriction fragments hybridized to the whole mRNA; only one of these hybridized to the 3′-end probe. This result suggested that one of the fragments contained *V* sequences and the other *C* sequences. In plasmacytoma DNA, a single restriction fragment hybridized to the probe corresponding to whole mRNA, and also to the 3′-end probe; this fragment migrated faster than either of the hybridizing fragments from the embryo DNA, and hence was smaller (Hozumi and Tonegawa, 1976). The data were interpreted as showing that the *V* and *C* genes encoding the expressed κ chain are encoded separately in the germline and are brought together and expressed in the tumor. Alternative explanations were also discussed. For example, there might be a *Bam*HI site close to the *V-C* junction in embryo DNA and this site might be lost by mutation or base modification in the tumor. However, since the hybridizing *Bam*HI fragment in the tumor was smaller than either of the fragments in the embryo, one would have to suppose also that new flanking *Bam*HI sites were acquired in the tumor DNA.

It was shown subsequently that the rearranged gene encoding the V region of the MOPC-321 κ chain does indeed have a *Bam*HI site in CDR3, near the end of the *V* region, as well as one at the beginning of the third framework region (Matthyssens and Tonegawa, 1978; Sakano et al., 1979a; Zeelon et al., 1981). Since a specific *V* probe was not available, there was no direct evidence that the restriction fragment from the plasmacytoma, which hybridized to the whole mRNA probe and to its 3′-end fragment, carried V_κ as well as C_κ sequences. Indeed, as pointed out by Matthyssens and Tonegawa, the *Bam*HI site near the 3′ end of the MOPC-321 V_κ sequence means that *V* and *C* sequences are not expected to be present on the same restriction fragment.

The presence of the *Bam*HI site complicates the interpretation of the Hozumi-Tonegawa experiment. It cannot be inferred that DNA encoding the V region is necessarily near that encoding the C region in the plasmacytoma. Moreover, if a *Bam*HI restriction site near the end of the *V* region is also present in the embryo DNA that was detected in this experiment, V_κ and C_κ sequences that are near each other in the germline might have been separated by the digestion. It was, however, clear that some somatic process affecting the C_κ gene must have occurred since the C_κ-containing *Bam*HI fragment from the tumor was smaller than that in the embryo.

Additional experiments using the same technique with different restriction enzymes and with a variety of plasmacytomas supported the hypothesis that *V* and *C* sequences are separate in embryo DNA and that rearrangements of these sequences have occurred in immunoglobulin-producing cells (Tonegawa et al., 1977b; Tonegawa et al., 1976). It was also shown that *V* genes that were not expressed in a plasmacytoma remained in the embryo configuration, as judged by restriction analysis, suggesting that there is a correlation between *V/C* joining and expression of the rearranged gene (Tonegawa et al., 1977a).

At about this time, Southern (1975) developed a method for transferring fragments of DNA from agarose gels to cellulose nitrate filters. The transferred fragments, bound to the filters, could be hybridized to radiolabeled probes for detection of sequences related to the probe. The sensitivity and high resolution of this simple technique greatly facilitated the detection and discrimination of fragments containing different immunoglobulin genes in embryo DNA and in a variety of κ and λ-producing plasmacytoma cell lines. Detection of specific restriction fragments was also facilitated by the availability of techniques for preparing cDNA probes. It was soon confirmed that *V* and *C* genes are separate in embryo DNA and in close proximity in plasmacytoma DNA (e.g., Brack et al., 1978; Lenhard-Schuller et al., 1978; Seidman and Leder, 1978). An example of such an experiment is shown in Figure 1. This conclusion was soon to be subjected to a rigorous test: the unrearranged as well as the rearranged genes would be isolated and characterized by cloning, using the recently developed methods of recombinant DNA.

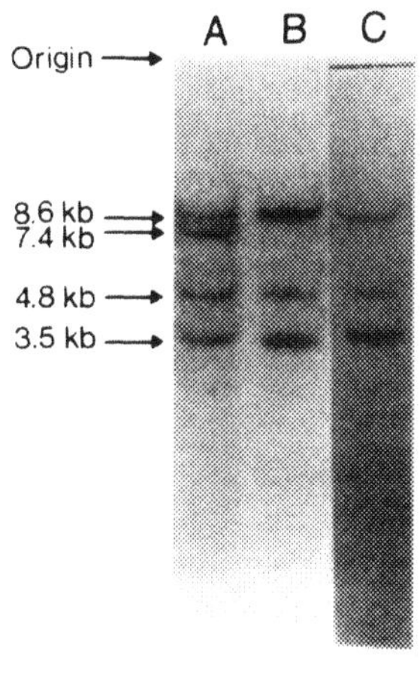

Figure 1. Rearrangement of gene segments in a plasmacytoma producing an immunoglobulin λ chain. DNA from a λ-producing plasmacytoma (A), 13-day BALB/c embryos (B), and a κ-producing plasmacytoma (C) were digested with a restriction endonuclease, *Eco*RI, electrophoresed through an agarose gel, transferred to a nitrocellulose filter, and hybridized to a ^{32}P-labeled cDNA probe encoding the plasmacytoma λ chain. Three fragments were found in all three lanes: one of 8.6 kb containing Cλ-sequences and two at 4.8 kb and 3.5 kb containing Vλ sequences. The 7.4 kb fragment was found only in the λ-producing plasmacytoma (A) and contained both Cλ and Vλ sequences, presumably a result of Vλ to Jλ rearrangement leading to expression of the λ chain in this tumor. (Reproduced with permission from Brack et al., 1978.)

In the original experiment of Hozumi and Tonegawa, fragments containing embryo *V* or *C* genes, such as might have been derived from an unrearranged chromosome, were not detected in the *Bam*HI digest of DNA from plasmacytoma MOPC-321. Possible explanations for the failure to find such fragments were suggested (Tonegawa et al., 1977a): 1) loss from the plasmacytoma cell of any chromosome bearing only unrearranged κ genes; this loss could be followed by duplication of the chromosome carrying the rearrangement; 2) the same joining event occurs on both chromosomes; however, there are many V_κ genes and there was no known reason why the same one should rearrange on both chromosomes.

In contrast, in DNA from another plasmacytoma, TEPC-124, unrearranged as well as rearranged genes were detected by the same method that had been used with MOPC-321 (Tonegawa et al., 1977a). When MOPC-321 tumor DNA was subsequently reexamined by the more sensitive Southern blotting technique, fragments corresponding to unrearranged *C* and/or *V* genes were detected after digestion with *Bam*HI (Seidman and Leder, 1978) or *Eco*RI (Lenhard-Schuller et al., 1978). The presence of unrearranged and rearranged *V* and *C* genes in plasmacytomas, as well as in normal B cells[d], will be discussed in the section Allelic and Isotypic Exclusion.

The next objective was to isolate and characterize an embryo *V* gene and to compare it to related expressed V regions. For this experiment, a gene encoding λ rather than κ V regions was chosen (Tonegawa et al., 1977c). This choice was based on evidence that most V_λ regions appeared to be encoded by very few germline genes, perhaps only one or two (see section Counting Genes). Therefore, relating the sequence of a germline gene to its expressed counterpart should be easier for λ than for κ chains.

Two fragments carrying V_λ sequences were identified in an *Eco*RI digest of embryo DNA (Tonegawa et al., 1977c; see also Figure 1). One of these, corresponding to the 4.8 kb fragment in Figure 1, was cloned. The insert hybridized to whole mRNA from a λ-producing plasmacytoma, HOPC-2020, but not to the segment of the mRNA corresponding only to the C region; therefore, it presumably contained only *V* sequences. The complementarity between the whole mRNA and the insert in the clone was confirmed by observing formation of an R loop by electron microscopy. (Under appropriate annealing conditions, RNA-DNA hybrids are more stable than DNA duplexes; an R loop is the structure formed when the sense DNA strand is displaced by the mRNA and loops out from the RNA-DNA duplex (Thomas et al., 1976).) Only the 5′-portion of the whole HOPC-2020 mRNA hybridized to the insert, consistent with the absence of *C*-region sequence.

To determine the sequence of the cloned insert, Tonegawa collaborated with Maxam and Gilbert, who had recently developed a method for determining DNA sequences (Maxam and Gilbert, 1977). Although the clone had been identified by hybridization to mRNA obtained from a λ1 plasmacytoma, HOPC-2020, its DNA sequence corresponded more closely to the V region of the λ2 chain of MOPC-315. Evidently, the clone carried the $V_{\lambda 2}$, not the $V_{\lambda 1}$, germline gene. The results also confirmed that no sequence corresponding to C region was present on the cloned fragment (Tonegawa et al., 1978).

The DNA sequence of $V_{\lambda 2}$ revealed a completely unanticipated feature: DNA 3′ to the codon for amino acid position 98 (or no. 96 according to "Kabat numbering" (Kabat et al., 1991)) did not correspond to the known amino acid sequence of that portion of the $V_{\lambda 2}$ region. Soon thereafter, Seidman et al. (1978) reported that V_{κ} genes also do not encode the entire κ variable region. It was subsequently shown that a separate short DNA segment, called *J* (for joining), encodes the last 12 or 13 amino acid residues of the light chain variable region (Brack et al., 1978; Bernard et al., 1978; Seidman and Leder, 1978). The *J*-encoded residues encompass one or two residues of the third CDR and the fourth framework region. Results obtained by amino acid sequence analysis of the variable regions of a group of closely related κ chains also indicated that there are two segments, V and J, that appear to associate independently with each other, presumably corresponding to the *V* and *J* gene segments (Weigert et al., 1978).

It had previously been established that light chains synthesized in a cell-free system differ slightly from secreted light chains (Stavnezer and Huang, 1971); additional studies established that these chains are synthesized as precursors having an extension of about 20 amino acid residues at the amino-terminus relative to mature light chains (Milstein et al., 1972; Swan et al., 1972; Mach et al., 1973; Schechter, 1973; Tonegawa and Baldi, 1973). The $V_{\lambda 2}$ DNA sequence confirmed the presence of a 5′ "leader" extension of the coding region. Within this extension was a stretch of nucleotides that did not encode any part of the light chain. Comparison of the DNA sequence (Tonegawa et al., 1978) with the partial amino acid sequence of the MOPC-315 λ leader (Burstein and Schechter, 1977) allowed the precise length and position of this segment—an intron—to be determined.

Tonegawa et al. (1978) pointed out that knowledge of the *V* region sequence might be informative regarding the hypothesis, proposed by Kabat and colleagues (Wu and Kabat, 1970; Wu et al., 1975), that the CDRs result from the insertion of short segments of DNA into the structural gene for a light or heavy chain. Although a separately encoded *J* segment might appear to be consistent with this idea, the *J* segment encoded mostly framework residues. The first two and part of the third CDR were entirely encoded in the continuous germline *V* segment, not consistent with the Wu-Kabat hypothesis. However, later studies of heavy chain *V* region genes (see below) revealed the presence of yet another short gene segment, designated *D* (or D_H), which encodes most of the third CDR of the heavy chain V region.

Even after rearrangement, the coding sequences for the V^e and C regions are not contiguous. This was demonstrated by R-loop analysis of hybrids formed between a cloned rearranged gene from a plasmacytoma and mRNA from the same tumor. As shown in Figure 2, the *V* and *C* regions formed the expected R-loops but, in addition, a double-stranded segment was looped out between them, evidently containing DNA that was not complementary to any sequence in the mRNA, i.e., an intron (Brack and Tonegawa, 1977; Lenhard-Schuller et al., 1978; Seidman and

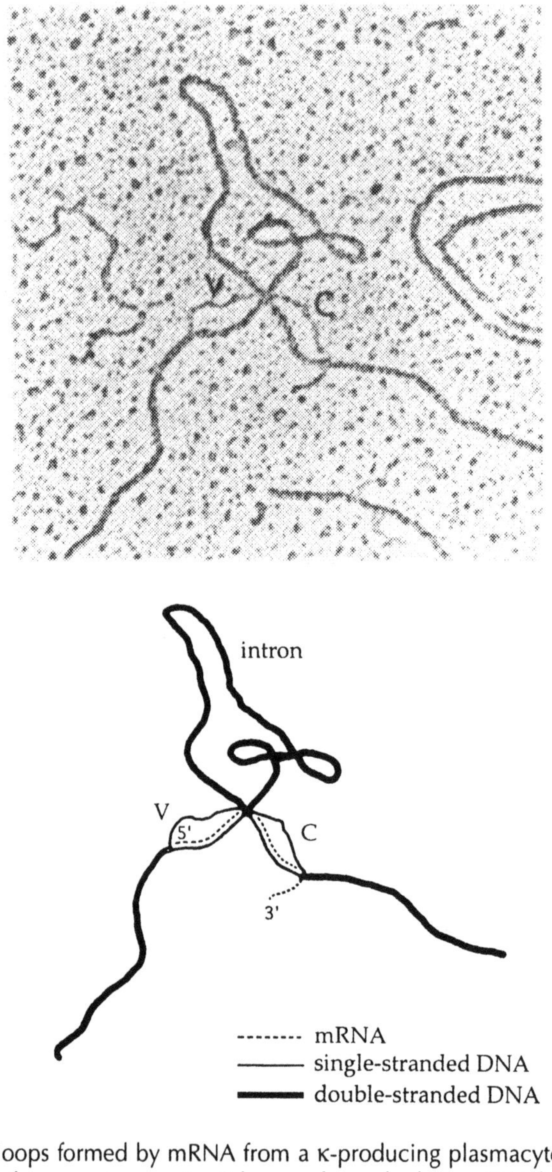

Figure 2. R loops formed by mRNA from a κ-producing plasmacytoma hybridized to DNA from the same tumor. An R loop is formed when mRNA hybridizes to the complementary DNA strand, displacing the sense DNA strand. In this example, two R loops are formed, one containing C and the other V region sequences. These two loops are separated by a double-stranded intron of ~3.7 kb. The tail extending from the C region R loop presumably consists of the poly(A) segment of the mRNA. (Photograph reproduced with permission from Seidman and Leder, 1978; diagram modified from same figure.)

Leder, 1978). The intron was shown to lie between the *J* and *C* gene segments. Similar conclusions were reached by the technique of S1 nuclease protection (Matthyssens and Tonegawa, 1978; Rabbitts and Forster, 1978). This intron and that near the 5′ end of the *V* gene segment (within the leader region) were among the first introns to be identified within the genes of eukaryotes. Thus, in addition to DNA translocation, RNA splicing is required for the expression of immunoglobulin genes.

Recombination Signal Sequences

That the V region of an immunoglobulin chain is encoded by two separate gene segments (*V* and *J*) was not only surprising but of profound relevance for antibody diversification. It was now clear that diversity can be achieved by a previously unknown mechanism operating in somatic cells: from a limited number of gene segments a far larger library of sequences can be generated by a combinatorial joining process (Weigert et al., 1978). But this discovery immediately raised a question: how do these segments become juxtaposed in a B cell to form a gene encoding a complete V region?

Examination of the sequences flanking V_κ and J_κ gene segments revealed the presence of conserved sequence elements: 3′ to each V_κ is a highly conserved palindromic heptamer and a somewhat less conserved A/T-rich nonamer, separated by a 12 bp spacer of unspecified sequence; 5′ to each J_κ is another conserved palindromic heptamer and an A/T-rich nonamer, separated by a 23 bp spacer (Max et al., 1979; Sakano et al., 1979a). The situation with V_λ and J_λ is basically similar except that the spacer lengths are reversed, i.e., the 23 bp spacer is 3′ to V_λ and the 12 bp spacer is 5′ to J_λ (see Figure 3). It was suspected that these sequence elements are important for the rearrangement events that bring the gene segments together; as evidence for this hypothesis accumulated, the heptamer-spacer-nonamer motifs came to be known as "recombination signal sequences (RSS)" (Akira et al., 1987).

The heptamers and nonamers 3′ to the *V* segments are largely complementary to the heptamers and nonamers 5′ to the *J* segments. This finding led to the hypothesis that these elements form a "stem-loop" structure as an intermediate in the recombination reaction, with deletion of the DNA between the recombining segments (Max et al., 1979; Sakano et al., 1979a). In contrast, Early et al. (1980a) suggested that the rearrangement is mediated by proteins that recongnize the heptamer-nonamer pair when they are separated by one or two turns (11 or 22 bp) of the DNA helix. It was, indeed, subsequently shown that base-pairing between the RSS is not essential (Hesse et al., 1989), supporting the idea that the RSS serve primarily as protein recognition and binding sites. As will be discussed in the section, Mechanism of Rearrangement, a site-specific recombinase is targeted to these conserved sequences and initiates the recombination.

When the RSS were first found adjacent to the *V* and *J* gene segments encoding light chains, it was pointed out that their structure is reminiscent of inverted repeats

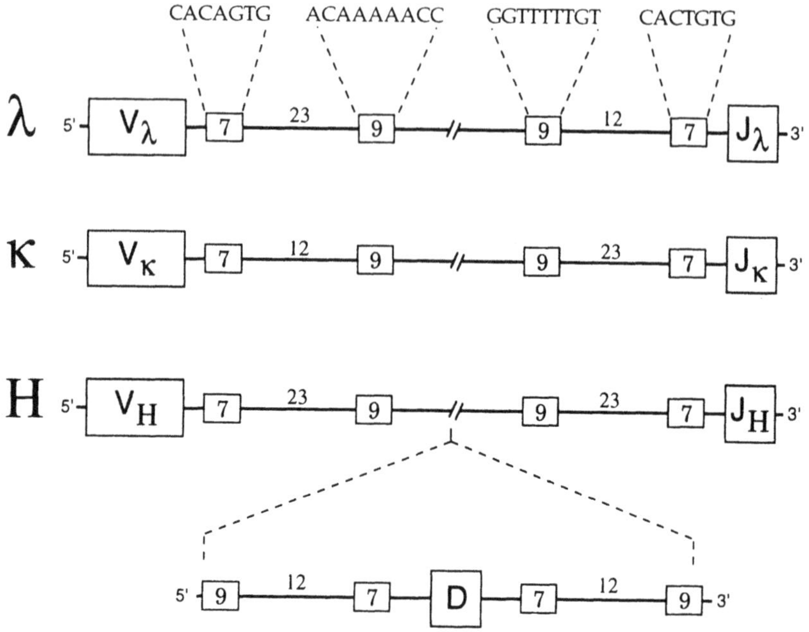

Figure 3. Arrangement of *V, D,* and *J* gene segments and adjacent recombination signal sequences in λ, κ, and heavy (H) chain loci, showing heptamer-spacer-nonamer motifs. The numbers designate bp. Recombination occurs between one signal having a 12 bp spacer and another having a 23 bp spacer (the 12/23-bp rule). (Redrawn from Tonegawa, 1983.)

at the ends of prokaryotic insertion elements. Perhaps, in the evolution of immunoglobulin genes, a DNA element was inserted into an ancestral *V* gene, splitting it into *V* and *J* segments; these gene segments then evolved independently, each retaining one of the two inverted repeat flanking sequences, thus facilitating subsequent excision in *V-J* recombination (Sakano et al., 1979a).

V and *J* gene segments were also shown to encode segments of the heavy chain V region. However, these two gene segments did not appear to encode the complete V region; part of the third CDR was missing (Early et al., 1980a; Sakano et al., 1980). It was also noted that the heptamer/nonamer pairs adjacent to V_H and J_H are both separated by spacers of 23 bp (Figure 3). In light chains, the spacers next to *V* and *J* are unequal in length, one is 12 bp and the other is 23 bp. These two observations (i.e., the missing sequence and the identical spacer lengths) led to the prediction that there must be a third gene segment encoding part of the third CDR of a heavy chain, and that its flanking RSS must have spacers of 12 bp to allow recombination with both *V* and *J*. It was proposed that this hypothetical gene segment be called *D* for diversity, since it presumably encodes a highly variable section of the heavy chain V region, part of the third CDR (Schilling et al., 1980). Such *D* (or D_H) gene segments with RSS having 12 bp spacers were soon identified, as described in the section Genes Encoding Human and Mouse Heavy Chains.

The discovery of *D* segments with 12 bp spacers confirmed that recombination generally requires one spacer to be 12 bp and the other to be 23 bp. This restriction became known as the "12/23-bp rule." Adherence to this rule prevents *V* segments from combining with each other or, in the case of the heavy chain locus, from combining directly with *J*, which would exclude interposition of a *D* segment. The rule also forbids combination of *J* with *J* or *D* with *D* (see next section for contrast with T cell receptor gene segments). However, exceptions to the 12/23 bp rule may occur, as will be discussed in the section Genes Encoding Human and Mouse Heavy Chains.

Junctional Variability

The nucleotide position at which *V-J* recombination actually takes place was found to vary by one, two, or even several bp, as illustrated in Figure 4. This introduces further diversity into the amino acid sequence of the light chain at the *V-J* junction (Max et al., 1979; Sakano et al., 1979a; Weigert et al., 1980). The recombination of *V* and *D* and that of *D* and *J* in the heavy chain locus is also imprecise (reviewed by Tonegawa, 1983), resulting in additional variability in the third CDR of the heavy-chain V region. Another source of diversity in this region of the heavy chain is the translation of *D* segments in all three reading frames (e.g., Kaartinen and Makela, 1985, Sanz, 1991; Corbett et al., 1997).

A feature noted in early studies of gene recombination at the heavy chain locus was the frequent deletion and/or addition of a few nucleotides at the *V-D* and *D-J* junctions (Sakano et al., 1981; Kurosawa and Tonegawa, 1982). (See Figure 5). It was proposed that the template-independent addition of so-called N (for "nucleotide") regions is catalyzed by the enzyme terminal deoxynucleotidyltransferase (TdT) (Alt and Baltimore, 1982; Desiderio et al., 1984); dG is preferentially added and is then complemented by dC, so that N regions tend to be rich in G/C bp. The role of TdT in N addition was supported by showing that mice lacking this enzyme are deficient in N nucleotides (Gilfillan et al., 1993; Komori et al., 1993). N region addition is developmentally regulated, being low or absent in fetal or neonatal animals, presumably reflecting the level of TdT expression (see references in Lewis, 1994a and Okada and Alt, 1995).

Although originally thought to be confined to *V-D* and *D-J* junctions in heavy chain gene rearrangements, there have been a number of reports of N nucleotides at the *V-J* junctions of light chain gene segments, where they may contribute to the diversity in CDR3. Short insertions consistent with N addition have been noted infrequently at mouse V_κ-J_κ junctions (Max et al., 1980; Milstein et al., 1992; Victor et al., 1994); they are found more frequently and are longer at human V_κ-J_κ junctions (e.g., Klobeck et al., 1987; Lee et al., 1992; Martin et al., 1992; Victor and Capra,

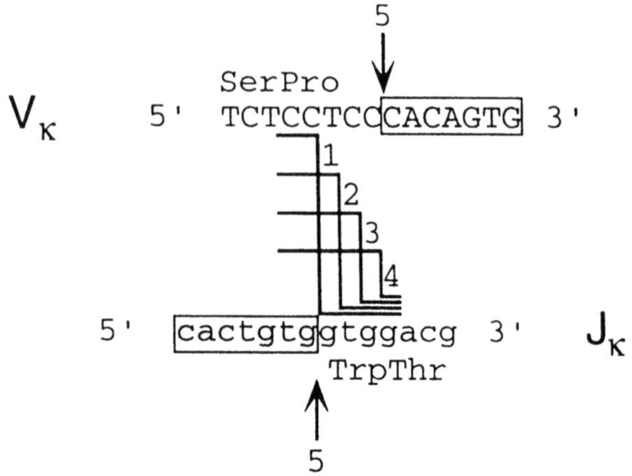

1. 5' TCTCCgtggacg 3'
   SerProTrpThr

2. 5' TCTCCTtggacg 3'
   SerProTrpThr

3. 5' TCTCCTCggacg 3'
   SerProArgThr

4. 5' TCTCCTCCgacg 3'
   SerProProThr

5. 5' TCTCCTCCgtggacg 3'
   SerProProTrpThr

Figure 4. Junctional diversity generated by varying the position of recombination between a V_κ and a J_κ gene segment. Nucleotides in *V* segment are in upper case, those in *J* segment are in lower case. Heptamers are shaded. The position of strand breakage and rejoining can vary (numerals 1 to 5), resulting in different codons at the recombination site, as indicated in the five sequences shown. The arrows marked 5 indicate sites of cleavage that would lead to a sequence with one additional codon, as shown. These sequences are examples and have not necessarily been found in κ light chains examined to date. Trp, Arg and Pro have all been found at the junctional position, as shown, and sequences consistent with recombination 5 have also been described (Weigert et al., 1980). (Based on Sakano et al., 1979a and Max et al., 1979.)

1994; Feeney et al., 1997; Foster et al., 1997). N addition has also been observed at human V_λ-J_λ junctions (Feeney et al., 1997). Sanz (1991) noted that long (>10 bp) junctional insertions are frequent in human but not mouse heavy chains. However, in both species, N insertions are much more prominent at heavy than at light chain junctions. TdT was found to be expressed mainly at the time of heavy chain gene rearrangement in both species (Loken et al., 1987; Li et al., 1993), but was also detected, albeit at very low levels, when light chain gene segments rearrange in humans (Billips et al., 1995). Extensive N nucleotide addition in a large collection of human heavy chain junctions was recently reported by Corbett et al. (1997) and Brezinschek et al. (1997).

The presence of one or two additional nucleotides palindromic to the terminal nucleotides of nontruncated coding ends was demonstrated at *V(D)J* junctions of chicken light chain genes (McCormack et al., 1989), as well as of T cell receptor γ and δ chain genes (Lafaille et al., 1989). The latter authors also surveyed immunoglobulin and T cell receptor junctional sequences in the literature and found many other examples of such insertions, which they called P (for "palindromic") nucleotides. It was suggested that P insertions are a consequence of the cleavage of a hairpin intermediate in the recombination event (Lafaille et al., 1989; Lieber, 1991; Roth et al., 1992b; Lewis, 1994b), as illustrated in Figure 5.

Meier and Lewis (1993) analyzed the process of P nucleotide insertion by examining junctions of an extrachromosomal recombination plasmid substrate that undergoes rearrangement in a pre-B cell line. Their data indicated that the frequency of P addition is variable, according to the sequence of the joined coding ends. P nucleotides were, however, absent from signal ends, confirming earlier observations (Lafaille et al., 1989; McCormack et al., 1989). Meier and Lewis also carried out a systematic survey of endogenous junctions at mouse immunoglobulin and T cell receptor loci. They noted that, with rare exceptions (e.g., Milstein et al., 1992), P nucleotides were not found at junctions of expressed light chain genes, but they appeared to be present at V_κ-J_κ junctions in the circular excision products analyzed by Harada and Yamagishi (1991), presumably nonfunctional byproducts of secondary rearrangement (see section Secondary Rearrangement). This observation suggested that P nucleotides are generated in light chain rearrangements but may not be selected in the expressed repertoire. P insertions were found with variable frequency (2 to 15%) at heavy-chain junctions and sometimes coexisted with N nucleotides. Most P inserts were one or two nucleotides in length, but a few were longer. Recent analysis of a large number of productive and nonproductive human $V_\kappa J_\kappa$ junctions yielded evidence for P as well as N nucleotides (Foster et al., 1997).

Removal of a few base pairs at *V(D)J* junctions occurs frequently. An early suggestion to account for this truncation was that an exonuclease trims the coding ends before they are joined (Alt and Baltimore, 1982), but the enzyme responsible has not yet been identified. Lewis (1994a) has pointed out that it is even possible that trimming results from endonucleolytic rather than exonucleolytic activity. The imprecision in junctional joining, as well as the removal or addition of bases not

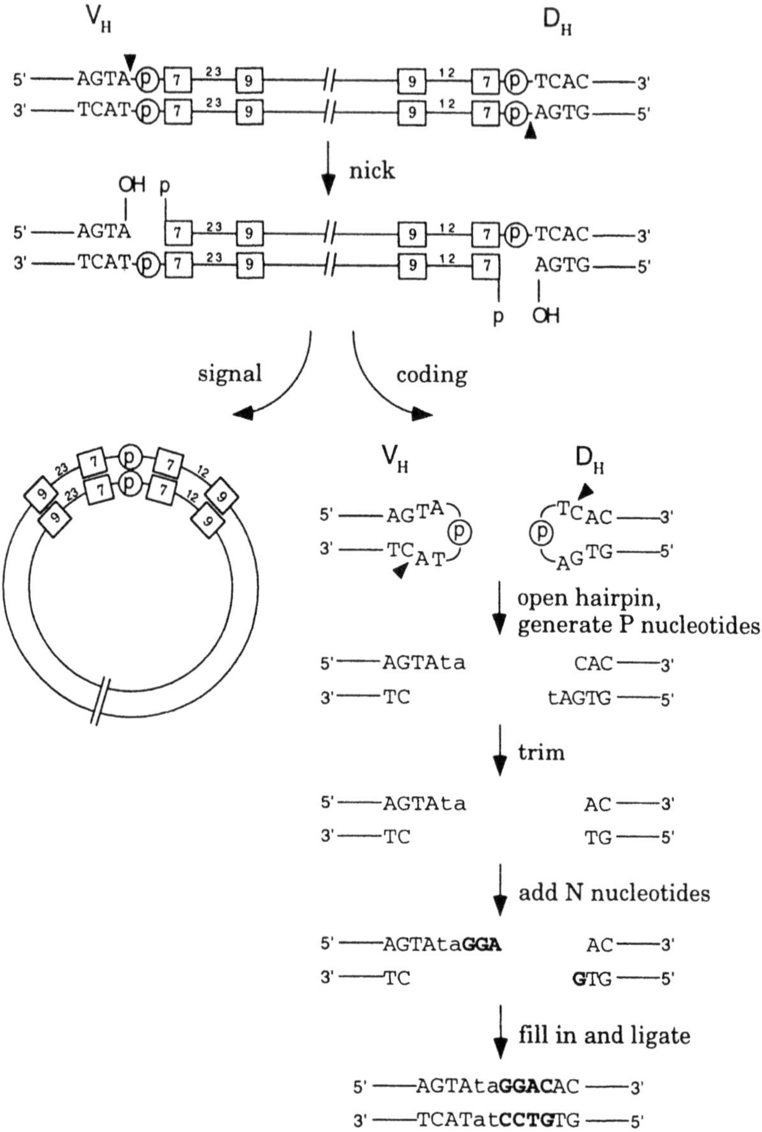

Figure 5. Outline of steps in *V(D)J* recombination, shown for formation of a $V_H D_H$ junction. A possible scheme for adding P and N nucleotides and trimming at the junctions is illustrated. Heptamers and nonamers are indicated by rectangles. If p is within a circle, it indicates a phosphodiester linkage; these are shown occasionally for clarity and not between all nucleotides in the continuous chains; p not in a circle indicates a terminal phosphate group. An initial cleavage (arrowhead) occurs at the 5′ end of the signal heptamer in one strand, leaving a 5′-phosphate group on the signal end and a 3-hydroxyl on the coding end. The 3′-hydroxyl forms a phosphodiester

Figure 5. (continued) linkage with the complementary nucleotide in the opposite strand, resulting in a hairpin at the 3′ end of the coding region. The blunt signal ends are fused, usually without loss or gain of nucleotides at the junctions, forming a closed circle (but see also Figure 6). The hairpins are nicked, either symmetrically or 5′ or 3′ to the midpoint. In this example, the nicks in V_H and D_H generate 3′ palindromic overhangs; the complementary nucleotides between the nick and the midpoint are designated P nucleotides, shown here in lower case. Trimming of terminal nucleotide(s) may occur. Here, nucleotides are removed only from the *D* coding end. In addition, N nucleotides may be added, shown here in bold. After filling in complementary nucleotides, the chains are ligated. The final coding junction in this example contains two P nucleotides, derived from the *V* coding end, and three N nucleotides. There is firm evidence for the steps leading to the formation of the hairpin; the later steps are conjectural (see text for references).

only introduces additional variability into the V region, but may also lead to frame shifts, rendering the rearranged gene nonfunctional (e.g., Max et al., 1980).

V(D)J recombination, as well as associated junctional modifications (truncation, insertion of N and P nucleotides), also occurs in the joining of T cell receptor gene segments. *D* gene segments, flanked by RSS with one-turn and two-turn spacers, encode part of the V region of T cell receptor β and δ chains; the spacer lengths in the RSS are such as to allow joining of *V* (two-turn spacer) directly to *J* (one-turn spacer) as well as the insertion of one or two *D*'s between *V* and *J* (Kronenberg et al., 1986; Chien et al., 1987).

Chromosomal Events in Rearrangement

In the initial report on the somatic rearrangement of *V* and *C* genes, before *J* segments were identified or the sequence of the rearranging gene segments was known, Hozumi and Tonegawa (1976) proposed several models to account for the integration of V and C sequences. 1) Copy-insertion: a *V* gene might be copied and then inserted next to the *C* gene; this seemed unlikely, however, as the expressed *V* gene should then be retained in its germline position, which at that time did not appear to be the case with the MOPC-321 plasmacytoma. 2) Copy excision: a *V* gene is excised and then integrated adjacent to a *C* gene. 3) Deletion: the DNA between the joining *V* and *C* genes, including the two flanking RSS, is deleted and ultimately lost from the cell. 4) Inversion: provided that the *V* and *C* genes are in opposite transcriptional orientation, inversion of a segment of chromosome would result in the juxtaposition of these gene segments in the same transcriptional direction. In all of these models, rearrangement occurs between gene segments lying on the same chromosome. This is compatible with older findings that V_H and C_H regions usually express allotypic determinants associated with genes present on a single parental chromosome (Kindt et al., 1970; Landucci Tosi et al., 1970). (See the section, Class Switching for exceptions to this general rule.)

Early analyses of κ light chain rearrangements yielded results consistent with the excision-deletion model in which DNA between the rearranging *V* and *J* gene segments is lost from the cell (Sakano et al., 1979a; Seidman et al., 1980). It was subsequently found, however, that in some cell lines this DNA, including two precisely fused RSS heptamers (a "signal joint"), may be retained, arguing against a simple deletion model (Steinmetz et al., 1980). Other studies also provided evidence for retention of DNA between *V* and *J* (Selsing and Storb, 1981a; Höchtl et al., 1982; Lewis et al., 1982; Van Ness et al., 1982). Although this retained DNA appeared to be a byproduct of *VJ* joining, it was not necessarily the reciprocal product expected from the recombination of the particular *VJ* join found in the cell line, suggesting that more than a single rearrangement had occurred. An aberrant recombinant product involving D_H and J_H gene segments, having a fusion of heptamer elements and an inversion of J sequence within the J locus, was also described in a B cell tumor line (Alt and Baltimore, 1982).

Mechanisms initially proposed to account for the results seen with κ rearrangements were excision and reintegration (Steinmetz et al., 1980) and unequal sister chromatid exchange (Höchtl et al., 1982; van Ness et al., 1982). However, it was also noted that the results could be explained if some, but not all, *V* genes are in inverse transcriptional polarity relative to *J* (Lewis et al., 1982). As illustrated in Figure 6, if *V* and *J* are in the same polarity, rearrangement results in deletion of the DNA separating them; the deleted DNA might be recovered as a circular product, as was shown for DJ_H recombination by Toda et al. (1989). If *V* and *J* are in opposite polarity, rearrangement results in inversion of the intervening DNA, which remains on the chromosome (Figure 6). The orientation of some *V* and *J* segments may be reversed as a result of the inversion. Successive inversion and deletion could generate non-reciprocal products.

To test the hypothesis that gene segments in opposite polarity rearrange by inversion, Lewis et al. (1984) prepared a retroviral construct that contained oppositely oriented V_κ and J_κ gene segments, each with their RSS. A selectable marker was introduced to allow detection of rearrangement. Following infection into a cell line capable of effecting *V-J* rearrangement, joining of the appropriate sequences in the construct occurred by inversion, with retention of intermediate sequence and the two RSS. With a plasmid construct that contained signals allowing recombination by either inversion or deletion, it was shown that inversions occur about one-third as frequently as deletions (Hesse et al., 1987).

As will be discussed in the section, Genes Encoding Human and Mouse Kappa Chains, many V_κ in both mouse and human are in inverse orientation; accordingly, recombination by inversion as well as deletion is a prominent feature of these loci. Recombination of immunoglobulin gene segments by reinsertion of DNA or by sister chromatid exchange has not been documented. (See also section, Secondary Rearrangements.)

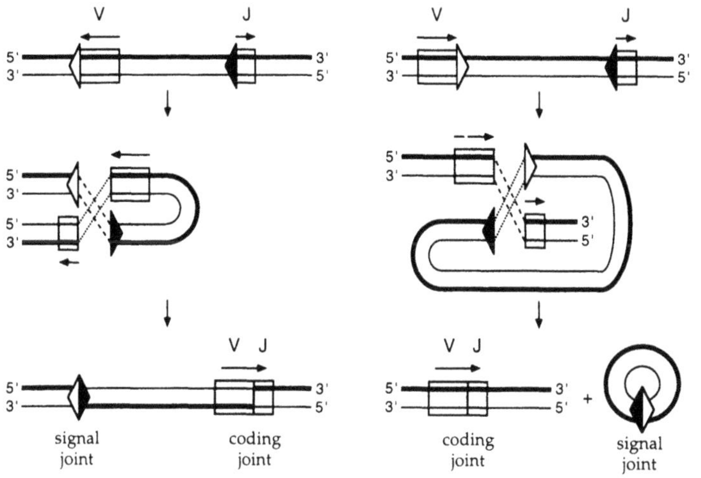

Figure 6. Rearrangement at the *VJ* junction by inversion (left) or deletion (right). To distinguish between the two DNA strands, one is shown thicker. Coding segments are represented by shaded rectangles and signal sequences by white and black triangles. Positions of recombination are denoted by the dotted and dashed lines. Horizontal arrows indicate direction of transcription.

The Mechanism of Rearrangement

An important step toward unraveling the molecular mechanism of *V(D)J* recombination was the discovery of two closely linked genes, designated recombination activating genes 1 and 2 (*RAG1* and *RAG2*), which turned out to be the only lymphoid-specific genes required for this process. The identification of these genes relied on the earlier demonstration, by Schatz and Baltimore (1988), that transfection of sheared genomic DNA into a fibroblast cell induced *V(D)J* recombinase activity. The activity was detected with a modified version of the retroviral construct developed by Lewis et al. (1984), mentioned above. Tagging the genomic DNA with an oligonucleotide allowed the DNA fragment responsible for the recombinase activity to be followed through several rounds of transfection. A tagged DNA fragment that cosegregated with the activity was identified and served as the start of a chromosomal walk that led to the identification of *RAG1* (Schatz et al., 1989) and then *RAG2* (Oettinger et al., 1990).

The two *RAG* genes are closely linked and are convergently transcribed. The encoded polypeptides are quite large, consisting of 1040 amino acid residues in the case of mouse RAG1 and 527 amino acid residues for mouse RAG2. Their indispensable role in *V(D)J* recombination was demonstrated by the introduction of targeted deletions into either germline gene. The resulting mice contained no mature B or T lymphocytes and were profoundly immunodeficient (Mombaerts et al., 1992; Shinkai et al., 1992). More recently, several cases of human severe combined immunodeficiency (SCID) have been shown to be correlated with mutations in *RAG1* or *RAG2* (Schwarz et al., 1996).

The *RAG* genes appear to be expressed together only in cells of the lymphocytic lineage. Genes encoding immunoglobulin V regions are completely assembled only in B lineage cells and genes encoding T cell receptor V regions are completely assembled only in T lineage cells (see Lewis, 1994a). What determines accessibility to the recombinase (e.g., gene methylation or transcription) in certain cell types and at particular developmental stages is not yet known. The expression is strictly regulated and can be correlated with specific rearrangement events; it is up-regulated in immature bone marrow or fetal liver B lineage cells when heavy chain genes rearrange, and again later, when light chain genes rearrange (see section Allelic and Isotypic Exclusion) (Grawunder et al., 1995).

Although it was originally thought that *RAG* expression in B lineage cells occurs only during their early differentiation in bone marrow and fetal liver, encounter with self-antigen was shown to induce reexpression of *RAG* in bone marrow, accompanied by secondary rearrangements of light chain genes (Tiegs et al., 1993). Moreover, it has recently been shown that *RAG* expression can also occur in B lymphocytes within germinal centers in spleen or lymph nodes (Han et al., 1996; Hikada et al., 1996). (See section Secondary Rearrangements: Receptor Editing.)

The *V(D)J* recombination reaction can be divided into two stages: 1) double-strand breaks in the DNA are introduced between the RSS heptamer and the adjacent coding segment, and a hairpin is formed at the coding end; 2) opening of the hairpin, followed by processing and joining of the cleaved ends. A current view of the steps involved, including the subtraction and addition of nucleotides at junctions, is shown in Figure 5. (See reviews by Lewis 1994a; Okada and Alt 1995; Gellert 1997; Schatz 1997; Smider and Chu 1997.)

In the first stage of the reaction, a nick is introduced into one DNA strand at the 5′ end of the RSS, yielding a 5′ phosphate group at the signal end and a 3′-hydroxyl at the coding end. The 3′-hydroxyl then forms a bond with the phosphate of the complementary nucleotide in the opposite strand, resulting in a hairpin structure on the coding end and a blunt 5′-phosphorylated signal end. For details about these steps, see Roth et al. (1992a, 1992b; 1993), Schlissel et al. (1993), and Ramsden and Gellert (1995). This phase of the reaction was shown to occur in a cell-free system (van Gent et al., 1995). Recombinant RAG1 and RAG2 were found to be sufficient for the early site-specific steps in the *in vitro* reaction: recognition of RSS, cleavage at the signal/coding border, and converting the coding ends into hairpins (McBlane et al., 1995). Optimal cleavage, *in vitro*, of substrates containing a pair of RSS follows the 12/23-bp rule (Eastman et al., 1996; van Gent et al., 1996a).

The second stage in *V(D)J* recombination (i.e., opening of the hairpin followed by joining of the two coding ends) is not yet understood in as much detail, but it is clear that it requires proteins involved in repair of double-strand breaks in DNA. This was first recognized when mice with severe combined immunodeficiency (scid) were found to be defective in *V(D)J* recombination (Lieber et al., 1988) and in general double-strand break repair (Fulop and Phillips, 1990; Biederman et al., 1991; Hendrickson et al., 1991). Cell-free systems that perform this phase of the reaction have also been developed (Cortes et al., 1996; Leu et al., 1997; Ramsden et al., 1997; Weis-Garcia et al., 1997). Results to date suggest that the RAG proteins are not only involved in the initial cleavage, as described above, but also influence the resolution of coding and signal ends (Leu et al., 1997; Ramsden et al., 1997).

Speculations about the origin of the *RAG* genes in the vertebrate genome have drawn on analogies between the details of *V(D)J* rearrangement and two different types of recombination in prokaryotes: site-specific recombination and transposition (reviewed by Lewis and Wu, 1997). Shortly after the discovery of the RSS adjacent to gene segments encoding immunoglobulin *V* regions, Simon et al. (1980) noted a similarity in sequence between the RSS and DNA motifs flanking the DNA segment that is inverted in phase variation in *Salmonella*, a site-specific recombinational event. Recently, it has been shown that recognition of the RSS nonamer site is mediated by a conserved region of RAG1 that has some similarity in sequence to the helix-turn-helix DNA-binding domain of the *Salmonella* Hin recombinase, which binds to the similar nonamer-like motif in *Salmonella* (Feng et al., 1994; Difilippantonio et al., 1996; Spanopoulou et al., 1996). Indeed, this domain in Hin, which also resembles eukaryotic homeodomains, can functionally replace the corresponding region of RAG1 in *V(D)J* recombination (Spanopoulou et al., 1996).

In support of a transposase rather than a Hin-like origin for the RAG proteins, van Gent et al. (1996b) have shown that the formation of the hairpin intermediates in *V(D)J* recombination is by direct transesterification, like strand transfer reactions of bacteriophage Mu and the integrases of retroviruses. A similarity between the RSS and the termini of transposable elements in *Ceanorhabditis elegans* was noted by Dreyfus (1992). Thompson (1995) has suggested that the *RAG*s and the RSS represent a disassembled transposon. Presumably, after its initial entry, the hypothetical "RAG transposon" lost its ability to move to another location. No canonical RSS-like sequences have been identified in or near the *RAG* locus. Although *V(D)J* recombination shares some features with site specific recombination and others with transposition, Lewis and Wu (1997) have pointed out that this process has unique features that do not allow straightforward classification with either of these mechanisms.

The discovery that the mouse *RAG1* and *RAG2* genes are closely linked and lack introns in their coding regions prompted Oettinger et al. (1990) to propose that they may have evolved as part of a viral or fungal recombination system. These structural features are also consistent with an origin as a transposon. Indeed, the *RAG* genes in several other vertebrate species, mouse (Oettinger et al., 1990), rabbit (Fuschiotti et al., 1993), chicken (Carlson et al., 1991), and *Xenopus laevis* (Greenhalgh et al., 1993) were also found to be devoid of introns in their coding regions. However, one or two introns have been found within the *RAG1* coding regions of two teleosts, trout (Hansen and Kaattari, 1995) and zebrafish (Willett et al., 1997). *RAG1* has also been identified in sharks, the most primitive vertebrate group known to have an adaptive immune system (Bernstein et al., 1994; Greenhalgh and Steiner, 1995; Bernstein et al., 1996), but the complete genomic sequence has not been reported. Preliminary comparison of the genomic sequence with the corresponding cDNA suggests that the shark gene lacks introns (R.M. Bernstein and J.J. Marchalonis, personal communication). Perhaps, introns were not present when the *RAG* genes were acquired by vertebrates, but were introduced subsequently into the teleost lineage.

ORGANIZATION OF IMMUNOGLOBULIN LOCI

In parallel with studies of the rearrangement of immunoglobulin genes, indeed a necessary accompaniment, were investigations of the number and arrangement of the various *V* and *C* segments in the genomes of different species, especially the human and mouse. These studies have been ongoing until the present and have led to a complete picture of some of the immunoglobulin loci (see Table I and Figure 7).

As noted in Part I, The Four-Chain Model, inheritance patterns of allotypic markers in the rabbit had indicated that the genes encoding κ, λ and heavy chains are unlinked. Suitable genetic markers for all three types of chains were not available in other species, but in the human and mouse, genes encoding these three major classes of immunoglobulin chains were shown to reside on different chromosomes. All three loci (κ, λ and heavy) have been described in considerable detail in humans and BALB/c mice. Despite some differences, the principles of the gene organization for each chain type are quite similar in these two species.

It has long been noted that V region sequences fall naturally into distinct sets, rather than forming a continuum. Initially, these sets or subgroups, as they were called, were classified mainly according to the amino acid sequence of the amino-terminal ~20–25 residues (see Part I, Two Genes, One Polypeptide). Subsequently, examination of complete V region sequences as well as discovery of new *V* gene segments led to some modification of these classifications (e.g., Dildrop, 1984). More recently, classification has generally been based on the degree of identity in nucleotide sequence. Strongly related sequences are said to belong to the same "family." Sequences within a family are generally more than 80% identical in nucleotide sequence, whereas those in different families are generally less than 70% identical. As a practical tool, cross-hybridization following Southern blotting has been used to differentiate among the families (e.g., Brodeur and Riblet, 1984). As

Table 1. Immunoglobulin Gene Loci in Human and Mouse

	Chromosome	*Locus Size (Mb)*	*Number of Total[1] (Functional[2]) Gene Segments*		
			V	*D*	*J*
Human					
κ	2p11–12	1.8	76(40)	—	5(5)
λ	22q11	0.9	69(31)	—	7(5)
H	14q32.3	~1.5	90(51)	27 (25)	9(6)
Balb/c					
κ	6	3–3.5	~140	—	5(4)
λ	16	0.2–0.3	3(3)	—	4(3)
H	12	?	~200(~150)	13	5(4)

Notes: [1]Due to polymorphism, the number of total and functional genes in different individuals (or mouse strains) can vary. The data shown here for the human loci were taken on 1/12/98 from a web site maintained by Ian Tomlinson at the MRC Centre for Protein Engineering: http://www.mrc-cpe.cam.ac.uk/imt-doc/. See text for specific references.

[2]"Functional" as used here for the human loci means rearranged *in vivo* with a sequence appropriate for transcription and translation into a stable protein (see Tomlinson et al., 1995). A gene segment may have an open reading frame and no obvious defect, but is not counted as functional unless it is detected as a *V(D)J* rearrangement that appears to encode a protein that can form a stable three-dimensional structure.

one might expect, classifications by different methods are, by and large, in fair agreement, but some reclassification may be necessary when all the germline sequences in each locus are determined.

A point of interest in the analysis of the immunoglobulin loci has been to determine whether or not members of the same family are located in physical proximity on the chromosome. Gene localization presumably reflects the pattern of gene duplication and divergence through which the *V* genes evolved. As will be discussed, the location of a *V* gene segment may have some bearing on its utilization, but the relation is not simple.

Genes Encoding Human and Mouse Kappa Chains

The *J* and *C* gene segments of the human κ locus were cloned and characterized in the early 1980s by Leder and colleagues (Hieter et al., 1980; Hieter et al., 1982; Emorine et al., 1983). The sequence of the entire region between *J* and *C* was completed only some 10 years later (Whitehurst et al., 1992). Five *J* segments, each about 40 bp in length, are distributed within about 1350 bp; the distance from the single C_κ to the nearest J_κ is 2.9 kb. An enhancer element is located in the *J-C* intron (Emorine et al., 1983; Picard and Schaffner, 1984; Queen and Stafford, 1984) and another is 9 kb 3′ to C_κ (Meyer and Neuberger, 1989).

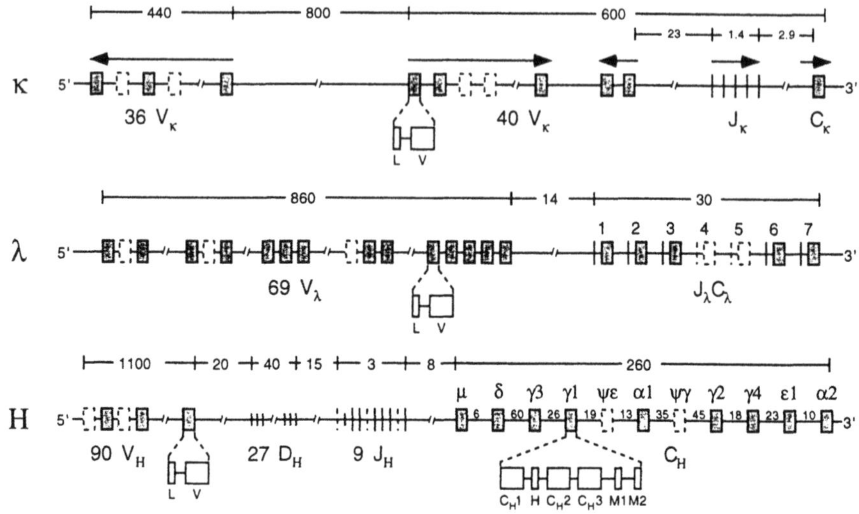

Figure 7. Human immunoglobulin gene loci. Approximate distances (small numbers) are in kb and are not drawn to scale. Shaded boxes denote *V* or *C* gene segments and open boxes show exons in representative segments. Longer vertical lines denote *J* segments and shorter lines *D* segments. Note one *D* segment within the J_H cluster. Dashed open boxes or lines represent pseudogenes. The exon/intron organization of *V* gene segments and of one C_H gene segment is shown: L and V are leader and *V* exons; C_H1, C_H2, C_H3, and H are exons encoding the three C domains and the hinge region of the γ1 chain; M1 and M2 are exons encoding the carboxyl-terminal segment of the membrane form of the γ1 chain (see Figure 8). Arrows above the κ locus show direction of transcription. The human V_κ genes are organized in two clusters in opposite orientation, presumably the result of gene duplication. The numbers of *V* segments are approximate and vary according to polymorphisms. See text and Table 1 for details and references.

The first studies of human light chain *V* genes were carried out by Bentley and Rabbitts (1980, 1981, 1983), who initially identified a V_κ gene segment by cross-hybridization with a cDNA clone encoding a κ chain from a mouse plasmacytoma. In the years since, extensive studies of the human V_κ locus, largely by Zachau and co-workers, have led to a comprehensive description of this locus (reviewed by Schäble and Zachau, 1993; Zachau, 1995, 1996; also Brensing-Küppers et al., 1997). These and other recent studies of immunoglobulin gene organization have relied heavily on techniques for cloning and separating large fragments of DNA (e.g., cloning in cosmids (Collins and Hohn, 1978; Royal et al., 1979) or yeast artificial chromosomes (Burke et al., 1987) and pulsed-field gel electrophoresis (Schwartz and Cantor, 1984)).

The human V_κ gene segments lie near the centromere on the short arm of chromosome 2 (2p11-12) (Malcolm et al., 1982). Two large regions of DNA contain 76 V_κ gene segments, including pseudogenes (Schäble and Zachau, 1993). Forty of these were classified as potentially functional, and all of these are now known to be expressed (Tomlinson et al., 1995; see also Table 1). The "proximal" region of 600 kb contains 40 V_κ as well as the five J_κ and single C_κ; except for the two V_κ nearest J, all of these V gene segments are in the same transcriptional orientation as the J_κ and C_κ, and presumably rearrange by deletion. The distance from the 3′-most V_κ to J_κ is 23 kb. About 800 kb from the proximal region is the "distal" region of 440 kb containing 36 V_κ, all in inverse orientation. That one of these distal V_κ rearranges by inversion was shown by pulsed-field gel electrophoresis; the size of the inverted segment was over a megabase (Weichold et al., 1990). Presumably, the other gene segments in the distal segment also rearrange by inversion.

Gene segments encoding different members of the major V_κ families are interspersed within each region, consistent with early observations of Pech and Zachau (1984). In one infrequent haplotype, the distal region is entirely absent. Cox et al. (1994) reported a strong bias towards use of the J_κ-proximal V_κ segments in the adult repertoire; 97% of the expressed V_κ they analyzed were in the J_κ-proximal region. The possible relevance of distance from the J_κ and transcriptional orientation to gene segment usage is not clear. The data indicate bias toward use of particular V_κ, as only 11 of these segments are used frequently.

The human κ locus appears to have undergone a recent gene duplication since 33 gene segments in the proximal region have nearly identical counterparts in the distal region. The κ locus in chimpanzees does not appear to be duplicated, implying that the duplication occurred after the divergence of the human and chimpanzee lineages (~4 to 5 million years ago). In addition to the V_κ in the main locus, there are at least two dozen V_κ scattered in other locations, some at distant locations on chromosome 2 and some on other chromosomes; most, but not all, of these "orphons" have defects in their sequences (Zachau, 1996).

The mouse κ locus, which is on chromosome 6 (Swan et al., 1979), is generally similar to the human κ locus, but has not yet been described in as much detail. The J_κ and C_κ gene segments were among the first immunoglobulin germline gene segments to be sequenced (Max et al., 1979; Sakano et al., 1979a; Max et al., 1981). As in the human κ locus, there are five J_κ segments, spaced at ~0.3 kb intervals, but the middle J element is not functional. The distance from the single C_κ to the nearest J_κ is 2.5 kb.

Recently, a continuous physical map of the entire mouse κ locus has been produced (George et al., 1995; Zocher et al., 1995; Kirschbaum et al., 1996; Schupp et al., 1997). The size of the locus is between 3 and 3.5 Mb. There is evidence for about 140 V_κ gene segments in the locus and a number of others are outside the locus. Individual members of a mouse V_κ family are frequently clustered in the same region (Heinrich et al., 1984; D'Hoostelaere et al., 1988), but interspersion with members of different families is not uncommon (Zocher et al., 1995). Some mouse V_κ rearrange by inversion implying that they are in inverse transcriptional orientation with respect to J_κ and C_κ.

Genes Encoding Human and Mouse Lambda Chains

About 40% of human immunoglobulin light chains are λ (Hood et al., 1967). Early studies of λ chains produced by patients with multiple myeloma identified four types of *C* regions that were distinguished serologically and/or by amino acid sequence analysis; they appeared not to be allelic and were presumed to be the products of distinct *C* genes (Appella and Ein, 1967; Ein and Fahey, 1967; Ein, 1968; Hess et al., 1971; Gibson et al., 1971; Fett and Deutsch, 1975).

The first investigation of genes in the human λ locus was by Leder and co-workers who screened a human genomic library with a mouse $C_{\lambda 1}$ probe (Hieter et al., 1981a). Six closely spaced gene segments were identified and the sequences of three of these ($C_{\lambda 1}$, $C_{\lambda 2}$, and $C_{\lambda 3}$) were determined. These sequences appeared to encode three of the four known protein C_λ regions. A *J* gene segment was located 1.3 to 1.6 kb 5′ to each of the *C* gene segments (Udey and Blomberg, 1987).

Vasicek and Leder (1990) extended the analysis to include the sequence of the remaining three human C_λ genes ($C_{\lambda 4}$, $C_{\lambda 5}$, and $C_{\lambda 6}$), as well as the identification of another C_λ segment ($C_{\lambda 7}$) at the 3′ end of the complex. Other investigators also cloned and sequenced parts of the C_λ complex (Dariavach et al., 1987; Bauer and Blomberg, 1991; Combriato and Klobeck, 1991). $C_{\lambda 4}$ and $C_{\lambda 5}$ were found to be pseudogenes. It was suggested initially that $C_{\lambda 6}$ (Dariavach et al., 1987) and then $C_{\lambda 7}$ (Vasicek and Leder, 1990; Combriato and Klobeck, 1991) might encode the fourth type of C region. However, reexamination of $C_{\lambda 6}$ indicated that it had a frame shift, although it can be expressed as a truncated protein (Stiernholm et al., 1995). $C_{\lambda 7}$ was shown to be expressed in a human lymphoma cell line and, at a low level, in peripheral blood lymphocytes (Berinstein et al., 1989; Bauer and Blomberg, 1991).

Bauer and Blomberg pointed out that $C_{\lambda 7}$ was unlikely to encode the fourth orginally identified C region because its translated sequence differs from that C region by five amino acids. Instead, they suggested that the fourth C region might be the product of a rare allelic variant or a somatically mutated form of $C_{\lambda 2}$, as it differs from the translated sequence of $C_{\lambda 2}$ at only one position. Recently, Niewold et al. (1996) identified a Bence-Jones protein (light chain secreted by myeloma cells) that differs from the sequence encoded by $J_{\lambda 7}C_{\lambda 7}$ in three positions, one of which could be explained by imprecise V_λ-J_λ joining. They suggested that this protein is the first identified representative of a family of λ chains encoded by the $C_{\lambda 7}$ gene segment; no additional chains of this type have been reported to date.

The human immunoglobulin C_λ gene segments were localized to chromosome 22q11 (de la Chapelle et al., 1983). The seven *J-C* gene segments are spread over about 30 kb of DNA (Vasicek and Leder, 1990) and the closest V_λ is 14 kb from the nearest *J* (Combriato and Klobeck, 1991). In some individuals, there are duplications of the region containing the $C_{\lambda 2}$ and $C_{\lambda 3}$ gene segments; these polymorphisms may contribute to variability among the λ chains (Taub et al., 1983).

Recently, the entire human λ locus has been described in great detail and the germline genes have been compared to expressed V_λ sequences (Frippiat et al., 1995; Williams et al., 1996; Ignatovich et al., 1997; Kawasaki et al., 1997). Indeed, Kawasaki et al. have reported a sequence of 1,025,415 bases containing this locus, at the time of publication the longest known continuous DNA sequence. A total of 69 V_λ gene segments are distributed in several gene-rich clusters within about 860 kb; 31 of these gene segments are actually expressed. In contrast to the human V_κ, all the V_λ have the same transcriptional polarity as the J_λ and C_λ. Ten families of related V_λ sequences had been defined previously (e.g., Chuchana et al., 1990). The largest families tend to be located in one or two regions, suggesting that these gene segments were generated by duplications. However, each region contains members of different families. In the more distal regions the *V* gene sequences are highly diverged. Ignatovich et al. (1997) analyzed a large cDNA library of unbiased V_λ sequences and found that only three families are used to any significant extent. Remarkably, only three V_λ gene segments in two of these families appeared to encode half the expressed V_λ repertoire. Although the *J*-proximal clusters were used most frequently, the usage of individual V_λ segments within each cluster did not seem to be correlated with position.

In BALB/c mice, only a few percent of mouse serum immunoglobulins bear λ chains (Hood et al., 1967; McIntyre and Rouse, 1970; Blaser and Eisen, 1978). The V and C regions of these chains are encoded by relatively few genes segments and the BALB/c λ locus was the first immunoglobulin gene complex to be described in detail. As noted previously (section Counting Genes), by 1971 two types of λ chain (λ1 and λ2) had been described; a third type, λ3, expressed even more rarely, was identified a decade later (Azuma et al., 1981).

The *J* and *C* gene segments encoding the BALB/c λ chains are arranged in two clusters, each ~5 kb: $J_{\lambda 2}C_{\lambda 2}$-$J_{\lambda 4}C_{\lambda 4}$ and $J_{\lambda 3}C_{\lambda 3}$-$J_{\lambda 1}C_{\lambda 1}$ (Blomberg et al., 1981; Miller et al., 1981). It seems likely that these clusters evolved by two successive rounds of gene duplication (Blomberg et al., 1981; Selsing et al., 1982). $J_{\lambda 4}C_{\lambda 4}$ is nonfunctional due to a defective splice site (Blomberg and Tonegawa, 1982; Miller et al., 1982). As in the human λ locus, but in contrast to the κ loci, there is a *J* segment 5′ to each *C*. Consequently, in a λ chain, each *C* gene segment (e.g., $C_{\lambda 1}$) is expressed with the nearby 5′ *J* (e.g., $J_{\lambda 1}$).

$V_{\lambda 1}$ and $V_{\lambda 2}$, the first *V* gene segments to be identified and sequenced (Bernard et al., 1978; Tonegawa et al., 1978), were the only two V_λ gene segments known for a number of years. Based on the V and C region combinations in the known λ chains, as well as gene rearrangement patterns, Blomberg et al. (1981) suggested that $V_{\lambda 1}$ is 5′ to the $J_{\lambda 3}C_{\lambda 3}J_{\lambda 1}C_{\lambda 1}$ cluster and $V_{\lambda 2}$ is 5′ to the $J_{\lambda 2}C_{\lambda 2}J_{\lambda 4}C_{\lambda 4}$ cluster. With the assumption that rearrangements to a nearby cluster are favored, this order would account for the most frequent types of λ chain expression: $V_{\lambda 1}J_{\lambda 1}C_{\lambda 1}$ (i.e., λ1 chains), $V_{\lambda 1}J_{\lambda 3}C_{\lambda 3}$ (λ3 chains), and $V_{\lambda 2}J_{\lambda 2}C_{\lambda 2}$ (λ2 chains). (The convention is to base nomenclature for immunoglobulin chains on the C region.) In addition to the more frequent expression with $C_{\lambda 2}$, $V_{\lambda 2}$ was occasionally found to be expressed with $C_{\lambda 3}$ or $C_{\lambda 1}$ (Elliott et al., 1982; Reilly et al., 1984). Since V to J rearrangements occur in the 5′ to 3′ direction, these findings allowed the ordering of the two clusters relative to each other: "2-4" 5′ to "3-1."

Subsequently, another V_{λ} region was identified in hybridomas from BALB/c (Sanchez et al., 1987) and C57BL/6 (Dildrop et al., 1987) mice. This V region, designated $V_{\lambda x}$, appears always to be associated with the λ2 C region (Sanchez et al., 1991) but is no more closely related in sequence to $V_{\lambda 1}$ or $V_{\lambda 2}$ than to V_{κ}. Interestingly, mouse $V_{\lambda 1}$, $V_{\lambda 2}$, and $V_{\lambda x}$ are most closely related in sequence to those human V_{λ} families that are expressed infrequently, and it has been suggested that the low level of expression of the mouse V_{λ} may be a consequence of the absence of gene segments resembling the more highly expressed human V_{λ} (Williams et al., 1996).

A significant advance was the mapping of the complete λ locus in BALB/c mice. The gene segments are all in the same transcriptional orientation and were estimated to occupy 200 to 300 kb. The gene order that had been deduced from expression patterns was confirmed and $V_{\lambda x}$ was found to lie between $V_{\lambda 2}$ and $C_{\lambda 2}$. Thus, the arrangement is $5'V_{\lambda 2}$-$V_{\lambda x}$-$C_{\lambda 2}$-$C_{\lambda 4}$-$V_{\lambda 1}$-$C_{\lambda 3}$-$C_{\lambda 1}$ (Miller et al., 1988; Storb et al., 1989; Carson and Wu, 1989). This order presumably accounts for the preferential rearrangement of $V_{\lambda 2}$ and $V_{\lambda x}$ to $C_{\lambda 2}$ and of $V_{\lambda 1}$ to $C_{\lambda 3}$ or $C_{\lambda 1}$. The locus has been assigned to chromosome 16 (D'Eustachio et al., 1981).

BALB/c and most laboratory mouse strains are derived from *Mus musculus domesticus*. Surprisingly, the λ loci in wild mice (e.g. *Mus musculus musculus*) were observed to contain substantially more *C* and *V* gene segments than had been found in the common inbred strains. Southern blot analysis of genomic DNA showed that in some wild mice as many as a dozen restriction fragments cross-hybridize to C_{λ} probes derived from BALB/c (Scott and Potter, 1984; Kindt et al., 1985). In addition, hybridomas made in wild mice were found to express light chains with V regions encoded by genes that did not cross-hybridize with any of the known V_{λ} from BALB/c; cDNAs encoding such chains were detected only because the C segments did cross-hybridize with $C_{\lambda 1}$ (Reidl et al., 1992).

Two of the wild mouse V_{λ} were found to be only ~40% identical in amino acid sequence to $V_{\lambda 1}$, $V_{\lambda 2}$, or $V_{\lambda x}$, and were closer in sequence to a number of human V_{λ}. Indeed, one of these wild mouse sequences was 64% identical to a human $V_{\lambda}6$. No genes cross-hybridizing to these wild mouse V_{λ} were present in BALB/c mice, in other common laboratory strains, or in mice derived recently by inbreeding from *Mus musculus domesticus* (Reidl et al., 1992; C. M. Kinoshita and L. A. Steiner, unpublished observations). However, cross-hybridizing V_{λ} were detected in *Mus musculus musculus*, *Mus musculus castaneus*, *Mus musculus molossimus*, *Mus cookii, Mus pahari*, and almost all other wild-derived mice examined, as well as in inbred lines derived from these subspecies (C. M. K. and L. A. S., unpublished). It seems likely that gene expansions and contractions of the λ locus are responsible for these variations in gene number and that the inbred mice that happen to have become laboratory standards are atypical in having a contracted λ locus.

Genes Encoding Human and Mouse Heavy Chains

Heavy chain immunoglobulin loci are more complex than light chain loci. This is mainly because: 1) there are more heavy- than light-chain C regions (e.g., μ, α, γ1, etc.), each encoded by a different gene segment; 2) each *C* gene segment consists of a number of exons including some that allow insertion of the heavy chain into the B cell membrane; 3) heavy chain V regions are encoded by three gene segments rather than two (*D* in addition to *V* and *J*); and 4) features of the heavy chain locus are presumably required to regulate which *C* gene segment is expressed at any time, e.g. to explain the switch, during an immune response, from expression of IgM to expression of one of the other classes having a different heavy chain (see section The Class Switch).

By methods of classical genetics, utilizing allotypic markers, close linkage of genes encoding some of the different heavy chain classes of mice (Herzenberg, 1964; Lieberman and Potter, 1966) and humans (Kunkel et al., 1969; van Loghem et al., 1970) was demonstrated. In the case of rabbit heavy chains, allotypic markers are available for V as well as C regions, and it was shown that V_H and C_H are also closely linked (Dubiski, 1969; Mandy and Todd, 1970; Mage et al., 1971; Kindt and Mandy, 1972). Genetic linkage of V_H and C_H was also shown in mice using idiotypic determinants[f] as markers for germline genes encoding certain V regions (e.g., Blomberg et al., 1972; Pawlak et al., 1973). The human heavy chain gene complex has been mapped to chromosome 14q32.3 (Cox et al., 1982; Kirsch et al., 1982) and the mouse complex to chromosome 12 (D'Eustachio et al., 1980).

The nucleotide sequences of all heavy chain *C* gene segments in humans and mice have been determined (see Honjo and Matsuda, 1995, Table III for references). Generally, each domain of the heavy chain, as well as the hinge region, is encoded by a discrete exon (e.g., Sakano et al., 1979b). The long hinge region of human γ3 chains is actually encoded by four exons (Takahashi et al., 1982). The hinge region of human and mouse α chains is encoded as a 5′ extension of the $C_{\alpha 2}$ exon; Tucker et al. (1981) have speculated that this variation results from the incorporation of an acceptor RNA splice site into the coding portion of the $C_{\alpha 2}$ exon. Additional exons encode membrane forms of the heavy chains (see section Identical Specificity of Receptor and Secreted Antibody).

The entire mouse J_H-C_H region of 200 kb was cloned by Honjo and co-workers. The gene order was shown to be: 5′-J_H-C_μ-C_δ-$C_{\gamma 3}$-$C_{\gamma 1}$-$C_{\gamma 2b}$-$C_{\gamma 2a}$-C_ε-C_α (Shimizu et al., 1982). This was consistent with earlier predictions of the order of the C_μ and C_γ gene segments based on estimates of gene loss after different rearrangements (Honjo and Kataoka, 1978). There are four functional J_H segments and one J_H pseudogene, separated by several hundred bp (Early et al., 1980a; Newell et al.,

1980; Sakano et al., 1980; Gough and Bernard, 1981); these are located ~6.5 kb 5′ to the C_μ gene segment (Shimizu et al., 1982).

A notable feature of the human C_H gene complex is the presence of two clusters of related γ, ε and α *C* gene segments, thought to have arisen by gene duplication: $C_{\gamma 3}$-$C_{\gamma 1}$-ψ C_ε-$C_{\alpha 1}$ and $C_{\gamma 2}$-$C_{\gamma 4}$-C_ε-$C_{\alpha 2}$ (Flanagan and Rabbitts, 1982). Based on data from individuals having deletions in this locus, the former cluster was proposed to be 5′ to the latter and 3′ to the single C_μ and C_δ gene segments (Lefranc et al., 1982). This order was confirmed in a subsequent study and additional distances were determined, resulting in a complete physical map of the human C_H locus, which occupies about 260 kb, from C_μ to $C_{\alpha 2}$ (Milstein et al., 1984; Bottaro et al., 1989; Hofker et al., 1989; see also Honjo and Matsuda, 1995). In addition to the nine functional heavy chain *C* genes and the single ψε *C* gene, a ψγ *C* gene is located between the duplicated clusters. There are nine human J_H gene segments, distributed within 3.2 kb. These are located about 8 kb from the human C_μ gene segment; three of the J_H are not functional (Ravetch et al., 1981).

In both the heavy chain and λ chain loci of mice and humans, there are a number of *C* as well as *J* gene segments. However, in the heavy chain loci, all the *J*'s are clustered separate from the *C*'s whereas in the λ loci, each *C* segment is preceded by a *J* segment. In the case of heavy chains, any of the J_H can, at least in principle, be expressed in association with any of the C_H, whereas in λ chains, J_λ-C_λ pairing is fixed.

As noted in the section Recombination Signal Sequences, the existence of *D* segments encoding much of the third CDR of heavy chains had been predicted; however, the identification of the corresponding gene segment in the germline was problematic because the *D* segments themselves are too short to be detected by hybridization. This difficulty was overcome by Tonegawa and co-workers who found that incomplete rearrangements involving J_H but not V_H frequently occur in plasmacytomas and T cell lines. Cloning and analysis of such incomplete rearrangements revealed short nucleotide segments 5′ to the rearranged J_H having sequence consistent with the predicted *D* segment and flanked by a conserved palindromic heptamer and a nonamer separated by a 12-bp spacer, as had been predicted for the *D* segments (Sakano et al., 1981; Kurosawa et al., 1981). The 5′ flanking sequence to the incompletely rearranged J_H provided a suitable hybridization probe for the identification of germline *D* segments.

This approach soon led to the identification of 12 D_H segments in the mouse heavy-chain gene complex (Sakano et al., 1981; Kurosawa et al., 1981; Kurosawa and Tonegawa, 1982). As predicted, these segments are flanked on both sides by RSS separated by 12 bp spacers. The D_H were classified, by sequence similarity, into three families. One of the families consists of only one member located 700 bp 5′ to the J_H cluster. The other two families contain nine and two members and are 10 to 80 kb from the *J* cluster (Wood and Tonegawa, 1983). Another D_H segment was identified more recently (Feeney and Riblet, 1993). All the D_H are located between the V_H and J_H.

Comparison of the sequence of the human J_H region with that of the mouse allowed the identification of a D_H segment between the most 5′ functional J_H and a pseudo-*J* segment at the very 5′ end of the human J_H cluster. This D_H was similar in sequence to the D_H that had been located 700 bp from the mouse J_H cluster (Ravetch et al., 1981). A family of four additional human D_H segments was identified with probes derived from an incomplete rearrangement involving J_H, as described above (Siebenslist et al., 1981). These D_H were spaced at about 9 kb and appeared to be embedded in repetitive units of DNA, suggesting that they had been generated by gene duplication. The segments of heavy-chain V regions corresponding to the *D* segment (i.e., CDR3) appeared to be much more diverse in the human than in the mouse and did not correspond to the germline *D* segments that were initially identified. Subsequently, additional *D* segments were identified (Buluwela et al., 1988; Ichihara et al., 1988a, b) and, by the late 1980s, 18 of an estimated ~30 *D* segments had been sequenced. In addition, a number of unusual "*DIR*" (*D* segments with irregular spacer lengths) were described (Ichihara et al., 1988b) and may be utilized in some rearrangements (e.g. Sanz, 1991).

The possible occurrence of D_H–D_H joining was suggested by Kurosawa et al. (1981) and Siebenlist et al. (1981). Kurosawa and Tonegawa (1982) suggested that heptamer-like elements embedded within the coding sequence of certain D_H elements might provide the required signal, while preserving the 12/23 bp rule, since these heptamers are separated from the flanking nonamer by 24 bp. Rearrangements consistent with D_H-D_H deletional, or inversional joining, were reported by Meek et al. (1989), who analyzed partially rearranged gene segments in bone marrow; evidence was obtained for fusion of two 12-spacer signals in a signal joint, indicative of direct D_H-D_H joining in violation of the 12/23 bp rule. (*D–D* joining occurs frequently in the T cell receptor loci since the *D* segments in this locus are flanked by RSS with a 23 bp spacer on one flank and a 12 bp spacer on the other flank.) However, since the sequence of a number of *D* segments remained undetermined, understanding of the utilization of these segments in *VDJ* rearrangements was necessarily incomplete.

Recently, Corbett et al. (1997) determined the complete nucleotide sequence of the human immunoglobulin *D* locus on chromosome 14. The sequence of nine additional segments was obtained, bringing the total to 27. The *D* segments have potential coding sequences of 11 to 37 bp and are flanked by RSS with 12-bp spacers. They can be grouped into six families of at least four members and one single-member family, the segment within the J_H cluster. Except for the latter, the *D*'s are located about 15 to 55 kb from the *J*'s. Comparison with nearly 900 heavy-chain sequences indicated that 25 of the 27 *D* segments are utilized and that there is extensive variability at the *V-D* and *D-J* junctions. More than one reading frame is generally used. These authors proposed that conventional *VDJ* recombination, obeying the 12/23-bp rule, without utilization of *DIR* segments, is sufficient to account for the diversity of CDR3 in human heavy chains. Thus, at present, there does not appear to be a consensus about *D* and *DIR* segment utilization in heavy chain variable regions. See Lewis (1994a) and Corbett et al. (1997) for further discussion and references.

Large-scale cosmid cloning and physical mapping of the human V_H locus were initiated by Honjo and co-workers (Kodaira et al., 1986). These studies showed that members of V_H families are frequently interspersed on the chromosome, in contrast to the situation with mouse V_H families (see below). Interspersion of members of human V_H families was also observed by Berman et al. (1988). Because of difficulties in linking cosmid clones, cloning into yeast artificial chromosomes was carried out by Shin et al. (1991) who mapped the J_H-proximal ~200 kb region and suggested a systematic nomenclature for the V_H gene segments according to family number and order from the 3′ end. The J_H-proximal map was extended to ~800 kb by Matsuda et al. (1993). Part of the J_H-proximal region, on an alternative haplotype, was analyzed by Walter et al. (1993). Subsequently, Cook et al. (1994) reported a map covering ~200 kb of the distal portion of the V_H complex. The V_H region was found to extend to within a few kb of the 14q telomere (the end of the long arm of this chromosome), indicating that no large segment was missing from the cloned region. The distance from the most 3′ V_H to J_H is about 75 kb (Schroeder et al., 1988; Corbett et al., 1997).

The properties of the human V_H locus have been summarized by Cook and Tomlinson (1985) and Matsuda and Honjo (1996). There is considerable polymorphism affecting both the overall size and the number of V_H gene segments. However, the amount of sequence polymorphism of the V_H segments is low, most allelic segments having relatively few nucleotide substitutions. The total size of the locus is 1.0 to 1.1 Mb. There are about 90 V_H, but the number varies in different individuals; somewhat over half of these are functional. Cook and Tomlinson pointed out that the V_H locus is, in fact, substantially smaller than originally estimated by pulsed-field electrophoresis. As in the light chain loci, some V segments appear to be used more frequently than others. Most—possibly all—of the V_H are in the same transcriptional orientation as the J_H. Early work by several investigators revealed that pseudogenes are abundant among human and mouse V_H gene segments; many have only minor defects such as point mutations (e.g., Bothwell et al., 1981; Givol et al., 1981; Huang et al., 1981; Kodaira et al., 1986; Blankenstein et al., 1987).

Like the other immunoglobulin gene loci, it is thought that the human V_H locus has evolved through repeated gene duplication, deletion and translocation. This has led to the many polymorphisms as well as to the interspersion of related gene segments along the locus. The same processes may have led to clusters of "orphon" V_H gene segments on chromosomes 15 and 16; some of these have no obvious defects and are similar in sequence to V_H in the authentic locus on chromosome 14. Orphon D_H gene segments are also found on chromosome 15, sometimes near V_H (Cook and Tomlinson, 1985; Matsuda and Honjo, 1996).

The physical mapping and sequencing of the mouse V_H locus is much less advanced than that of the human. Current estimates of the number of mouse V_H

gene segments are largely based on determinations of the number of restriction fragments hybridizing to different V_H probes by Southern blotting, and are therefore likely to be overestimates since pseudogenes would be included. Such approaches led initially to estimates of ~100 mouse V_H (Brodeur and Riblet, 1984); this number was expanded as several additional V_H families were found (summarized by Kofler et al., 1992). Solution hybridization studies in probe excess suggested that one family (called J558) in BALB/c mice may be considerably larger, with perhaps 500 to 1000 members (Livant et al., 1986). However, it has been argued that the probe excess hybridization analysis might have included highly diverged pseudogenes, which are unlikely to have been detected by Southern blotting (Blankenstein et al., 1987). Indeed, the number of functional genes in the J558 family appears to be only about 100 in the CB.20 strain (Gu et al., 1991), as well as in BALB/c (P. Brodeur, personal communication). Establishing the precise number of mouse V_H genes will require detailed sequence analysis of the locus. In the meantime, a reasonable estimate is approximately 200 V_H genes in total, of which about 150 are functional.

Extensive analyses of the gene organization of the mouse V_H locus by deletion mapping have indicated that members of the same family tend to be physically segregated, although there may be some interspersion (e.g., Kemp et al., 1981; Reth et al., 1986a; Rathbun et al., 1987; Blankenstein and Krawinkel, 1987; Brodeur et al., 1988; Meek et al., 1990; Kofler et al., 1992; Mainville et al., 1996). There may also be variation in number and placement of *V* gene segments among different strains (e.g., Meek et al., 1990).

A summary of information for each of the immunoglobulin loci is presented in Table 1. For the human loci, the estimates of gene number should be reasonably firm. The total number of functional human *V*, *D* and *J* gene segments is a substantial but not a vast number. However, assortment of these elements plus diversity at the junctions, as described above, as well as different combinations of heavy and light chains, can generate an enormous collection of antibody combining sites. These constitute the repertoire before introduction of antigen. However, use of particular gene segments or certain combinations may be favored and others occur rarely, thereby introducing some restriction. The repertoire is subject to substantial additional diversification after antigen challenge, as will be discussed in the section Tuning the Antibody Response.

REGULATION OF GENE REARRANGEMENT

As described in the previous section, investigations of the arrangement of the gene segments within the immunoglobulin loci have been ongoing for the last 20 years. Concurrently, the rearrangement and subsequent expression of these gene segments have been explored. Certain guiding principles that control immunoglobulin gene utilization have emerged, and these will now be considered.

Allelic and Isotypic Exclusion

A central tenet of the clonal selection theory is that an antibody-producing cell or clone expresses only a single specificity (see Part I). In molecular terms, this restriction can be achieved by allowing synthesis of only one heavy chain V region and one light chain V region, each such V region resulting from a single *V(D)J* rearrangement. Genes encoding heavy, κ and λ chains are found on three distinct chromosomes. In each of the three loci there are many *V* gene segments, a number of *J* gene segments and, in the case of the heavy chain locus, a number of *D* segments. If productive *V(D)J* rearrangements were to occur without any restriction, each B cell might produce several, perhaps many, different light and heavy chain V regions, and therefore antibodies having a number of different specificities.

A variety of evidence was presented in Part I (see section, Clonal Selection Prevails) in support of the supposition that a single B cell does indeed produce antibody of only a single specificity. This restriction implies that a productive rearrangement occurs at only one of the two alleles of an immunoglobulin heavy or light chain locus and at one allele of either the κ or λ light chain locus. The restriction in expression to one of a pair of alleles is known as allelic (or haplotype) exclusion. The restriction in expression to a locus bearing either κ or λ genes is known as isotypic exclusion.

Allelic exclusion, as manifest in the expression of immunoglobulin (and T cell receptor) genes, is an exception to the general finding that in a heterozygote, both alleles of autosomal genes are expressed. Thus, erythrocytes in individuals heterozygous for the normal and sickle allele of the hemoglobin β chain contain both normal and sickle hemoglobin. However, there are other exceptions, such as genomic imprinting in mammals (see Bartolomei and Tilghman, 1997 and Jaenisch, 1997 for review), expression of odorant receptors on neurons of the olfactory system (Chess et al., 1994), and expression of MHC-specific receptors on certain natural killer (NK) cells (Held et al., 1995). In some cases, as discussed in the section, One V Region, Many C Regions, a B cell may synthesize antibodies having heavy chains of two different classes (e.g., IgM and IgD or IgM and IgG), but these always have the same V region.

Early evidence pointing to a restriction in immunoglobulin expression was the demonstration that, in individuals who are heterozygous for allotype, myeloma proteins carry one or the other, but never both, of the allelic determinants found in the pooled immunoglobulin of that person (Mårtensson, 1961; Harboe et al., 1962). This finding was interpreted to indicate that the alternate allotypic determinants are present on different immunoglobulin molecules, which are produced by different cells. Oudin (1960) had observed a presumably related phenomenon in rabbits.

In 1963, shortly after the four-chain model for immunoglobulins was proposed, Dray and Nisonoff showed that immunoglobulin molecules from rabbits heterozygous for allotypic determinants at a light chain locus contain either light chains of one allotype, or of the other, but not both. These authors did not interpret the results in terms of a restriction in cellular synthesis, but rather they proposed that within each molecule there is preferential association of pairs of the same chain. Soon thereafter it was shown that B cells from mice or rabbits heterozygous for allotypic determinants form two distinct populations, one containing antibody of one allotype and the other antibody of the alternative allotype (Pernis et al., 1965; Weiler, 1965; Cebra et al., 1966), establishing the fact of allelic exclusion. It was also shown that a B cell produces either a κ or a λ chain, but not both, i.e. isotypic exclusion (Bernier and Cebra, 1964).

What is the molecular basis for allelic and isotypic exclusion? When somatic rearrangements of gene segments encoding immunoglobulin κ chains were first described, it was noted that rearrangement appeared to be restricted to one of a pair of chromosomes, the κ genes on the second chromosome retaining their germline configuration; it was proposed that this might provide the explanation for allelic exclusion (Hozumi and Tonegawa, 1976; Seidman and Leder, 1978). It soon became evident, however, that rearrangement at more than a single immunoglobulin allele occurs frequently (Lenhard-Schuller et al., 1978; Steinmetz et al., 1979; Wilson et al., 1979; Alt et al., 1980a; Cory et al., 1980; Perry et al., 1980; Seidman and Leder, 1980); consequently such a restriction was unlikely to be the basis for allelic exclusion.

Plasmacytoma cells were utilized in the initial experiments on gene rearrangements and there was some uncertainty that the results would necessarily be representative of rearrangements in normal B lymphocytes. Examination of mouse splenic B cells expressing κ light chains again suggested that the C_κ gene had rearranged on only one chromosome (Joho and Weissman, 1980), but another study indicated that about one-third of B cells contain two rearrangements at the κ locus, one of which was presumed to be nonproductive with respect to κ synthesis (Coleclough et al., 1981). It was noted that λ1 genes are infrequently rearranged in κ-producing B cell lines, whereas both κ alleles are often rearranged, non-productively, in λ-producing cell lines (Coleclough et al., 1981; Hieter et al., 1981b). In some λ-producers, both C_κ gene segments were found to be deleted by specific recombination, an event that was proposed to activate V_λ to J_λ rearrangement (Durdik et al., 1984). It was also observed that aberrant heavy chain gene rearrangements are particularly frequent, presumably because of the complexity of these rearrangements, which involve three rather than two gene segments (Coleclough et al., 1981).

These observations about immunoglobulin gene rearrangements formed the basis for additional hypotheses to explain allelic exclusion. In the "stochastic" model, it was proposed that the rearrangement process operates independently at each immunoglobulin locus and at each allele (Adams, 1980; Perry et al., 1980; Walfield et al., 1980; Coleclough et al., 1981; Langman and Cohn, 1987). Whether or not a particular allele is rearranged productively depends on the rearrangement frequency as well as the probability of productive versus nonproductive rearrangement. Since mice have many more V_κ than V_λ gene segments, the observation that rearrangement at the κ locus generally precedes that at the λ locus could be attributed to a statistical effect. The frequent presence of an unrearranged κ locus in κ producers, together with the rarity of an unrearranged κ locus in λ producers, could be explained by assuming that the rate of gene recombination drops substantially upon formation of a productive light chain gene. A higher intrinsic rearrangement frequency at the heavy chain locus could explain why heavy chain genes generally rearrange before light chain genes.

A consequence of a strict stochastic model is that not all B lymphocytes will exhibit allelic exclusion. If the rate of productive rearrangement is sufficiently high, there will be occasional "double producers" (see related discussion in Part I, Instruction and Selection). On the other hand, if the rate of productive rearrangement is low, there will be few "doubles," but many cells will have only non-functional rearrangements.

Alternatively, the rearrangement order might be strictly determined rather than relying on probabilistic events. It had been shown that B cell precursors (i.e., pre-B cells) synthesize μ chains but not light chains (Burrows et al., 1979; Levitt and Cooper, 1980; Siden et al., 1981), thereby setting a precedent for the idea that rearrangement might follow a fixed order. A model, proposed by Alt et al. (1980a), focussed on the events that turn off rearrangement. It was suggested that when a functional light chain is produced, it combines with μ chain in the pre-B cell to form an intact IgM molecule, which is inserted into the membrane, marking the transition from a pre-B to a B cell. Once IgM has appeared on the cell surface, further light chain rearrangement ceases. If the rearrangement is not productive, additional light chain genes rearrange until a productive rearrangement is achieved. If the cell exhausts the possibilities without producing a functional light chain, it dies at the pre-B cell stage. The restriction that in any single cell only one light chain allele, whether κ or λ, can rearrange productively leads to isotypic as well as allelic exclusion. It was considered either that some unspecified mechanism dictates the order of light chain gene rearrangement (κ, then λ) or that this order is a consequence of the greater number of κ than of λ genes.

Only synthesis of functional protein, not gene rearrangement and not mRNA production, can provide the negative feedback signal. Occasional B cells have been found to produce two light chains, or a light chain and a light chain fragment resulting from a different gene rearrangement, but only one of these combines with heavy chain to form an IgM molecule[g] (e.g., Rose et al., 1977; Alt et al., 1980a; Bernard et al., 1981; Kwan et al., 1981). Indeed, as discussed below, to achieve allelic exclusion, it is thought that the light chain must combine with a membrane form of the heavy chain produced by the same cell.

A similar proposal was subsequently advanced for heavy chain rearrangements (Alt et al., 1981): synthesis of a functional heavy chain in some way signals the cell to cease further *V(D)J* rearrangement. An alternate explanation for the presence in a B cell of no more than one heavy chain, proposed by Wabl and Steinberg (1982), was based on the supposition that free heavy chains are intrinsically toxic to the cell. As heavy chains are ordinarily produced before light chains, these authors suggested that their potential toxicity is averted by the binding of a special "heavy chain binding protein," which is subsequently displaced by light chain. If heavy chain is produced from both alleles there is insufficient binding protein and toxic heavy chain accumulates. However, the recent finding that mice can express two different heavy chains from distinct transgenic V_HDJ_H constructs inserted into each heavy chain locus has cast some doubt on the toxicity model (Sonoda et al., 1997).

A number of studies were directed at determining the order of rearrangement of the *V*, *D* and *J* gene segments within the heavy-chain locus. It had been observed that most B cells have rearrangements of J_H genes on both chromosomes (Cory et al., 1980; Alt et al., 1981; Coleclough et al., 1981; Nottenburg and Weissman, 1981; Sakano et al., 1981). Studies of Alt et al. (1984) showed that most immature B cell lines derived from fetal liver cells have two DJ_H but no V_HD rearrangements; in cell lines representing more mature B cells one or two V_HDJ_H rearrangements were seen. Based on these results it was proposed that the initial gene rearrangements are *D* to J_H on both chromosomes, followed by rearrangement of one V_H. If, and only if, the first *VDJ* rearrangement is not productive, a second V_H rearranges. The synthesis of μ chain marks the transition from a pro-B cell (*D* to J_H joined) to a pre-B cell (productive V_H to DJ_H rearrangement).

Additional investigation showed that it is the membrane, not the secreted, form of the μ chain that provides a signal for inhibiting further heavy-chain gene rearrangement (Nussenzweig et al., 1987; Reth et al., 1987; Manz et al., 1988; Kitamura and Rajewsky, 1992). Evidence that a productive V_HDJ_H rearrangement signals rearrangement to begin at the κ locus was provided by Reth et al. (1985). In this case also, the membrane form of the μ chain appears to provide the necessary signal (Reth et al., 1987). Similar results were reported by Iglesias et al. (1991) and Shapiro et al. (1993). Thus, both positive as well as negative signals are provided by the membrane μ chain.

In the last ten years it has become evident that the membrane μ chain is complexed with a moiety called "surrogate light chain," as well as with two additional polypeptides, Ig-α and Ig-β. The surrogate light chain consists of yet two more polypeptides, the products of genes, V_{preB} and λ5; no rearrangement is involved in the expression of these genes and their products associate with each other noncovalently. This complex (membrane μ plus Ig-α/Ig-β plus surrogate light chain) forms a structure, expressed on the surface of pre-B cells, called the pre-B cell receptor. The importance of membrane μ chain in allelic exclusion suggested that the critical signaling molecule might, in fact, be the pre-B cell receptor complex. In support of this hypothesis, Löffert et al. (1996) demonstrated that targeted disruption of the λ5 gene abrogates allelic exclusion.

The involvement of the pre-B cell receptor in allelic exclusion at the heavy chain locus is reminiscent of the original suggestion of Alt et al. (1980a) that surface IgM on a B lymphocyte is the molecule that transmits a signal inhibiting further light chain gene rearrangement. Surface IgM is part of the complex known as the B cell receptor, which is identical to the pre-B cell receptor except that either κ or λ light chain replaces the surrogate light chain. See Reth et al. (1991) and Reth (1992) for review of B cell and pre-B cell receptors and Melchers et al. (1993) for review of surrogate light chain. Two recent reviews discuss the structure and signaling functions of the pre-B cell receptor and the corresponding receptor on precursor T lymphocytes, the pre-T cell receptor (Borst et al., 1996; Owen and Venkitaraman, 1996).

It has been difficult to obtain direct evidence for an analogous role for membrane μ chain and the B cell receptor in regulating light chain rearrangement in B cells; if membrane μ is not expressed, B cell development is arrested at an early stage, before light chain rearrangement commences. However, Ma et al. (1992) showed that in a transformed B cell line, cross-linking of surface immunoglobulin with antibodies to μ chain downregulated the expression of *RAG*-1 and *RAG*-2, implying a link between a signal transduced by the B cell receptor and immunoglobulin gene rearrangement in these cells and perhaps also in normal B cells. Similarly, Turka et al. (1991) showed that cross-linking of the T cell receptor on thymocytes by antibodies led to termination of the expression of both *RAG* genes.

In the years since models to explain allelic exclusion were introduced, numerous experiments have been devised to test their predictions. A powerful approach has been to determine whether the introduction of transgenes expressing either rearranged heavy- or light-chain genes will inhibit rearrangement of endogenous genes. Indeed, the first immunoglobulin transgenic mice were produced for this purpose (Brinster et al., 1983; Ritchie et al., 1984). Results of such experiments, while not entirely definitive, have generally supported the idea that there is an ordered progression of immunoglobulin gene rearrangement (see review by Storb, 1995). However, exceptions to the supposition that heavy chain rearrangement must precede light chain rearrangement have been noted (e.g., Blackwell et al., 1989; Schlissel and Baltimore, 1989; Kubagawa et al., 1989; Kitamura and Rajewsky, 1992; Ehlich et al., 1993; Chen et al., 1993a), as have exceptions to the proposition that κ rearrangement is a necessary prelude to λ rearrangement (e.g., Chen et al., 1993b; Takeda et al., 1993; Zou et al., 1993). The mechanism of allelic and isotypic exclusion has continued to stimulate experimentation and discussion (see Nature, Oct. 1, 1992, Vol. 359, pp. 370–372, for correspondence by Wabl and Steinberg, Oancea and Shulman, and Rajewsky et al. in response to Kitamura and Rajewsky (1992)).

Bias in Rearrangement

It was noted in the section Organization of Immunoglobulin Loci that utilization of the gene segments encoding heavy- and light-chain V regions is uneven, certain segments being used frequently and other rarely or not at all. A number of factors are likely to be important in determining the observed frequency of rearrangement of these gene segments. These include chromosomal location, for example whether a particular *V* gene segment is at the *J*-proximal or *J*-distal end of the *V* cluster; amino acid sequence encoded by the gene segment, which determines whether the V region will be able to participate in a productive and selectable binding site; and other features of the particular gene segment, such as the nature of its promoter and RSS (e.g., Hesse et al., 1989).

Utilization of *V* gene segments in the heavy chain locus of pre-B and B cells of mice of diverse ages has been explored in considerable detail.[h] Cells from fetal liver were used in several studies because it was supposed that the effects of antigen selection on such cells would be minimal. Yancopoulos et al. (1984) observed that in transformed pre-B cell lines derived from fetal liver, those V_H gene segments physically closest to the J_H region (the "J_H-proximal *V* segments") were preferentially rearranged. A similar bias was noted in a number of other studies of pre-B cells in fetal and neonatal mice (e.g., Perlmutter et al., 1985; 1986a; Lawler et al., 1987; Yancopoulos et al., 1988; Gu et al., 1991). In B lineage cells derived from adult bone marrow, J_H preferential utilization was also seen in most (Yancopoulos et al., 1984, Reth et al., 1986a; Freitas et al., 1990; Malynn et al., 1990) but not all (Jeong and Teale, 1988) studies. Differences among reported results may reflect differences in cell populations.

In BALB/c mice, members of a V_H family tend to lie near each other on the chromosome (see section, Genes Encoding Human and Mouse Heavy Chains). Therefore, rearrangement preference for *V* gene segments that are in physical proximity would tend to generate heavy chains with related *V* regions. Preferential rearrangement has been invoked to explain a number of early observations that the capability of animals to respond to different antigenic stimuli is not random but often follows a characteristic program during ontogeny (e.g., Sterzl and Silverstein, 1967; Klinman and Press, 1975).

In contrast to the J_H-proximal utilization characteristic of pre-B cells, the utilization of V_H families in peripheral B cells was generally found to correspond not to the location, but to the size of the family, as would be expected if utilization of individual segments were random (Dildrop et al., 1985; Wu and Paige, 1986; Shulze and Kelsoe, 1987; Yancopoulos et al., 1988). Gu et al. (1991) observed strong bias in utilization of *V* gene segments within one large V_H family. Although many of the V_H segments in the family were found to be rearranged in pre-B cells, only a limited number were rearranged in splenic B cells. The results were interpreted as consistent with ligand selection of individual family members, either by foreign antigens or, possibly, through recognition by internal ligands such as anti-idiotypic antibodies. Selection presumably occurs as the newly generated B cells enter the peripheral B cell pool. It has also been proposed that selection may occur at the pre-B cells stage (Decker et al., 1991; Huetz et al., 1993; Kraj et al., 1997). A pre-B cell might be selected for expression of a heavy chain that can associate at the cell surface with the surrogate light chain (see previous section).

The recent identification of virtually all the *V*, *D* and *J* gene segments in the human immunoglobulin loci has allowed systematic comparison of germline and expressed genes. In the human fetal repertoire, preferential usage of J_H-proximal *V* segments was observed initially (e.g., Schroeder et al., 1988; Cuisinier et al., 1989), but it was later realized that there is no strict relation between chromosomal location and utilization of V_H segments (e.g, Schroeder and Wang, 1990; Willems van Dijk et al., 1992). As was mentioned in the section Organization of Immunoglobulin Loci, the usage of gene segments (*J* and *D* as well as *V*) in the adult repertoire is not evenly distributed; relatively few germline segments are expressed frequently and others are used rarely, if at all (Cox et al., 1994; Brezinschek et al., 1997; Foster et al., 1997; Corbett et al., 1997; Ignatovich et al., 1997).

A striking feature of the fetal repertoire in both mice and humans is that it consists to a large extent of antibodies that are polyreactive, including reactivity against self. Many of these antibodies are produced by a distinct set of B lymphocytes, known as B1 cells. First identified by virtue of a cell surface antigen, Ly-1, that had previously been detected only on T lymphocytes (Hayakawa et al., 1983), these special B cells have a number of distinctive properties (reviewed by Casali and Notkins, 1989; Wortis, 1992; Haughton et al., 1993; Hardy and Hayakawa, 1994). They constitute a large fraction of the B cell population in the fetus and newborn, but are also present in adults and are frequently committed to the production of antibodies that bind to a variety of antigens, including self-antigens, with relatively low affinity.

Why do so many of the antibodies produced early in life display weak reactivity with self-antigens? It has been proposed that the germline genes encoding these V regions have been selected in evolution because their polyreactivity provides a useful baseline defense against a variety of pathogens. A low degree of self-reactivity may assure that these cells will be selected from the germline pool to be available when needed. A requirement for reactivity against self antigens may also be a conservative force in maintaining the germline pool of *V* genes (Stollar, 1991). A possible role for "network" interactions (see part I, Anti-Antibodies), whereby germline-encoded antibodies directed against epitopes on other antibodies regulate the development of the B cell repertoire, has also been considered (Kearney and Vakil, 1986).

Preferential V_H gene usage has been invoked to explain a phenomenon that puzzled immunologists for many years: the allelic behavior of rabbit *a*-locus allotypic determinants. As discussed in Part I (Two Genes, One Polypeptide and Many Germline Genes or Few?), these determinants, which are associated with the heavy chains of all the rabbit immunoglobulin classes, behave like products of classical Mendelian alleles at a single genetic locus. However, the C regions of the different heavy chains are the products of non-allelic gene segments. Therefore, it was reasoned that the allotypic determinants must be associated with the V regions. This conclusion led to an apparent paradox, for the V regions were believed to be the products of many V_H gene segments, which are not expected to segregate as if they are alleles of a single gene. Indeed the rabbit was shown to have at least 100 V_H gene segments (Gallarda et al., 1985; Currier et al., 1988). Unless some special mechanism is invoked to prevent meiotic recombination among these genes, one would expect multiple crossovers to scramble the genes encoding different allotypic determinants, with loss of simple Mendelian behavior on breeding.

The conundrum was resolved when Knight and Becker (1990) showed that one of the rabbit V_H genes, that nearest the J_H locus, is preferentially rearranged in leukemic rabbit B cells. The most J_H-proximal V_H gene is also preferentially rearranged in normal adult (Becker and Knight, 1990; Raman et al., 1994) and neonatal (Friedman et al., 1994) rabbit B lymphocytes. Thus, although there are indeed many *V* gene segments in the rabbit heavy chain locus, the preferential rearrangement of one of these means that, in effect, the inheritance pattern conforms to that of a single gene.

The bias in rearrangement and expression of a single rabbit V_H gene segment neatly explained the inheritance pattern of the V_H allotypes, but it raised two new questions: 1) How are rabbit V_H regions diversified? and 2) If only one V_H is rearranged in most B cells of rabbits, why have all the other V_H gene segments been preserved? Answers to both questions were provided by Becker and Knight (1990) who presented data indicating that the predominantly rearranged V_H gene is diversified by gene conversion from more *J*-distal V_H gene segments. Presumably, the allotypic determinants are located on the relatively invariant framework regions, which are shared by most of the diversified V regions. Gene conversion as the principal means of generating antibody diversity had previously been demonstrated only in chickens and related species (reviewed by McCormack et al., 1991 and Reynaud et al., 1994).

TUNING THE ANTIBODY RESPONSE

The repertoire that exists before encounter with antigen is generated by gene rearrangement and diversification at the junctions of gene segments. This preimmune repertoire may be modified by somatic mutation of the rearranged genes or by the emergence of antibodies resulting from new gene rearrangements. These changes frequently lead to improvement in the affinity of the antibody-antigen interaction. Recently, the possibility that the repertoire may also be modified by the further rearrangement of already-rearranged gene segments has also been proposed.

Somatic Hypermutation: Hypothesis and Evidence

A key feature of the clonal selection theory is that each antibody-forming cell or clone is committed to the expression of one antibody. The diversity in the response is a consequence of the great variety of distinct clones. Initially, Burnet (1957) assumed that antibody diversification occurs exclusively early in life, by a random and unknown process, so that all possible antibody specificities are present in the animal before initial encounter with antigen. In this formulation, antigen serves as a selective agent, stimulating the amplification of cells that produce complementary antibody. The selective theories proposed by Jerne (1955) and Talmage (1957) also assign to antigen a similar role.

Whereas Burnet proposed that antibody "randomization" occurs only in perinatal life and is unrelated to encounter with antigen, Lederberg (1959) put forward the idea that the "diversity of the precursors of antibody-forming cells arises from a high rate of spontaneous mutation during their lifelong proliferation." This suggestion implied that antigen also plays an indirect role in the diversification process since mutations occur as cells proliferate, a process that is driven by antigen reacting with cellular receptors. The possibility that diversification might be an ongoing process was incorporated by Burnet (1964) into clonal selection in what he referred to as the "Darwinian" view that lymphoid cells in the body constitute a "microcosm continually subject to evolutionary change . . . from early embryonic life until death."

The possible contribution of somatic mutation to antibody diversification has been the subject of speculation and experimentation from the time of Lederberg's proposal until the present. Indeed, before it was understood that the rearrangement of genetic elements is a major factor in generating V-region diversity, somatic mutation was the most obvious candidate for generating the required variability, unless one adopted the view that all V-region variation is encoded in the germline (see Part I, Many Germline Genes or Few?).

In 1966, Brenner and Milstein proposed a mechanism for somatic mutation that included a means for restricting mutations to a specific region of a gene. This was an attractive feature of the proposal as it had recently been shown that antibody polypeptide chains are composed of variable (V) and constant (C) regions. They suggested that DNA encoding the invariant (C) region contains a recognition site for a DNA-cleaving enzyme, which cuts one of the DNA strands within the *V* segment, exposing a free 3′ end. The broken strand is degraded by an exonuclease and the gap is repaired by a polymerase that introduces errors such as base-pairing mistakes, deletions or insertions. The end of the region of variation is specified by the point of cleavage. It was further postulated that the error-prone repair process is expressed only at specific stages in the differentiation of antibody precursor cells. As has been pointed out by Bachl and Wabl (1996), this model bears much resemblance to the process of N region addition (see section, Junctional Variability). Brenner and Milstein commented that identification of the responsible enzymes would "not be easy to do." In the years since, the process of somatic mutation has been described in considerable detail, but the enzymatic mechanism has remained elusive.

The first persuasive, albeit indirect, evidence that antibodies are diversified by somatic mutation came from the analysis of mouse λ light chains by Weigert et al. (1970). In the initial experiments, V regions of ten λ chains, obtained from mouse plasmacytomas, were examined by peptide mapping and partial sequence analysis.

Six of the V regions examined seemed to be identical, two differed from these in a distinct single amino acid position (compatible with one base change), one differed by two residues (two base changes) and another by three residues (four base changes). All the amino acid replacements were in the CDRs. The authors suggested that the six identical sequences are encoded by a single germline gene and that the four variant sequences result from spontaneous somatic mutation and sequential selection by antigen of single-step mutants. Additional λ chains examined subsequently fell into the same pattern, strengthening this conclusion (Weigert and Riblet, 1976). Following the discovery of a second, substantially different BALB/c λ chain (Eisen et al., 1968; Schulenburg et al., 1971), the group of λ chains studied by Weigert and collaborators was designated λ1, and the second type of mouse λ chain was designated λ2.

The prediction of a single germline gene ($V_{\lambda 1}$) that encodes all the V regions of λ1 chains was ultimately confirmed by molecular cloning and DNA sequencing, which yielded results consistent with the data obtained by protein analysis (Brack et al., 1978; Bernard et al., 1978). The V region of the second type of BALB/c λ chain was also shown to be encoded by a single germline gene ($V_{\lambda 2}$) that also can undergo somatic mutation (Dugan et al., 1973; Brack et al., 1978; Tonegawa et al., 1978; Wu et al., 1982; Elliott et al., 1984).

The mouse BALB/c λ system is particularly favorable for identifying somatic mutants because of the very small number of germline genes. It was straightforward to compare the sequences of V regions of expressed λ chains with those of germline $V_{\lambda 1}$ and $V_{\lambda 2}$. (A third BALB/c V_{λ} gene, designated $V_{\lambda x}$, which was discovered later, appears to be expressed much less frequently than are $V_{\lambda 1}$ or $V_{\lambda 2}$ (Dildrop et al., 1987; Sanchez et al., 1987).)

The existence of somatic mutation in mouse (or human) V_{κ} and V_{H} was more difficult to evaluate because there are many germline genes in each of these loci. Nevertheless, evidence was obtained to suggest that somatic mutation also diversifies germline genes encoding mouse κ chains (Gershenfeld et al., 1981; Pech et al., 1981; Selsing and Storb, 1981b; Heinrich et al., 1984). This conclusion was based on comparison of expressed sequences to those germline genes that seemed, by criteria of cross-hybridization and sequence analysis, to be most closely related. As noted in the section Counting Genes, most of the evidence obtained by hybridization kinetic analyses also indicated that there are probably fewer V_{κ} genes in the germline than related expressed sequences, implying that some somatic mechanism operates to diversify these genes.

Before methods of molecular cloning permitted the isolation and direct analysis of germline and rearranged antibody genes, it was possible to draw some inferences about these genes from idiotypic analysis. In occasional immune responses, a substantial proportion of the antibody molecules produced by different individuals share idiotypic determinants, implying that the combining regions are very similar. Such a shared or recurrent idiotype is called a "public idiotype," in contrast to determinants that are specific to one antibody and one individual, which are called "private idiotypes." Responses in which the antibodies share a public idiotype are usually found in related individuals, such as the members of an inbred population, and it was suggested that the *V* regions of such antibodies may be encoded by one or a few germline genes (Eichmann and Kindt, 1971; Blomberg et al., 1972; Kuettner et al., 1972; Sher and Cohn, 1972).

Evidence was subsequently obtained by sequence analysis that certain sets of antibodies expressing a public idiotype contain V_H regions that are derived, presumably by somatic mutation, from a single germline *V* gene (Bothwell et al., 1981; Crews et al., 1981; Gearhart et al., 1981). For example, Siekevitz et al. (1983) demonstrated that all hybridoma cell lines from A/J mice that produce antibody specific for the azophenylarsonate determinant and that express a particular public idiotype contain the same germline V_H gene rearranged to the same J_H segment. Related V_H sequences expressing the same idiotypic determinant have presumably diverged by somatic hypermutation. Although in many cases, idiotypic determinants appear to reflect the products of *V* genes, the other gene segments that encode the V region, *D* and *J*, or the junctions of these segments, may also contribute to idiotype (e.g., Berek, 1984).

In the human, the recent identification of all or nearly all the germline *V* gene segments in the three immunoglobulin loci (see sections in Organization of Immunoglobulin Loci) has made it possible to evaluate the contribution of somatic mutation (and antigen selection) toward diversifying the germline repertoire. Except for the λ locus (see above), this is not yet possible in the mouse. Tomlinson et al. (1996) and Ignatovich et al. (1997) compared ~2500 expressed V_H, V_κ and V_λ sequences to the known germline sequences. Somatic mutation was found especially to alter certain CDR residues that are conserved in the germline sequences and has the effect of spreading the diversity away from the center of the binding site. (The greatest diversity in the *V* region is actually not in the part encoded by the *V* gene segment but in CDR3 of V_H and the end of CDR3 of V_κ, due to *D* gene segment usage and variability at *V(D)J* junctions. In the three dimensional structure, these highly diverse regions form the center of the antigen binding site.)

Somatic Hypermutation in Clonally Related B Cell Products: Changes in Antibody Affinity

One approach to obtain information about the progression of somatic mutation in one or a few germline genes is to examine V regions expressed by hybridomas obtained after immunization of a single mouse. In one such study, the sequences of a number of V_κ regions belonging to the same V_κ subgroup were examined and shown to be consistent with stepwise descent, by mutation, from a single V_κ germline gene (McKean et al., 1984). Genealogical relationships among hybridoma products were also demonstrated in other studies (e.g., Rudikoff et al., 1984; Sablitzky et al., 1985).

It was recognized that antibody diversification cannot rely solely on antigen selection, but requires a specially efficient mutational mechanism. A somatic mutation rate for *V* genes of approximately 10^{-3} per base pair per generation was estimated (McKean et al., 1984; Sablitzky et al., 1985), many orders of magnitude greater than other mutation rates, hence the term "hypermutation." Such a rate, which will lead to approximately one mutation in the *V* gene at each cell division, is necessary so that cells expressing mutated antibodies will become a major fraction of the expanding clone.

As described in Part I, increases in affinity or "avidity" for antigen are characteristic features of the immune response. In antibody responses to well-defined haptenic determinants, the affinity of the antibodies for the hapten had been shown to increase during the period after immunization; this was interpreted in terms of clonal selection, cells with antibody-receptors of higher affinity successfully competing for antigen against cells with receptors of lower affinity.

The analysis of clonal B cell products was now extended to determine the relation of somatic hypermutation to changes in antibody affinity. In one study, hybridomas were obtained at various times during the course of an immune response to a haptenic determinant (Griffiths et al., 1984; Berek et al., 1985). Early in the response (one week), many of the hybridoma *V* gene segments were identical in sequence to a particular pair of germline V_H and V_κ gene segments. After another week, all hybridomas expressing these same basic sequences showed nucleotide changes leading to amino acid replacements; most of the changes in V_κ were clustered around residue 34 in CDR1. The modified antibodies generally displayed increased affinity for hapten. Sequence changes were found in IgM and IgG antibodies expressed relatively late in the response, but not in antibodies of either class expressed early in the response. These results indicated that mutations can occur before, as well as after, class switching. In the secondary response, only a minority of the hybridomas expressed the original germline gene segments, and these were even more extensively mutated. Heavy and light chains in the secondary response were largely encoded by new combinations of germline gene segments. The resulting antibodies generally displayed high affinity for hapten.

The finding that V regions of IgM antibodies may be altered by somatic mutation was at variance with earlier impressions. Karush (1978), noting that IgM, in contrast to IgG, antibodies are generally low in affinity for antigen, had proposed that the V regions of both heavy and light chains of IgM antibodies are encoded by unmutated germline genes, whereas IgG antibodies include somatic variants of these genes. Consistent with this hypothesis, Karush and co-workers observed that V_H segments from purified IgM antibodies from different individuals were identical in their isoelectric focussing patterns, suggesting that they were similar, perhaps even identical in sequence (Rodwell and Karush, 1980). In contrast IgG antibodies displayed gross differences (Ghose and Karush, 1973). Sequence analysis of heavy and light chain V regions of monoclonal IgM antibodies appeared to support the idea that IgM antibodies are products of unmutated germline genes (Bothwell et al., 1981; Gearhardt et al., 1981). However, in most immune responses, IgM antibodies are produced before IgG antibodies (see section, The Class Switch) and the frequency of somatic mutation is now thought to be unrelated to any intrinsic difference in immunoglobulin class, but to be a function mainly of time after immunization.

A combination of "mutational drift"[i] and "repertoire shift" was invoked to explain the ongoing changes in antibody sequence and affinity (Berek et al., 1985; Berek and Milstein, 1987). The early primary response is dominated by the expression of genes that rearrange at a reasonably high frequency and that also have good affinity for hapten. In some cases a single light chain *VJ* rearrangement and a single heavy chain *VDJ* rearrangement satisfy these criteria and encode most antibodies formed during the early part of the response. Next, hypermutation of the germline genes leads to a drift toward higher affinity. In the later stages of the primary or in the secondary response, repertoire shift may occur: other gene combinations, which were initially at lower frequency, are eventually selected and begin to appear. Some of these may be somatic mutants of different genes, which happen to display favorable affinities. Thus, an initial adequate response is improved up to a limit imposed by the nature of the initially rearranged gene segments and is then gradually superseded by a more optimal response.

Other scenarios are also possible. There may be no initial single favorable heavy or light chain rearrangement or the first rearrangement may prove to be so good that it is retained throughout the response. Thus, Wysocki et al. (1986) found that antibodies with a particular germline encoded V_H were a minority in an early primary response, but mutated versions of this germline V_H encoded high-affinity antibodies that dominated the secondary response.

Limited substitutions due to somatic mutation can have major effects on antibody affinity, as was demonstrated by sequence analysis of antibodies generated during a response, in some cases supplemented by site-specific mutagenesis followed by expression of the modified antibody. Rajewsky and colleagues had observed that a tryptophan residue in CDR1encoded by a particular germline V_H gene segment is frequently replaced by leucine in secondary response antibodies to a haptenic determinant (Allen et al., 1987). It was then shown that this change alone was sufficient to cause a 10-fold increase in affinity (Allen et al., 1988). The role of specific amino acid residues in modulating affinity of anti-hapten antibodies was also explored in a number of other studies (e.g., Sharon, 1990; Denzin et al., 1993).

In a different approach, Azuma et al. (1984) recombined the same heavy chain with two different light chains. The reconstituted immunoglobulins, which were identical except for a single Phe/Tyr replacement at the $V_\lambda J_\lambda$ junction, differed in affinity for a haptenic determinant by a factor of about 100. This sequence difference is probably a result of variation in the position of *VJ* recombination and not of somatic mutation. Chain recombination was also used by Berek and Milstein (1987) to show that a single residue in CDR1 of a V_κ region can have a substantial effect on affinity.

That minor changes in sequence can have major effects on binding was already known from earlier studies carried out with polyclonal antibodies that bind particular haptens. In an extensive series of studies in the late 1960s and 1970s, the laboratory of David Pressman and colleagues showed that chemical modification of residues in the active site could abolish hapten binding activity or greatly reduce the affinity (e.g., Roholt et al., 1973; Grossberg et al., 1976).

Somatic mutation plus selection frequently leads to improvement in affinity. However, some antibody responses have not shown such an increase, possibly because antibodies early in the response were already effective in binding antigen (Newman et al., 1992; Roost et al., 1995; see also commentary by Foote and Eisen, 1995). Perhaps, genes specifying such antibodies happened to be present in the germline and any mutations might have been deleterious. Hengartner, Zinkernagel and co-workers have argued that to achieve adequate protection against potentially lethal infection with cytopathic viruses, neutralizing antibodies that are highly effective in binding virus must be produced early in the response, without the delay that would be required if progressive somatic mutation were needed to generate efficient binding capacity (Roost et al., 1995; Kalinke et al., 1996). One possible explanation for the rapid production of such antibodies is that the mice had already been exposed to the same or a related antigen, so that what was considered to be a primary response was actually a secondary response, in which case high affinity antibodies would be expected. The mice used in these experiments had been kept under pathogen-free conditions, which would mitigate against, but not rule out, previous exposure to a related antigen.

Non-binding or otherwise nonfunctional variants may arise within a clone during the late primary (Manser et al., 1987) or secondary (Rada et al., 1991) response. Presumably, such variants represent mutants that have not yet been eliminated by selection. Direct evidence for the loss of hapten-binding in a cell line cloned in tissue culture was obtained by Scharff and collaborators (Rudikoff et al., 1982; Giusti et al., 1987). A single base change resulted in an amino acid replacement in V_H CDR1 and complete loss of binding activity for the original ligand, phosphorylcholine. Unexpectedly, the mutated antibody acquired activity for a variety of phosphorylated macromolecules, including double-stranded DNA (Diamond and Scharff, 1984).

As mentioned above, mutations are frequently not found early in a response, but occur stepwise as the response progresses, allowing the sequence variations to accumulate. These observations imply that the process is initiated only after introduction of antigen. However, an alternative possibility is that rare preexisting variants become detectable only after stringent antigen selection. Evidence against the latter possibility was provided by Manser and Gefter (1986) who showed that anti-idiotypic suppression of B cells expressing the direct product of a germline gene suppressed the emergence of mutants derived from that gene.

The occurrence of mutation only after exposure to antigen probably accounts for early reports that high affinity antibodies were observed only late during immune responses to haptenic determinants (see Part I, Clonal Selection Prevails). If an appreciable number of cells having high-affinity receptors had been present early in the response, it should have been possible to detect the presence of antibodies of correspondingly high affinity (see discussion by Eisen, 1989).

MacLennan and Gray (1986) suggested that the B-cell rich germinal center, which appears only after antigenic stimulation (Thorbecke, 1959; Nieuwenhuis and Opstelten, 1984), is a site for hypermutation. Direct evidence for this hypothesis was provided by Jacob et al. (1991), who amplified, by the polymerase chain reaction, rearranged V_H genes from genomic DNA of cells picked from individual germinal centers in the spleens of immunized mice. Genealogical relationships could be demonstrated among most of the sequences obtained from the amplified *V* regions, suggesting that each germinal center contains a variety of related B cell clones with *V* gene segments that have undergone stepwise somatic mutation.

Küppers et al. (1993) analyzed single cells picked from frozen sections of individual germinal centers from human lymph nodes. Analysis of the products of individual cells, rather than populations of cells, allows clonal relationships to be deduced from rearrangements at both heavy and light chain loci. The data confirmed that a germinal center contains several B cell clones, the individual members of which have been diversified by somatic mutation. In contrast, the rearranged genes in proliferating B cells surrounding the germinal center were mostly unmutated. Evidence that somatic mutation occurs in germinal center B cells was also provided by Berek et al. (1991). For details about the germinal center, see reviews by MacLennan (1994), Thorbecke et al. (1994), and Liu and Banchereau (1996).

Characteristics of Somatic Hypermutation

The molecular mechanism of somatic hypermutation is not yet understood, nor have the enzyme(s) involved been identified. Nevertheless, a considerable amount of descriptive information about the process has accumulated (see reviews by Milstein and Neuberger, 1996; Storb, 1996).

A number of early studies indicated that somatic mutations occur predominantly within and near rearranged *V* genes, whether or not the rearrangement is productive (e.g., Kim et al., 1981; Pech et al., 1981; Selsing and Storb, 1981b; Gearhart and Bogenhagen, 1983). Unrearranged genes are usually not mutated (Gorski et al., 1983), although rare somatic mutations have been reported in unrearranged V_λ genes found in plasmacytoma or hybridoma cell lines (Weiss and Wu, 1987; Motoyama et al., 1994). Mutations were also found in nonproductive DJ_H rearrangements, but far fewer than in rearrangements in which a V_H had rearranged to a DJ_H in the same cell line (Roes et al., 1989). The requirement for *V* rearrangement would be compatible with errors in a repair process linked to transcription, as will be discussed below.

Most of the somatic mutations are single nucleotide substitutions; less frequently, insertions or deletions are found. The mutations are not confined to coding regions. Gearhart and Bogenhagen (1983) detected as many mutations in $V_\kappa J_\kappa$ flanking sequences as in the coding region itself. Subsequent studies revealed that almost all of the mutations occur in a segment of about 1.5 kb, beginning usually 3′ to the promoter (Rada et al., 1994; Rogerson, 1994) and extending for about 1 kb 3′ to *V(D)J*, no matter which *J* segment is rearranged (Lebecque and Gearhart, 1990; Weber et al., 1991). That mutations extend over a uniform distance from some initiation point, independent of sequence, may account for a few nucleotide replacements in the *C* gene segment of a rearranged mouse λ1 gene; presumably, the relatively short *JλCλ* intron (1.2 kb) allows mutation to extend into the *C* gene (Motoyama et al., 1991). The border of the mutation domain is relatively sharp at the 5′ end and trails off at the 3′ end.

A useful tool for gaining insight into the mutation process has been the analysis of mutations in transgenes. O'Brien et al. (1987) demonstrated that specific chromosomal location is not required for mutation since a pre-rearranged κ-transgene was mutated. Subsequent studies showed that specific sequences are not required as targets; mutations occurred in a variety of non-immunoglobulin sequences introduced on transgenes, provided that appropriate control elements were present (Azuma et al., 1993; Yèlamos et al., 1995).

Certain *V* gene positions tend to be favored sites for mutation. This could be the result of selection by antigen or of intrinsic mutability at these positions. To distinguish between these possibilities, mutation patterns were analyzed in rearranged genes that are not subject to selection by antigen (Sharpe et al., 1991; Betz et al., 1993). The pattern of mutations in a passenger κ transgene, which is not expressed in an antibody, revealed a preference for transitions rather than transversions. Mutations were found to occur preferentially in and near CDR1, with a major "hotspot" in the second base of the codon (AGT) for Ser-31. Two other serine codons of sequence AGY (Y = pyrimidine) at positions 26 and 77 were also found to be favored sites for mutation. The sequence RGYW (R = purine, W = A or T) had been identified as a consensus sequence for mutation by Rogozin and Kolchanov (1992). Preference for transitions as well as for certain sequences previously identified as hotspots was also revealed by analyzing non-immunoglobulin transgenes (Yèlamos et al., 1995).

Wagner et al. (1995) noted a difference between CDR and framework regions in usage of codons for serine. In both V_κ and V_H, AGY codons are used more frequently in CDRs whereas TCN (N = any nucleotide) codons are used more frequently in framework regions, suggesting that the CDR segments have been selected, through evolution, for mutability. In contrast, this bias was not seen in the *V* gene segments encoding the α and β chains of T cell receptors, which are generally considered not to be diversified by somatic mutation. Biased serine codon usage, AGC being favored, was also observed in CDRs of V_H genes of the amphibian, *Xenopus laevis* (Schwager et al., 1989).

Bachl and Wabl (1996) observed a strong bias toward mutation of GC base pairs in a μ gene construct introduced into a pre-B cell line in culture. They argue that this preference is revealed in this situation because antigen selection does not play a role. Further, they point out that preference for mutation of GC bp was observed in *Xenopus laevis* (Wilson et al., 1992) and in the shark, *Heterodontus*, (Hinds-Frey et al., 1993). In both of these species, *V* gene segments have been found to undergo somatic mutation, but it has been suggested that antigen-driven selection may be weak. It may be significant that germinal centers, sites for hypermutation and antigen selection, do not appear to be present in cold-blooded vertebrates (reviewed by Nahm et al., 1992).

Observations that the 5′ end of the mutation domain is near the transcriptional promoter stimulated speculation that the initiation of transcription is in some way involved in the generation of mutations. Betz et al. (1994) showed that the two transcriptional enhancer elements in the κ locus, one in the *JC* intron and the other 9 kb 3′ to C_κ, are important for effective hypermutation. The intron enhancer appeared to be absolutely required, whereas deletion of the 3′ enhancer reduced but did not abolish mutation. The promoter 5′ of the V_κ transcription start site (Falkner and Zachau, 1984; Parslow et al., 1984) was replaced by the human β-globin promoter without deleterious effect on mutation, indicating that specific promoter elements may not be required. Heavy chain transgenes with a heterologous promoter can also undergo mutation (Tumas-Brundage and Manser, 1997).

Peters and Storb (1996) provided additional support for the role of transcription initiation in somatic hypermutation. A κ transgene was constructed with two transcriptional promoters. One of these was in the normal position 5′ to the *V* gene segment. An exact duplicate of the promoter region was inserted into the intron 5′ to the *C* gene segment. Transcripts from the two promoters were equally abundant. Mutations were found in two domains, one centered over the *VJ* segment and the other over the *C* segment; a region of ~500 bp between the two domains was not mutated, indicating that mutations in the *C* segment had not spilled over from V_H but were newly initiated. A role for transcription-coupled repair in somatic hypermutation was proposed. However, patients with defects in nucleotide excision repair (e.g., xeroderma pigmentosum, Cockayne syndrome) are fully capable of somatic hypermutation of immunoglobulin genes (Wagner et al., 1996; Kim et al., 1997). Therefore, if the transcription-coupled repair model is correct, another type of repair mechanism, not defective in these patients, must be involved.

The importance of the position of the promoter was also demonstrated by Tumas-Brundage and Manser (1997): moving the V_H promoter 750 bp 5′ to its normal location resulted in a 5′ shift in the mutation domain. However, as Scharff et al. (1997) have pointed out, it is not yet clear whether it is transcription itself that is critical or some process dependent on or related to transcription, such as enhanced accessibility of the DNA in the transcribed gene segment. Moreover, antibodies produced in response to antigens that do not require collaboration of B and T cells (so-called "T-independent antigens such as polysaccharides") are not mutated despite high rates of transcription.

Somatic hypermutation has been studied in detail in mice and humans, but no doubt operates in a similar fashion in a variety of other species, modifying the repertoire that is initially established by *V(D)J* recombination before the introduction of antigen. In the case of V regions of λ light chains in sheep, somatic hypermutation also contributes toward the generation of the preimmune repertoire in an antigen-independent process (Reynaud et al., 1991, 1995).

Secondary Rearrangements: Receptor Editing

An initial *VJ* rearrangement at the κ locus does not necessarily preclude further rearrangement on the same chromosome. In a number of B-cell lines, secondary rearrangements involving deletion and/or inversion were observed (e.g., Lewis et al., 1982; Selsing et al., 1984; Fedderson and Van Ness, 1985; Shapiro and Weigert, 1987). Evidence consistent with secondary rearrangement was also obtained by identifying coding and signal joints on circular DNA excision products resulting from deletional $V_\kappa J_\kappa$ joining (Hirama et al., 1991). It is efficient for the primary rearrangement to be inversional as this leaves all the *V* and *J* gene segments on the chromosome, and therefore maximizes the options for gene utilization in a secondary rearrangement (see Figure 6). As mentioned previously, some *V* and *J* segments may be reversed as a consequence of inversional joining so that a particular pair that might originally have rearranged by deletion, will now rearrange by inversion (or conversely). In principle, rearrangement may continue until the last available *J* segment has been utilized.

A secondary rearrangement may replace a nonfunctional $V_\kappa J_\kappa$ joint with a functional one, possibly rescuing a cell that did not make a usable receptor (Feddersen and Van Ness, 1990). However, an apparently functional joint may also be replaced (Harada and Yamagishi, 1991), even with a nonfunctional product (Levy et al., 1989; Huber et al., 1992). Evidently, generation of a productive coding joint does not in all cases prevent another rearrangement on the same chromosome.

Analogous secondary rearrangements occur between *D* and *J* segments in the heavy chain locus. The first gene segments to rearrange at this locus are *D* and J_H. The initial DJ_H complex can be replaced by joining a more 5′ *D* to a more 3′ J_H (Reth et al., 1986a; Maeda et al., 1987; Toda et al., 1989). These rearrangements, as well as the ones at the κ light chain locus, depend on conventional recognition of RSS and obey the 12/23-bp rule.

Heavy chain *D* segments are located between V_H and J_H segments (see section, Arrangement of Immunoglobulin Genes). Therefore, secondary *V* to *DJ* rearrangements are not ordinarily possible in the heavy chain locus since the initial joining steps will have deleted all unrearranged *D* segments. Nonetheless, another type of secondary rearrangement, "V_H gene replacement" has been described (Kleinfeld et al., 1986; Reth et al., 1986b); this departs from the standard mode in that a V_H gene, with its 3′ heptamer/spacer/nonamer signal, rearranges into an already-assembled $V_H DJ_H$ complex by virtue of site-specific recognition of a cryptic heptamer found near the 3′-end of most V_H gene segments. Initially a nonamer was not thought to be involved, but subsequent re-examination of V_H sequences showed that many of them have a conserved nonamer exactly 12 bp 5′ to the heptamer (Chen et al., 1995). The consensus sequence of this nonamer differs from that of the usual nonamer found in RSS and it is not clear whether it has a role in the rearrangement. The original $V_H D$ junction is retained, including its N nucleotides. By providing a correct reading frame, the replacement may rescue a cell that has non-functional *V(D)J* junctions on both alleles or it may provide for diversity to compensate for rearrangement bias, which favors usage of certain V_H segments.

Evidence supporting the proposed mechanism of *V* gene replacement was obtained by showing inversional rearrangement, albeit at a low frequency, between a *V* gene segment and a *VDJ* sequence in a retroviral recombination substrate, with production of the expected signal and coding joints (Covey et al., 1990). That *V* gene replacement can occur by a deletional mechanism in a transformed cell line was reported by Usuda et al. (1992).

The original examples of *V* gene replacement were documented in transformed cell lines. More recently, such rearrangements have also been shown to occur *in vivo* (Chen et al., 1995; Taki et al., 1995). These experiments utilized mutant mouse strains in which a rearranged heavy chain gene ($V_H DJ_H$) was inserted into the heavy chain locus (Taki et al., 1993). Junctional diversity due to N addition was noted in the rearrangements studied by Chen et al., which utilized the cryptic heptamer near the 3′ end of V_H. This observation implies either that V_H replacement occurs early in the B cell lineage or that TdT, ordinarily not expressed in mature B lymphocytes, is reactivated during the secondary rearrangement.

That a heptamer alone may mediate rearrangement had been suggested in earlier studies showing rearrangement of either V_κ (Seidman and Leder, 1980) or J_κ (Kelley et al., 1985) to an isolated heptamer in the $J\kappa$-$C\kappa$ intron. Rearrangements to this heptamer have also been implicated in deletions of C_κ (Durdik et al., 1984; Siminovitch et al., 1985). Another aberrant rearrangement of a J_κ segment to a sequence unrelated to any *V* gene, but flanked by a canonical heptamer without a nonamer was reported by Höchtl and Zachau (1983). In an assay of the effect of sequence variation in the RSS on recombination efficiency, utilizing a plasmid recombination substrate, Hesse et al. (1989) found a low level of recombination when one of the two RSS consisted of a heptamer without a nonamer, but in the absence of a heptamer there was no recombination.

A question of current interest is the possible role of these secondary rearrangements in altering the specificity of antigen receptors that display anti-self reactivity, a process designated "receptor editing" (Gay et al., 1993; Radic et al., 1993; Tiegs et al., 1993). Much of the evidence for receptor editing comes from experiments with mice carrying transgenes for autoantibodies. Immature B cells in the bone marrow bearing IgM receptors reactive with a self-antigen are ordinarily rendered tolerant by encounter with that antigen (see recent reviews by Nemazee, 1995 and Klinman, 1996). However, such cells may alter their specificity by appropriate editing, for example by rearranging and expressing endogenous light chain genes, thereby escaping tolerance induction and developing into mature B cells.

Weigert and co-workers investigated the editing process by inserting rearranged heavy and light chain genes, from hybridomas with anti-DNA activity, into their respective loci, simulating normal rearrangements (Chen et al., 1995; Luning Prak, and Weigert, 1995; Chen et al., 1997). Secondary rearrangements (editing) were observed in this model system, more frequently at the κ than at the heavy chain locus. Editing of an autoreactive specificity at the κ locus was also recently reported by Pelanda et al. (1977).

Evidence that secondary rearrangements can occur not only in the bone marrow, but also in germinal centers, has recently been obtained (Han et al., 1997; Papavasiliou et al., 1997). That such rearrangements might occur had previously been suggested when *RAG* genes were found to be expressed in germinal center B cells (see section, The Mechanism of Rearrangement). Receptor editing, either in the bone marrow or in germinal centers, may rescue B cells that would otherwise have been lost through autoreactivity or inability to bind foreign antigen, thereby contributing to antibody diversification.

ONE V REGION, MANY C REGIONS

According to clonal selection, a cell synthesizing and secreting antibody of a particular specificity has surface receptors of exactly the same specificity. This requirement implies that the V regions of the secreted and receptor antibodies must be identical. When antigen binds to its surface receptors, the cell is stimulated to produce antibody and to proliferate, forming or expanding a clone. During the lifetime of such a clone, antibodies of several different classes may be produced. Indeed, at any time, more than one class may be expressed, even in an individual cell. The V regions of all of these antibodies, regardless of the class, maintain their original identity, except for the possible presence of somatic mutations or secondary rearrangements. In this final section, the means by which the B cell usually retains the original V region on its heavy chain, yet varies the C region to allow expression of both secreted and receptor forms, as well as the expression of different immunoglobulin classes, is considered.

Identical Specificity of Receptor and Secreted Antibody: One Gene, Two Polypeptides

In Part I, some of the early evidence for the existence of the receptor and for its identity in specificity to the antibody produced by the cell on which it resides was presented (see section, Clonal Selection Prevails). However, it was realized that the C region of the receptor is unlikely to be completely identical to antibody since secreted antibodies do not contain segments of hydrophobic amino acids, which would be needed for insertion into a membrane.

Following the description of the four-chain immunoglobulin molecule in the early 1960s, a number of experiments demonstrated that the receptor on B lymphocytes had properties expected of an immunoglobulin. Treatment of rabbit lymphocytes with antisera to allotypic determinants induced "blast transformation," i.e. DNA synthesis and morphological changes characteristic of activated dividing cells (Sell and Gell, 1965). This observation also suggested that signals can be transmitted to the interior of the cell following a specific interaction at the surface, activity expected for a receptor. Since $F(ab')_2$[j] but not Fab′ fragments of antibodies induced blast transformation, it was proposed that cross-linking the receptors was necessary (Woodruff et al., 1967; Fanger et al., 1970). It was subsequently shown that either bivalent antibodies directed against immunoglobulin or multivalent antigen can induce a redistribution of their respective ligands on the cell surface, a phenomenon known as "capping" (Taylor et al., 1971; Loor et al., 1972).

The presence of immunoglobulin epitopes on the surface of lymphocytes was also demonstrated by the fluorescent antibody technique, an enormously useful method that had been developed in the 1940s by Coons and co-workers (Coons et al., 1942; Coons and Kaplan, 1950). Cell suspensions were treated with fluorescent-labeled antibodies directed against a variety of immunoglobulin epitopes. The first such report was actually a control in a study by Möller (1961) of H-2 (major histocompatability) antigens on the surface of mouse lymphoid cells; additional evidence was obtained by Pernis et al. (1970), Raff et al. (1970), and Rabellino et al. (1971). Similar results were obtained by autoradiography, electron microscopy, and other methods (reviewed by Warner, 1974). A potential hazard of techniques in which cells are reacted with labeled antibodies is that many cells, including lymphocytes, have membrane receptors for the Fc fragment of IgG, and this can lead to non-specific binding; the use of bivalent $F(ab')_2$ fragments instead of complete antibody can prevent such undesired reactions (Vitetta and Uhr, 1975; Winchester et al., 1975).

Other experiments provided evidence that anti-immunoglobulin and antigen may compete in reacting with the receptor. The binding of antigen to nonimmune spleen cells was inhibited by pretreatment with antibodies directed against immunoglobulins (Byrt and Ada, 1969; Raff et al., 1973). Further, treatment of suspensions of unprimed spleen cells with antisera to immunoglobulins before transfer to irradiated recipients suppressed a subsequent antibody response (Warner et al., 1970). Moreover, antibodies directed against an idiotypic determinant specifically suppressed the synthesis of antibodies expressing that determinant *in vitro* (Cosenza and Köhler, 1972; Sher and Cohn, 1972) or *in vivo* (Hart et al., 1972; Eichmann, 1974).

By the early 1970s, an impressive body of evidence had accumulated pointing to the identity in specificity of the receptor on a B lymphocyte and the antibody produced by that cell. Nonetheless, final proof had to await the characterization of the receptor at the molecular level. Considerable information about the receptor was obtained by the application of methods of classical biochemistry. However, its detailed structure was not clarified until the genes encoding secreted and membrane immunoglobulins were cloned. Even then it was not clear how the receptor transduces signals across the plasma membrane, and information about this process has only recently begun to emerge (reviewed by Cambier et al., 1994).

The first step in the biochemical analysis of the receptor, isolation from the cell membrane, was difficult; the membrane-associated protein is present in only small amounts and is insoluble in aqueous solution. Sensitivity and selectivity would be improved by radiolabeling under conditions that maintain cell viability so that only proteins associated with the surface and not those within the cell become labeled. A procedure that met these conditions was iodination catalyzed by lactoperoxidase in the presence of hydrogen peroxide (Marchalonis, 1969; Phillips and Morrison, 1970; Marchalonis et al., 1971). It was also necessary to find conditions for solubilizing membrane proteins that would not interfere with subsequent specific immunoprecipitation.

Mouse splenic lymphocytes were lysed in nonionic detergent and the extracted proteins were treated with antibodies to mouse immunoglobulins; the specific precipitate was analyzed by polyacrylamide electrophoresis in sodium dodecyl sulfate (Vitetta et al., 1971). The major peak corresponded in mobility to "monomers" of IgM (units consisting of two light chains and two μ heavy chains). After reduction, components corresponding in mobility to μ and light chain were seen. Similar results were obtained when the immunoprecipitation was with class-specific antibodies to IgM. It was concluded that the major immunoglobulin present on lymphocyte surfaces is monomeric IgM. Secreted IgM in mammals, as found in serum, usually consists of five four-chain units (reviewed by Green, 1969; Metzger, 1970).

A variety of indirect evidence indicated that the heavy chain of membrane IgM (designated μ_m) and the heavy chain of secreted IgM (μ_s) differ slightly in structure, and that these differences account for the membrane attachment of the one immunoglobulin and the solubility in aqueous solution of the other. Small amounts of detergent appeared to be bound by membrane IgM and by isolated μ_m, but not by secreted IgM or μ_s (Melcher and Uhr, 1977; Vassalli et al., 1979; Parkhouse et al., 1980). The apparent molecular weight of μ_m was found to be greater than that of μ_s (Melcher and Uhr, 1976; Bergman and Haimovich, 1978) and this seemed to be related to differences in peptide, not carbohydrate (Vassalli et al., 1979; Singer et al., 1980; Sibley et al., 1980). At least part of the difference in peptide structure between μ_m and μ_s appeared to be localized to the carboxy-terminal region (Williams et al., 1978; Jaton and Vassalli, 1980; McCune et al., 1980; Singer et al., 1980). As judged by peptide mapping and protein sequence analysis, the two types of chain are very similar in the V_H and the four C_μ domains, differing only in their carboxy-terminal segments (Kehry et al., 1980).

DNA cloning and sequencing clarified the molecular basis for the distinction between μ_s and μ_m. Mapping studies indicated that mRNAs encoding the μ_s and μ_m forms of mouse μ chain differ only at their 3′ ends (Alt et al., 1980b; Rogers et al., 1980; Singer et al., 1980). Hood and co-workers then determined the nucleotide sequences of these RNAs in the 3′ segments where they appeared to differ and compared them to the sequence of the corresponding genomic DNA. The remarkable conclusion was that the two mRNAs appeared to be derived from a single germline gene by alternative processing pathways (Early et al., 1980b; Rogers et al., 1980).

The membrane form of the mouse μ chain was found to have a terminal segment of 41 amino acid residues that is entirely different from the terminal 20 amino acid residues in the secreted chain; elsewhere, the two forms of the μ chain are identical (Rogers et al., 1980). The terminal segment in μ_m is encoded by two "membrane exons" located about 1850 bp 3′ to $C_\mu 4$ (Early et al., 1980b). The 5′ membrane exon encodes the first 39 residues of the extra segment of μ_m and the 3′ exon encodes the remaining two residues and a 3′-untranslated region. This 41 amino acid region consists of a highly negatively charged segment of 12 residues, presumably outside the cell, a hydrophobic segment of 26 residues, presumably spanning the membrane, and a three-residue segment that contains two lysine residues and is thought to be located within the cytoplasm. The coding sequence of the human membrane exons differs in only 9 bp (three amino acid replacements) (Milstein et al., 1984).

It was puzzling how mIgM could perform its presumed function as a receptor, since the short intracytoplasmic segment was not considered adequate for transmitting a signal. The discovery of a transmembrane heterodimer, Ig-α/Ig-β, which associates non-covalently with mIgM and appears to have the properties required of a signaling receptor, apparently provided the missing link (see reviews by Reth et al., 1991; Reth, 1992; Cambier et al., 1994). It had already been shown that the T cell receptor polypeptide chains are associated, non-covalently, with a group of transmembrane polypeptides known as the CD3 complex, which transduce signals from the receptor to the cell interior (reviewed by Clevers et al., 1988). Thus, the receptor for antigen on B cells was found to resemble the antigen-specific receptor on T cells.[k]

The 20 amino acid terminal segment specific to the secreted form of the μ chain contains the half-cystine residue that forms a disulfide bridge to the J (joining) chain, a polypeptide that is found exclusively in polymeric immunoglobulins (reviewed by Koshland, 1985; not to be confused with the *J* gene segment with which it has nothing in common, except one letter of the alphabet).

The processing pathways that lead to the two forms of mRNA are illustrated in Figure 8. To produce mRNA encoding μ_s, a primary transcript is cleaved and polyadenylated at a "proximal" poly(A) site located 3′ to the exon encoding $C_\mu 4$ and the terminal segment of the secreted chain. To produce mRNA encoding μ_m, the transcript is cleaved and polyadenylated at a "distal" poly(A) site 3′ to the membrane exons. The primary transcript for μ_m is spliced from the border between

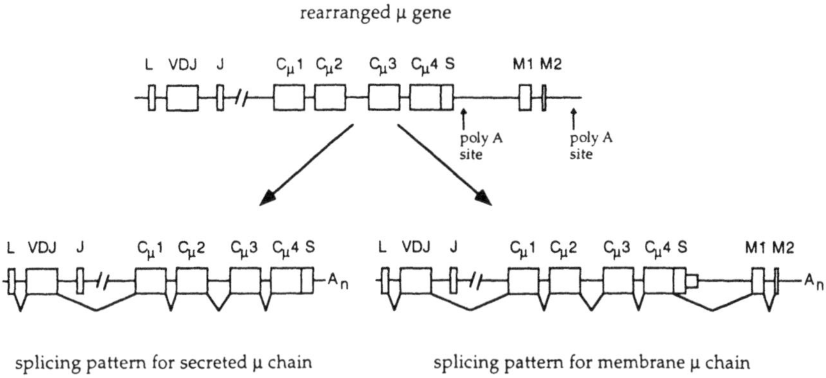

Figure 8. Alternative splicing patterns for generating mRNAs encoding secreted and membrane forms of mouse μ chain. Boxes represent exons. L, leader exon; VDJ, rearranged exon encoding a V region; $C_\mu 1$-$C_\mu 4$, exons encoding domains of μ chain; S, segment of $C_\mu 4$ exon encoding the carboxyl-terminal 20 amino acid residues of the secreted μ chain; M1 and M2, exons encoding the carboxyl-terminal 41 amino acid residues of the membrane form of the μ chain. Bent lines indicate splicing patterns generating mRNA for secreted and membrane forms of the μ chain. (Based on Early et al., 1980b.)

$C_\mu 4$ and the secreted segment to the 5′ border of the first membrane exon; the donor splice site lies within a GGU glycine codon that is retained in μ_s mRNA. Which form of mRNA is produced depends on competition between splicing and polyadenylation at the μ_s and μ_m poly(A) sites (Peterson and Perry, 1986; Tsurushita et al., 1987; Galli et al., 1988; Peterson and Perry, 1989). The balance between the two mRNA products determines the proportion of membrane and secreted IgM. Activation of the B lymphocyte appears to tip the balance in favor of secretion, but the regulatory signals are not known.

Rogers et al. (1980) pointed out that the sequence Gly-Lys had been found 20 and 19 residues from the carboxy-terminus of all heavy chains that had been sequenced at that time. These two residues could be encoded by the same sequence of nucleotides (GGUAAA) that form the RNA donor splice site used to generate μ_m. It seemed likely, therefore, that the other heavy chains also exist in secreted and membrane forms. In the next several years, mRNA encoding membrane and secreted forms of heavy chain, as well as one or two membrane exons, were identified for a number of the other immunoglobulin classes of mice and humans, e.g., subclasses of IgG (e.g., Rogers et al., 1981; Tyler et al., 1982; Yamawaki-Kataoka et al., 1982; Komaromy et al., 1983; Bensmana and Lefranc, 1990), IgA (Word et al., 1983; Bensmana and Lefranc, 1990; Yu et al., 1990), IgE (Ishida et al., 1982; Zhang et al., 1992; Peng et al., 1992), and IgD (see next section). The segments encoded by the membrane exons of the γ, α, and ε chains are longer than those of the mouse and human μ chains; the additional length appears to be both in the extracellular and cytoplasmic portions.

IgM and IgD

Studies of membrane immunoglobulins of mouse lymphocytes revealed that in addition to mIgM, a second immunoglobulin was present. Because of the similarity of its heavy chain to that of human IgD, this immunoglobulin was thought to be a membrane form of IgD (mIgD) (Abney and Parkhouse, 1974; Melcher et al., 1974). IgD had been identified originally as an unusual human myeloma protein (Rowe and Fahey, 1965a). It was shown to be present in low concentration in human serum, ~30 μg/ml (Rowe and Fahey, 1965b) and also on the surface of lymphocytes (Van Boxel et al., 1992; Rowe et al., 1973a).

Unexpectedly, most human lymphocytes were found to express IgM as well as IgD on their membranes (Rowe et al., 1973b), thereby providing precedence for finding the two immunoglobulins on mouse lymphocytes. The two classes on the same cell express the same idiotypic determinants (Salsano et al., 1974; Fu et al., 1975) and specificity for antigen (Pernis et al., 1974; Goding and Layton, 1976; Stern and McConnell, 1976), suggesting that they share the same heavy chain *V* region. Thus, the one cell-one specificity rule is maintained despite the existence of immunoglobulins belonging to two different classes in a single cell.

Further work elucidated the basis for the simultaneous expression, in a single cell, of two immunoglobulins, IgM and IgD, which differ only in the C regions of their heavy chains (Liu et al., 1980; Maki et al., 1981; Moore et al., 1981; Cheng et al., 1982). Like the μ chain, secreted and membrane forms of the IgD heavy (δ) chain are encoded by shared and unique exons. However, the carboxy-terminal segment of the secreted δ chain is encoded in a separate exon 3′ from the last shared exon. In two particular cell lines expressing both IgM and IgD, there was only one copy of chromosome 12, indicating that both heavy chains must be expressed from the same allele (Wabl et al., 1980; Knapp et al., 1982). Moreover, only a single $V_H D J_H$ was detected in cells producing both immunoglobulins; the C_μ and C_δ genes and the DNA between them were in the germline configuration (Maki et al., 1981; Knapp et al., 1982). It was concluded that the secreted and membrane forms of the μ and δ chains are derived by a complex series of alternative processing pathways from a large (~25 kb) transcript spanning the μ-δ locus (see reviews by Blattner and Tucker, 1984; Rogers and Wall, 1984).

The presence of two immunoglobulins on the surface of most lymphocytes naturally led to speculation and experiment regarding their respective roles. Since IgM appears before IgD on the surface of differentiating lymphocytes (Yuan and Vitetta, 1978), Vitetta and Uhr (1975) proposed that cells expressing surface IgM only and cells expressing both immunoglobulins respond differentially to antigenic stimulation. According to this model, relatively immature B lymphocyte, bearing only IgM, are rendered tolerant by encounter with antigen. In contrast, more mature

B cells, bearing IgD as well as IgM, are triggered by antigen and T cell signals to differentiate into IgM-secreting cells or into cells that have lost surface IgM and are the progenitors of cells that will secrete other isotypes (e.g., IgG, IgA). Thus, stimulation of IgM leads to tolerance and stimulation of IgD to antibody synthesis. That early B cells expressing only IgM are highly sensitive to tolerance induction was also proposed by Nossal (1983).

Despite considerable effort on the part of many investigators, it has proven difficult to obtain definitive evidence either to support or disprove the proposition that the fate of a B cell differs according to whether the antigen binds to a cell with only IgM receptors or to one with both classes of receptor. Many of the earlier experimental approaches and results have been summarized by Brink et al. (1992) and Roes and Rajewsky (1993), in reports of their own investigations of the problem. Brink et al. showed that B cells expressing either transgene-encoded IgM or transgene-encoded IgD (and no endogenously encoded immunoglobulin) did not differ substantially in their responses to foreign or self antigens. Roes and Rajewsky (1993) and Nitschke et al. (1993) generated IgD-deficient mice by gene targeting. There was no impairment in the immune response except for some delay in the affinity increase of antibodies. Taken together, these experiments suggest that there may be no unique role for IgD antigen receptors on the cell surface.

The Class Switch

It has been known for many years that during the course of a typical immune response, different antibody classes are produced: IgM is found early in the response and IgG later (e.g. Bauer and Stavitsky, 1961; Uhr and Finkelstein, 1963). In a typical heterogeneous polyclonal response, the IgM and IgG antibodies appear to have similar specificity, but the structural relationship of the binding sites (in modern terms, the V regions) associated with these different classes cannot easily be evaluated. However, a crucial insight into the relationship among V regions of different antibody classes was provided by Todd's observation (1963) that rabbit IgG and IgM express the same *a*-locus allotypic determinants.

As discussed in Part I, Two Genes, One Polypeptide Chain, the *a*-locus allotypic determinants behave genetically as if they are products of alleles at a single locus; these determinants were subsequently shown to be located on the heavy chain. However, the C regions of the heavy chains in IgM and IgG are the products of different, non-allelic genes. How can the heavy chain be, simultaneously, the product of one gene and of several different genes? Before long, the identification of V and C regions of immunoglobulin heavy and light chains, and the idea that these regions might be encoded separately in the germline, led to a possible resolution. Thus, it was proposed that one gene (i.e., V_H) encodes the part of the chain expressing the *a*-locus allotypic determinants and another gene (C_μ or C_γ) encodes the C region (Todd and Inman, 1967). (For further discussion of the "Todd phenomenon," see the section, Bias in Rearrangement.)

Todd's observation about shared allotypy could not be interpreted in terms of individual molecules, i.e., that **exactly** the same V_H region can be associated with the C region of either a γ or a μ chain. However, a molecular interpretation became possible when Fudenberg, Nisonoff and co-workers showed that two myeloma proteins found in the serum of the same patient, one IgG and the other IgM, express the same idiotypic determinants. It was proposed and subsequently confirmed that the V regions of the heavy chains were identical (Wang et al., 1970; Wang et al., 1977). The light chains also appeared to be identical and the two proteins were synthesized by different plasma cells (Wang et al., 1969). Presumably, a clone originally producing one of the myeloma proteins (e.g., IgM) generated a subclone producing the other, the same V_H gene being expressed in association with either C_H gene. Also consistent with the presence of identical V regions on different classes of heavy chain, was the finding of Klinman and coworkers that the clonal progeny of a single precursor B lymphocyte produced, after antigenic stimulation, IgM, IgG, and IgA antibodies expressing the same idiotypic determinant (Gearhart et al., 1975).

The expression of the same V_H region, sequentially, with more than one C_H region is known as "class switching." Ordinarily, in the lifetime of a B cell or clone, the switch occurs from IgM to one or more of the other classes. The observation that germinal center B cells express mostly IgM after initial exposure to antigen and IgG after repeated exposure led to the suggestion that isotype switching occurs in germinal centers (Kraal et al., 1982). Much subsequent work has confirmed that somatic hypermutation and isotype switching, probably in that sequential order (Liu et al., 1996), occur in germinal centers after appropriate interactions of B cells with antigen and T cells.

Class switching usually results in the translocation of a rearranged *V(D)J* gene segment from a position 5′ to C_μ to a position 5′ to another C_H, which is then expressed. Switching generally occurs in a 5′ to 3′ direction and may occur more than once in a single cell line, indeed can in principle continue until the *V(D)J* segment is next to the most 3′ C_H segment, C_α. Thus, at least two distinct DNA rearrangements occur in the formation of functional heavy chain genes: *V(D)J* recombination and one or more class switches.

In 1978, Honjo and Kataoka, proposed a molecular model for class switching based on their analysis, by hybridization kinetics, of the number of different C_H genes present in various plasmacytomas. The significant finding was that this number varied by a factor of two, according to the class of immunoglobulin produced by the particular plasmacytoma. The data could be explained if a specific deletion of C_H genes on one of the two homologous chromosomes accompanies the recombination of *V(D)J* to a C_H gene on that same chromosome. Relating the number of C_H genes in the plasmacytoma to the immunoglobulin class produced allowed a tentative ordering of the C_H genes in the locus. Thus, finding that an IgG1-producing plasmacytoma retained one copy each of the $\gamma1$, $\gamma2a$, and $\gamma2b$ *C* genes per haploid genome, but only 0.5 copy of the $\gamma3$ *C* gene, suggested that the other C_γ genes are 3′ to the $C_{\gamma 3}$ gene. The model specifies that the recombination occurs on only one allele and that the chromosomal segment between the two joining segments is excised.

The deletion model was supported by results obtained by Southern blotting with cloned probes in a number of other studies (Coleclough et al., 1980; Cory and Adams, 1980; Hurwitz et al., 1980; Rabbitts et al., 1980; Yaoita and Honjo, 1980). The model was elaborated by Kataoka et al. (1980) who found that a rearranged γ1 gene appeared to be formed by a recombination between 5′ flanking regions of the $C_{\gamma 1}$ and C_μ germline genes; they proposed that the DNA segment involved in the class-switch recombination be called the switch region.

Not all data favored stepwise linear deletion as the basis for switching. Obata et al. (1981) found that the DNA organization in a particular plasmacytoma could be explained by unequal sister chromatid exchange, but observations inconsistent with this possibility were reported by Wabl et al. (1985). Alt et al. (1982) observed that a subclone of a transformed pre-B cell line had switched from synthesis of a μ chain to synthesis of a γ chain, with retention of the same V_H and J_H as well as both copies of the C_μ gene; these results suggested that the switch in heavy chain synthesis was a transcriptionally controlled event (see discussion below).

More recently, the intrachromosomal deletion model has received strong support from the identification of circles consisting of the DNA originally lying between the recombined switch regions (Iwasato et al., 1990; Matsuoka et al., 1990; von Schwedler et al., 1990). Analysis of the sequences within the circles indicated that switching usually occurs from C_μ to another C_H, but can also occur from one of the C_γ to a more 3′ C_H gene.

Switch regions contain G-rich tandem repeat sequences and are 1 to 10 kb long; they are located 5′ to each of the C_H genes except C_δ (reviewed by Gritzmacher, 1989). Unlike *V(D)J* joining, in which the site of recombination always occurs within a few bp near specific RSS, switch recombination is not site-specific and can occur at many positions within switch regions. Most of the recombination sites fall within the tandemly repeated units, but some have been detected 5′ to the tandem repeats associated with C_μ (Katzenberg and Birshtein, 1988; Dunnick et al., 1993). Although there are no switch regions 5′ to C_δ, rare B cells or B cell tumors have been found to express IgD but not IgM, presumably due to deletion of C_μ by some recombinational event (Moore et al., 1981; Yasui et al., 1989; Kluin et al., 1995).

The enzymes responsible for class switching have not been identified but it is known that they are not products of the *RAG* genes. There is considerable information on the regulation of the class specificity of the switch (reviewed by Stavnezer, 1996). Which immunoglobulin class is produced during an immune response depends on a number of factors including the type of cytokine released from activated T cells. For example, addition of interleukin-4 to mouse B cells that have been activated (e.g. by addition of lipopolysaccharide) induces transcription of the unrearranged $C_{\gamma 1}$ and C_ε genes and also promotes switching to IgG1 and IgE; the germline transcription may indicate that the chromatin structure is in an open configuration and accessible to the switch recombinase.

Switch recombination usually results in the deletion of C_μ, so that switched cells no longer produce IgM. However, dating from an early study by Nossal et al. (1964), there have been a number of reports that both IgM and a heavy chain class other than IgD are occasionally expressed simultaneously in single normal or neoplastic B cells (e.g., Yaoita et al., 1982; Perlmutter and Gilbert, 1984; Chen et al., 1986; Kinashi et al., 1987). In these studies, no evidence for reorganization or deletion of heavy chain *C* genes was found. In analogy with the simultaneous expression of μ and δ chains, the data were interpreted as being most easily explained by post-transcriptional processing of long transcripts, extending from the rearranged V_HDJ_H through the C_H locus. The possible existence of such "giant" RNA precursors to explain class switching had originally been suggested by Rabbitts (1978), but no evidence for their existence has been obtained.

An alternate explanation for double producers, discontinuous transcription, was proposed by Tucker and co-workers (Chen et al., 1986). Discontinuous transcription might occur via *trans*-splicing of mRNA—the intermolecular joining of two or more transcripts (reviewed by Sharp, 1987 and Bonen, 1993)—or by "polymerase hopping," as may occur in replication of some RNA viruses (reviewed by Lai, 1992). Evidence in favor of discontinuous transcription, probably by trans-splicing, was obtained by Nolan-Willard et al., 1992).

Shimizu and colleagues (1989, 1991) demonstrated the production of mRNA containing the rearranged V_HDJ_H segment of a μ transgene and the C region of an endogenous C_γ gene ("trans-mRNA") when cells were cultured in the presence of lipopolysaccharide and interleukin-4. They suggested that the germline transcript produced under these conditions (see above) provides one of the partners in a *trans*-splicing reaction. No switch recombination in the DNA could be detected. Since mRNA containing both V and C regions of the transgene was also produced, both μ chain and γ chain with the same V region were synthesized by these cells. Reminiscent of the suggestion by Nossal et al. (1964) that double producers represent a transitional phase in class switching, Shimizu et al. (1991) proposed that cells expressing more than a single class are intermediates in the switching process, marked by the synthesis of germline transcripts. Evidence for a role for germline transcription in *trans*-splicing was recently presented by Fujieda et al. (1996).

Results of a number of studies over the years have suggested that recombination may occasionally occur between, as well as within, chromosomes. Reports from the early 1970s indicated that in rabbits, V_H and C_H are occasionally (~1 to 7%) encoded by genes on different members of a chromosome pair (Landucci Tosi and Tosi, 1973; Pernis et al., 1973; Knight et al., 1974). Such studies were feasible in the rabbit because of the availability of genetic markers (allotypes) for V as well as C regions. These findings antedated the discovery of immunoglobulin gene rearrangement and eukaryotic gene splicing. Possible explanations for the results that were considered were recombination at either the DNA or RNA level, i.e., recombination between homologous chromosomes or joining of separate mRNAs for V and C regions.

Evidence for a low frequency of interchromosomal switching during culture of mouse hybridoma cells was obtained by Kipps and Herzenberg (1986). In an immunized transgenic mouse, interchromosomal recombination was found to occur at a high frequency between the *VDJ* segment of a transgene and an autologous C_γ gene (Durdik et al., 1989; Gerstein et al., 1990). In a subsequent study (Umar and Gearhart, 1995), it was shown that a crossover between a transgene containing *VDJ* and the endogenous heavy chain locus can occur by homologous recombination.

Knight and co-workers have recently obtained direct evidence that *trans*-chromosomal recombination between homologous parental chromosomes can occur in the expression of rabbit α chains. By analyzing cDNA from rabbits heterozygous for both V- and C-region allotypes, they showed that V_H and J_H sequences were sometimes derived from one chromosome and C_α sequences from the other (Knight et al., 1995). Although the results, taken alone, could be explained either by *trans*-chromosomal recombination or by *trans*-splicing, the former possibility was favored as more consistent with previous results indicating that single B cells produce antibody with C region derived from only one chromosome (Hanly and Knight, 1975). If *trans*-splicing were to account for the observed results, then mRNAs for V and C on the same chromosome should also be spliced together, resulting in the expression of two C regions by one cell (one the result of trans-splicing and one from normal splicing).

In a follow-up study, the chromosomal origin of V_HDJ_H and C_α gene segments expressed in the IgA-secreting hybridomas from rabbits heterozygous for allotyope was evaluated by examining DNA flanking the recombined gene segments. Recombination **between** the homologous chromosomes (rather than recombination **within** one chromosome) was observed at a high (~ 7%) frequency (K.L. Knight, personal communication). In addition, one of the hybridoma α chains was found to have an unusual sequence that could most easily be explained by transchromosomal gene conversion. Interchromosomal gene conversion from a transgene to the endogenous heavy chain locus had previously been proposed by Giusti and Manser (1994) to explain what appeared to be a nonreciprocal exchange in mice carrying rearranged V_HDJ_H constructs that lacked switch or C regions.

Despite the exceptions cited above, it is currently thought that the predominant mode of class switching is by intrachromosomal recombination leading to deletion of the segment of chromosome between the two switch regions. Less frequently, *trans*-chromosomal recombination may also occur, at least in the rabbit. The contribution to class switching in normal B cells of the other mechanisms that have been proposed, such as *trans*-splicing and gene conversion, remains to be explored.

CODA[l]

The unifying thread, running through both parts of this chapter on antibody specificity and diversity, has been the clonal selection hypothesis: **clonal**, because it proposes that the great diversity of antibodies is generated by an equally diverse array of clones, each committed to the synthesis of antibody of a single specificity; **selective**, because in contrast to instructive theories, antigen has no direct role in antibody synthesis, but merely stimulates the proliferation of cells having antibody-like receptors that bind to complementary antigen.[m]

The isolation and biochemical description of the antibody molecule clarified how specificity and diversity could be achieved at the protein level, but it did not explain how this enormous family, with its numerous similar yet subtly different members, is derived from the germline. Studies at the DNA level, described here, established that genes specifying antibody polypeptide chains are assembled in somatic cells from large arrays of gene segments. This unique rearrangement process enables a relatively limited number of genetic elements to generate a vastly greater number of distinct sequences. The resulting repertoire of antibody combining sites, which is adequate to provide a baseline defense against most foreign pathogens, is further refined by additional somatic processes that are set into motion after the initial encounter with antigen.

Once the central issue of somatic gene rearrangement as a principal means for achieving antibody diversification was established, other aspects of the clonal selection hypothesis could be explored. The resolution of some of these was relatively straightforward: the identical specificity of the cell-surface receptor and the antibody secreted by an individual cell was shown to be achieved by alternative splicing of exons of a complex gene. Other problems, such as the mechanism of allelic exclusion (the expression of only one parental chromosome to ensure that a cell is restricted to a single specificity), have been more difficult to resolve. Progress in identifying the enzymes needed for critical molecular events has been uneven. The *V*(*D*)*J* recombinases encoded by the *RAG* genes are now known and their action is understood in great, if not complete, detail. The identification of other enzymes and mechanisms, those involved in somatic hypermutation and in class switching, has been more difficult.

The focus of the chapter has been antibody diversification in humans and mice, the species that have been most thoroughly studied for the longest time. However, in the last several years it has become clear that, although the principal features of antibody diversification are shared by all vertebrates, the specifics can vary considerably from one species to another, even within the class of mammals. Evolution may achieve workable solutions in a variety of ways, none necessarily optimal.[n] Among the tasks of the next few years will be exploration of the limits of this variation.

ACKNOWLEDGMENTS

I have received invaluable help from a number of colleagues who read part or all of the text, corrected errors, made helpful suggestions, and provided much encouragement. These colleagues include Tania Baker, Peter Brodeur, Jianzhu Chen, Herman Eisen, Ann Feeney, Julian Fleischman, Valerie Hohman, John Kimball, Katherine Knight, Leonard Lerman, Susanna Lewis, Alfred Nisonoff, Eduardo Padlan, Jonathan Seidman, Christopher Roman, Matthew Scharff, Janet Stavnezer, David Stollar, Philip Tucker, and Martin Weigert. I also thank Anne Battis and Louisa Worthington of the MIT Libraries for their assistance in obtaining articles that are not in the MIT collection. I acknowledge with thanks three decades of support for my laboratory, via grant AI-08054, from the National Institute of Allergy and Infectious Diseases of the National Institutes of Health.

NOTES

a. An allotypic determinant is an epitope on an immunoglobulin molecule that reflects allelic variation within a species. (See Part I, Two Genes, One Polypeptide.)

b. The "gene segments" are separate in the germline and must be rearranged in antibody-forming cells to allow expression of the complete light or heavy chain. Traditional usage would reserve the term "gene" for the rearranged form. Frequently in the literature, and occasionally in this chapter, gene segment and gene are used interchangeably, but the meaning is usually clear from the context.

c. Iodination of purified MOPC-321 κ mRNA resulted in fragmentation of the molecule, a 3′ fragment containing sequences corresponding to the *C* but not the *V* region, was recovered and used as a *C*-region probe. No fragment containing *V* but not *C* sequences was recovered.

d. As in Part I, the terms B lymphocyte and B cell are used interchangeably.

e. There is ambiguity in the use of the term *V*, as it sometimes refers to the entire variable region and sometimes only to that portion encoded by the *V* gene segment; in context, the meaning is usually clear.

f. Idiotypic determinants are epitopes on *V* regions typically associated with the combining site for antigen; they are detected by reaction with antibodies called anti-idiotypes. See Part I, Anti-Antibodies (Idiotypy).

g. Also in the T cell receptor α locus, two potentially productive *V*α*J*α rearrangements may occur in a single cell but, usually, only one of these forms a functional heterodimer with the β chain on the cell surface (Malissen et al., 1992).

h. In discussions of B cell development, it is important to distinguish between maturity of the cell and maturity of the animal. A similar distinction should be made for the repertoire. For example, the repertoire expressed by newly-formed B cells, which have presumably not been subjected to selection by foreign antigen, may not be identical in the fetus and the adult and both will generally differ from the repertoire expressed by peripheral B cells.

i. This terminology is borrowed from descriptions of changes in the antigenicity of the surface proteins (hemagglutinin and neuraminidase) of the influenza virus (see Webster and Laver, 1975). Antigenic drift refers to minor changes in antigenicity, due to point mutation, which are responsible for influenza epidemics. Antigenic shift refers to major changes in antigenicity, due to reassortment of genes from different vertebrate hosts, which are responsible for pandemics. Antigenic shift occurs only in influenza A strains. The term "immunological drift," referring to mutations in influenza A, appears to have been introduced by Burnet (1955), in analogy to "genetic drift."

j. $F(ab')_2$ is a bivalent derivative of IgG produced by digestion with pepsin, which cleaves on the carboxy-terminal side of the disulfide bridge(s) joining the heavy chains. Fab′ is a fragment produced by reduction of the interchain disulfide bridge(s) between the heavy chain fragments of $F(ab')_2$; it resembles Fab, consisting of the entire light chain and a heavy chain piece that is slightly larger than Fd, the heavy chain piece found in Fab. (See Part I, Figure 1.)

k. Unlike antibodies, the T cell receptor does not recognize isolated antigen, but only a complex of a peptide derived from antigen bound to a cell-surface major histocompatabilty protein (for a review, see Eisen et al., 1996).

l. The aims of this chapter, set out at the beginning of Part I, were to describe the historical background for our current understanding of the basis of antibody specificity and diversity. This turned out to be a far larger task than anticipated and so the article was divided into two parts along a natural cleavage plane: proteins in Part I and genes in Part II. Even so, it has been necessary to omit many topics of relevance to the subjects that have been discussed. The differentiation of B lymphocytes has been mentioned only in passing and the interaction between B and T cells, vital for almost all antibody responses, hardly at all. That antigen binding can specifically suppress or kill, as well as activate a B cell via its antibody receptor, has received only brief mention. If time and space were unlimited, it would have been interesting to consider the history of the discovery of the T cell receptor and its similarity to, and differences from, the B cell receptor.

m. "Selection from preexisting diversity appears as the means most frequently used in the living world to face an unknown future" (Jacob, 1982, p. 66).

n. "Evolution proceeds like a tinkerer who, during millions of years, has slowly modified his products, retouching, cutting, lengthening, using all opportunities to transform and create . . . different tinkerers interested in the same problem will reach different solutions" (Jacob, 1982, p. 35).

REFERENCES

Abney, E.R. & Parkhouse, R.M.E. (1974). Candidate for immunoglobulin D present on murine B lymphocytes. Nature 252, 600–602.

Adams, J.M. (1980). The organization and expression of immunoglobulin genes. Immunology Today 1, 10–17.

Akira, S., Okazaki, K., & Sakano, H. (1987). Two pairs of recombination signals are sufficient to cause immunoglobulin V-(D)-J joining. Science 238, 1134–1138.

Allen, D., Cumano, A., Dildrop, R., Kocks, C., Rajewsky, K., Rajewsky, N., Roes, J., Sablitzky, F., & Siekevitz, M. (1987). Timing, genetic requirements and functional consequences of somatic hypermutation during B-cell development. Immunol. Revs. 96, 5–22.

Allen, D., Simon, T., Sablitzky, F., Rajewsky, K., & Cumano, A. (1988). Antibody engineering for the analysis of affinity maturation of an anti-hapten response. EMBO J. 7, 1995–2001.

Alt, F.W. & Baltimore, D. (1982). Joining of immunoglobulin heavy chain gene segments: implications from a chromosome with evidence of three D-J_H fusions. Proc. Natl. Acad. Sci. USA 79, 4118–4122.

Alt, F.W., Bothwell, A.L.M., Knapp, M., Siden, E., Mather, E., Koshland, M., & Baltimore, D. (1980b). Synthesis of secreted and membrane-bound immunoglobulin mu heavy chains is directed by mRNAs that differ at their 3′ ends. Cell 20, 293–301.

Alt, F.W., Enea, V., Bothwell, A.L.M., & Baltimore, D. (1980a). Activity of multiple light chain genes in murine myeloma cells producing a single, functional light chain. Cell 21, 1–12.

Alt, F.W., Rosenberg, N., Casanova, R.J., Thomas, E., & Baltimore, D. (1982). Immunoglobulin heavy chain class switching and inducible expression in an Abelson murine leukaemia virus transformed cell line. Nature 296, 325–331.

Alt, F., Rosenberg, N., Lewis, S., Thomas, E., & Baltimore, D. (1981). Organization and reorganization of immunoglobulin genes in A-MuLV-transformed cells: rearrangement of heavy but not light chain genes. Cell 27, 381–390.

Alt, F.W., Yancopoulos, G.D., Blackwell, T.K., Wood, C., Thomas, E., Boss, M., Coffman, R., Rosenberg, N., Tonegawa, S., & Baltimore, D. (1984). Ordered rearrangement of immunoglobulin heavy chain variable region segments. EMBO J. 3, 1209–1219.

Appella, E. (1971). Amino acid sequences of two mouse immunoglobulin lambda chains. Proc. Natl. Acad. Sci. USA 68, 590–594.

Appella, E. & Ein, D. (1967). Two types of lambda polypeptide chains in human immunoglobulins based on an amino acid substitution at position 190. Proc. Natl. Acad. Sci. USA 57, 1449–1454.

Appella, E., McIntyre, K.R., & Perham, R.N. (1967). Lambda Bence Jones proteins of the mouse: chemical and immunological characterization. J. Mol. Biol. 27, 391–394.

Appella, E. & Perham, R.N. (1968). Amino-terminal sequences of two mouse lambda chains. J. Mol. Biol. 33, 963–966.

Azuma, T., Igras, V., Reilly, E.B., & Eisen, H.N. (1984). Diversity at the variable-joining region boundary of λ light chains has a pronounced effect on immunoglobulin ligand-binding activity. Proc. Natl. Acad. Sci. USA 81, 6139–6143.

Azuma, T., Motoyama, N., Fields, L.E., & Loh, D.Y. (1993). Mutations of the chloramphenicol acetyl transferase transgene driven by the immunoglobulin promoter and intron enhancer. Int. Immunol. 5, 121–130.

Azuma, T., Steiner, L.A., & Eisen, H.N. (1981). Identification of a third type of λ light chain in mouse immunoglobulins. Proc. Natl. Acad. Sci. USA 78, 569–573.

Bachl, J. & Wabl, M. (1996). An immunoglobulin mutator that targets G·C base pairs. Proc. Natl. Acad. Sci. USA 93, 851–855.

Bartolomei, M.S. & Tilghman, S.M. (1997). Genomic imprinting in mammals. Annu. Rev. Genet. 31, 493–525.

Bauer, D.C. & Stavitsky, A.B. (1961). On the different molecular forms of antibody synthesized by rabbits during the early response to a single injection of protein and cellular antigens. Proc. Natl. Acad. Sci. USA 47, 1667–1680.

Bauer, T.R., Jr. & Blomberg, B. (1991). The human λ L chain Ig locus. Recharacterization of JCλ6 and identification of a functional JCλ7. J. Immunol. 146, 2813–2820.

Becker, R.S. & Knight, K.L. (1990). Somatic diversification of immunoglobulin heavy chain VDJ genes: evidence for somatic gene conversion in rabbits. Cell 63, 987–997.

Bensmana, M. & Lefranc, M.-P. (1990). Gene segments encoding membrane domains of the human immunoglobulin gamma 3 and alpha chains. Immunogenetics 32, 321–330.

Bentley, D.L. & Rabbitts, T.H. (1980). Human immunoglobulin variable region genes—DNA sequences of two V_κ genes and a pseudogene. Nature 288, 730–733.

Bentley, D.L. & Rabbitts, T.H. (1981). Human V_κ immunoglobulin gene number: implications for the origin of antibody diversity. Cell 24, 613–623.

Bentley, D.L. & Rabbitts, T.H. (1983). Evolution of immunoglobulin V genes: evidence indicating that recently duplicated human V_κ sequences have diverged by gene conversion. Cell 32, 181–189.

Berek, C. (1984). The D segment defines the T15 idiotype: the immunoresponse of A/J mice to *Pneumococcus pneumoniae*. Eur. J. Immunol. 14, 1043–1048.

Berek, C., Berger, A., & Apel, M. (1991). Maturation of the immune response in germinal centers. Cell 67, 1121–1129.

Berek, C., Griffiths, G.M., & Milstein, C. (1985). Molecular events during maturation of the immune response to oxazolone. Nature 316, 412–418.

Berek, C. & Milstein, C. (1987). Mutation drift and repertoire shift in the maturation of the immune response. Immunol. Revs. 96, 23–41.

Bergman, Y. & Haimovich, J. (1978). B lymphocytes contain three species of μ chains. Eur. J. Immunol. 8, 876–880.

Berinstein, N., Levy, S., & Levy, R. (1989). Activation of an excluded immunoglobulin allele in a human B lymphoma cell line. Science 244, 337–339.

Berman, J.E., Mellis, S.J., Pollock, R., Smith, C.L., Suh, H., Heinke, B., Kowal, C., Surti, U., Chess, L., Cantor, C.R., & Alt, F.W. (1988). Content and organization of the human Ig V_H locus: definition of three new V_H families and linkage to the Ig C_H locus. EMBO J. 7, 727–738.

Bernard, O., Gough, N.M., & Adams, J.M. (1981). Plasmacytomas with more than one immunoglobulin κ mRNA: implications for allelic exclusion. Proc. Natl. Acad. Sci. USA 78, 5812–5816.

Bernard, O., Hozumi, N., & Tonegawa, S. (1978). Sequences of mouse immunoglobulin light chain genes before and after somatic changes. Cell 15, 1133–1144.

Bernier, G.M. & Cebra, J.J. (1964). Polypeptide chains of human gamma-globulin: cellular localization by fluorescent antibody. Science 144, 1590–1591.

Bernstein, R.M., Schluter, S.F., Bernstein, H., & Marchalonis, J.J. (1996). Primordial emergence of the recombination activating gene 1 (RAG1): sequence of the complete shark gene indicates homology to microbial integrases. Proc. Natl. Acad. Sci. USA 93, 9454–9459.

Bernstein, R.M., Schluter, S.F., Lake, D.F., & Marchalonis, J.J. (1994). Evolutionary conservation and molecular cloning of the recombinase activating gene 1. Biochem. Biophys. Res. Comm. 205, 687–692.

Betz, A.G., Milstein, C., González-Fernández, A., Pannell, R., Larson, T., & Neuberger, M.S. (1994). Elements regulating somatic hypermutation of an immunoglobulin κ gene: critical role for the intron enhancer/matrix attachment region. Cell 77, 239–248.

Betz, A.G., Rada, C., Pannell, R., Milstein, C., & Neuberger, M.S. (1993). Passenger transgenes reveal intrinsic specificity of the antibody hypermutation mechanism: clustering, polarity, and specific hot spots. Proc. Natl. Acad. Sci. USA 90, 2385–2388.

Biedermann, K.A., Sun, J., Giaccia, A.J., Tosto, L.M., & Brown, J.M. (1991). *scid* mutation in mice confers hypersensitivity to ionizing radiation and a deficiency in DNA double-strand break repair. Proc. Natl. Acad. Sci. USA 88, 1394–1397.

Billips, L.G., Lasssoued, K., Nuñez, C., Wang, J., Kubagawa, H., Gartland, G.L., Burrows, P.D., & Cooper, M.D. (1995). Human B-cell development. Ann. NY Acad. Sci. 764, 1–8.

Blackwell, T.K., Malynn, B.A., Pollock, R.R., Ferrier, P., Covey, L.R., Fulop, G.M., Phillips, R.A., Yancopoulos, G.D., & Alt, F.W. (1989). Isolation of *scid* pre-B cells that rearrange kappa light chain genes: formation of normal signal and abnormal coding joins. EMBO J. 8, 735–742.

Blankenstein, T., Bonhomme, F., & Krawinkel, U. (1987). Evolution of pseudogenes in the immunoglobulin V_H-gene family of the mouse. Immunogenetics 26, 237–248.

Blankenstein, T. & Krawinkel, U. (1987). Immunoglobulin V_H region genes of the mouse are organized in overlapping clusters. Eur. J. Immunol. 17, 1351–1357.

Blaser, K. & Eisen, H.N. (1978). λ2 light chains in normal mouse immunoglobulins. Proc. Natl. Acad. Sci. USA 75, 1495–1499.

Blattner, F.R. & Tucker, P.W. (1984). The molecular biology of immunoglobulin D. Nature 307, 417–422.

Blomberg, B., Geckeler, W.R., & Weigert, M. (1972). Genetics of the antibody response to dextran in mice. Science 177, 178–180.

Blomberg, B. & Tonegawa, S. (1982). DNA sequences of the joining regions of mouse λ light chain immunoglobulin genes. Proc. Natl. Acad. Sci. USA 79, 530–533.

Blomberg, B., Traunecker, A., Eisen, H., & Tonegawa, S. (1981). Organization of four mouse λ light chain immunoglobulin genes. Proc. Natl. Acad. Sci. USA 78, 3765–3769.

Bonen, L. (1993). *Trans*-splicing of pre-mRNA in plants, animals, and protists. FASEB J. 7, 40–46.

Borst, J., Jacobs, H., & Brouns, G. (1996). Composition and function of T-cell receptor and B-cell receptor complexes on precursor lymphocytes. Curr. Op. Immunol. 8, 181–190.

Bothwell, A.L.M., Paskind, M., Reth, M., Imanishi-Kari, T., Rajewsky, K., & Baltimore, D. (1981). Heavy chain variable region contribution to the NP^b family of antibodies: somatic mutation evident in a γ2a variable region. Cell 24, 625–637.

Bottaro, A., de Marchi, M., Migone, N., & Carbonara, A.O. (1989). Pulsed-field gel analysis of human heavy-chain constant region deletions reveals the extent of unmapped regions within the locus. Genomics 4, 505–508.

Brack, C., Hirama, M., Lenhard-Schuller, R., & Tonegawa, S. (1978). A complete immunoglobulin gene is created by somatic recombination. Cell 15, 1–14.

Brack, C. & Tonegawa, S. (1977). Variable and constant parts of the immunoglobulin light chain gene of a mouse myeloma cell are 1250 nontranslated bases apart. Proc. Natl. Acad. Sci. USA 74, 5652–5656.

Brenner, S. & Milstein, C. (1966). Origin of antibody variation. Nature 211, 242–243.

Brensing-Küppers, J., Zocher, I., Thiebe, R., & Zachau, H.G. (1997). The human immunoglobulin kappa locus on yeast artificial chromosomes (YACs). Gene 191, 173–181.

Brezinschek, H.-P., Foster, S.J., Brezinschek, R.I., Dörner, T., Domiati-Saad, R., & Lipsky, P.E. (1997). Analysis of the human V_H gene repertoire. Differential effects of selection and somatic hypermutation on human peripheral $CD5^+/IgM^+$ and $CD5^-/IgM^+$ B cells. J. Clin. Invest. 99, 2488–2501.

Brink, R., Goodnow, C.C., Crosbie, J., Adams, E., Eris, J., Mason, D.Y., Hartley, S.B., & Basten, A. (1992). Immunoglobulin M and D antigen receptors are both capable of mediating B lymphocyte activation, deletion, or anergy after interaction with specific antigen. J. Exp. Med. 176, 991–1005.

Brinster, R.L., Ritchie, K.A., Hammer, R.E., O'Brien, R.L., Arp, B., & Storb, U. (1983). Expression of a microinjected immunoglobulin gene in the spleen of transgenic mice. Nature 306, 332–336.

Britten, R.J. & Kohne, D.E. (1968). Repeated sequences in DNA. Hundreds of thousands of copies of DNA sequences have been incorporated into the genomes of higher organisms. Science 161, 529–540.

Brodeur, P.H., Osman, G.E., Mackle, J.J., & Lalor, T.M. (1988). The organization of the mouse *Igh-V* locus. Dispersion, interspersion and the evolution of V_H gene family clusters. J. Exp. Med. 168, 2261–2278.

Brodeur, P.H. & Riblet, R. (1984). The immunoglobulin heavy chain variable region (Igh-V) locus in the mouse. I. One hundred Igh-V genes comprise seven families of homologous genes. Eur. J. Immunol. 14, 922–930.

Buluwela, L., Albertson, D.G., Sherrington, P., Rabbitts, P.H., Spurr, N., & Rabbitts, T.H. (1988). The use of chromosomal translocations to study human immunoglobulin gene organization: mapping D_H segments within 35 kb of the C_μ gene and identification of a new D_H locus. EMBO J. 7, 2003–2010.

Burke, D.T., Carle, G.F., & Olson, M.V. (1987). Cloning of large segments of exogenous DNA into yeast by means of artificial chromosome vectors. Science 236, 806–812.

Burnet, F.M. (1955). Principles of Animal Virology, 1st ed. p. 380, Academic Press, New York.

Burnet, F.M. (1957). A modification of Jerne's theory of antibody production using the concept of clonal selection. Aust. J. Sci. 20, 67–69.

Burnet, M. (1964). The clonal selection theory of immunity—A Darwinian modification. Aust. J. Sci. 27, 6–7.

Burrows, P., Lejeune, M., & Kearney, J.F. (1979). Evidence that murine pre-B cells synthesize μ heavy chains but no light chains. Nature 280, 838–841.

Burstein, Y. & Schechter, I. (1977). Glutamine as a precursor to N-terminal pyrrolid-2-one-5-carboxylic acid in mouse immunoglobulin λ-type light chains. Amino acid-sequence variability at the N-terminal extra piece of λ-type light-chain precursors. Biochem. J. 165, 347–354.

Byrt, P. & Ada, G.L. (1969). An *in vitro* reaction between labelled flagellin or haemocyanin and lymphocyte-like cells from normal animals. Immunology 17, 503–516.

Cambier, J.C., Pleiman, C.M., & Clark, M.R. (1994). Signal transduction by the B cell antigen receptor and its coreceptors. Annu. Rev. Immunol. 12, 457–486.

Carlson, L.M., Oettinger, M.A., Schatz, D.G., Masteller, E.L., Hurley, E.A., McCormack, W.T., Baltimore, D., & Thompson, C.B. (1991). Selective expression of RAG-2 in chicken B cells undergoing immunoglobulin gene conversion. Cell 64, 201–208.

Carson, S. & Wu, G.E. (1989). A linkage map of the mouse immunoglobulin lambda light chain locus. Immunogenetics 29, 173–179.

Casali, P. & Notkins, A.L. (1989). Probing the human B-cell repertoire with EBV: polyreactive antibodies and $CD5^+$ B lymphocytes. Annu. Rev. Immunol. 7, 513–535.

Cebra, J.J., Colberg, J.E., & Dray, S. (1966). Rabbit lymphoid cells differentiated with respect to α-, γ-, and μ-heavy polypeptide chains and to allotypic markers Aa1 and Aa2. J. Exp. Med. 123, 547–558.

Cesari, I.M. & Weigert, M. (1973). Mouse lambda-chain sequences. Proc. Natl. Acad. Sci. USA 70, 2112–2116.

Chen, C., Luning Prak, E., & Weigert, M. (1997). Editing disease-associated autoantibodies. Immunity 6, 97–105.

Chen, C., Nagy, Z., Luning Prak, E., & Weigert, M. (1995). Immunoglobulin heavy chain gene replacement: a mechanism of receptor editing. Immunity 3, 747–755.

Chen, J., Trounstine, M., Alt, F.W., Young, F., Kurahara, C., Loring, J.F., & Huszar, D. (1993a). Immunoglobulin gene rearrangement in B cell deficient mice generated by targeted deletion of the J_H locus. Int. Immunol. 5, 647–656.

Chen, J., Trounstine, M., Kurahara, C., Young, F., Kuo, C.-C., Xu, Y., Loring, J.F., Alt, F.W., & Huszar, D. (1993b). B cell development in mice that lack one or both immunoglobulin κ light chain genes. EMBO J. 12, 821–830.

Chen, Y.-W., Word, C., Dev, V., Uhr, J.W., Vitetta, E.S., & Tucker, P.W. (1986). Double isotype production by a neoplastic B cell line. II. Allelically excluded production of μ and γ1 heavy chains without C_H gene rearrangement. J. Exp. Med. 164, 562–579.

Cheng, H.-L., Blattner, F.R., Fitzmaurice, L., Mushinski, J.F., & Tucker, P. W. (1982). Structure of genes for membrane and secreted murine IgD heavy chains. Nature 296, 410–415.

Chess, A., Simon, I., Cedar, H., & Axel, R. (1994). Allelic inactivation regulates olfactory receptor gene expression. Cell 78, 823–834.

Chien, Y., Iwashima, M., Wettstein, D.A., Kaplan, K.B., Elliot, J.F., Born, W., & Davis, M.M. (1987). T-cell receptor δ gene rearrangements in early thymocytes. Nature 330, 722–727.

Chuchana, P., Blancher, A., Brockly, F., Alexandre, D., Lefranc, G., & Lefranc, M.-P. (1990). Definition of the human immunoglobulin variable lambda (IGLV) gene subgroups. Eur. J. Immunol. 20, 1317–1325.

Clevers, H., Alarcon, B., Wileman, T., & Terhorst, C. (1988). The T cell receptor/CD3 complex: a dynamic protein ensemble. Annu. Rev. Immunol. 6, 629–662.

Cohn, M., Blomberg, B., Geckeler, W., Raschke, W., Riblet, R., & Weigert, M. (1974). First order considerations in analyzing the generator of diversity. In: The immune system: genes, receptors, signals. (Sercarz, E.E., Williamson, A.R., Fox, C.F., eds.), pp. 89–117. Academic Press, New York.

Coleclough, C., Cooper, D., & Perry, R.P. (1980). Rearrangement of immunoglobulin heavy chain genes during B-lymphocyte development as revealed by studies of mouse plasmacytoma cells. Proc. Natl. Acad. Sci. USA 77, 1422–1426.

Coleclough, C., Perry, R.P., Karjalainen, K., & Weigert, M. (1981). Aberrant rearrangements contribute significantly to the allelic exclusion of immunoglobulin gene expression. Nature 290, 372–378.

Collins, J. & Hohn, B. (1978). Cosmids: a type of plasmid gene-cloning vector that is packageable *in vitro* in bacteriophage λ heads. Proc. Natl. Acad. Sci. USA 75, 4242–4246.

Combriato, G. & Klobeck, H.-G. (1991). V_λ and J_λ-C_λ gene segments of the human immunoglobulin λ light chain locus are separated by 14 kb and rearrange by a deletion mechanism. Eur. J. Immunol. 21, 1513–1522.

Cook, G.P. & Tomlinson, I.M. (1985). The human immunoglobulin V_H repertoire. Immunology Today 16, 237–242.

Cook, G.P., Tomlinson, I.M., Walter, G., Riethman, H., Carter, N.P., Buluwela, L., Winter, G., & Rabbitts, T.H. (1994). A map of the human immunoglobulin V_H locus completed by analysis of the telomeric region of chromosome 14q. Nature Genet. 7, 162–168.

Coons, A.H., Creech, H.J., Jones, R.N., & Berliner, E. (1942). The demonstration of pneumococcal antigen in tissues by the use of fluorescent antibody. J. Immunol. 45, 159–170.

Coons, A.H. & Kaplan, M.H. (1950). Localization of antigen in tissue cells. II. Improvements in a method for the detection of antigen by means of fluorescent antibody. J. Exp. Med. 91, 1–13.

Corbett, S.J., Tomlinson, I.M., Sonnhammer, E.L.L., Buck, D., & Winter, G. (1997). Sequence of the human immunoglobulin diversity (D) segment locus: a systematic analysis provides no evidence for the use of DIR segments, inverted D segments, "minor" D segments or D-D recombination. J. Mol. Biol. 270, 587–597.

Cortes, P., Weis-Garcia, F., Misulovin, Z., Nussenzweig, A., Lai, J.–S., Li, G., Nussenzweig, M.C., & Baltimore, D. (1996). *In vitro* V(D)J recombination: signal joint formation. Proc. Natl. Acad. Sci. USA 93, 14008–14013.

Cory, S. & Adams, J.M. (1980). Deletions are associated with somatic rearrangement of immunoglobulin heavy chain genes. Cell 19, 37–51.

Cory, S., Adams, J.M., & Kemp, D.J. (1980). Somatic rearrangements forming active immunoglobulin μ genes in B and T lymphoid cell lines. Proc. Natl. Acad. Sci. USA 77, 4943–4947.

Cosenza, H. & Köhler, H. (1972). Specific suppression of the antibody response by antibodies to receptors. Proc. Natl. Acad. Sci. USA 69, 2701–2705.

Covey, L.R., Ferrier, P., & Alt, F.W. (1990). V_H to V_HDJ_H rearrangement is mediated by the internal V_H heptamer. Int. Immunol. 2, 579–583.

Cox, D.W., Markovic, V.D., & Teshima, I.E. (1982). Genes for immunoglobulin heavy chains and for α_1-antitrypsin are localized to specific regions of chromosome 14q. Nature 297, 428–430.

Cox, J.P.L., Tomlinson, I.M., & Winter, G. (1994). A directory of human germ-line V_κ segments reveals a strong bias in their usage. Eur. J. Immunol. 24, 827–836.

Crews, S., Griffin, J., Huang, H., Calame, K., & Hood, L. (1981). A single V_H gene segment encodes the immune response to phosphorylcholine: somatic mutation is correlated with the class of the antibody. Cell 25, 59–66.

Cuisinier, A.-M., Guigou, V., Boubli, L., Fougereau, M., & Tonnelle, C. (1989). Preferential expression of V_H5 and V_H6 immunoglobulin genes in early human B-cell ontogeny. Scand. J. Immunol. 30, 493–497.

Currier, S.J., Gallarda, J.L., & Knight, K.L. (1988). Partial molecular genetic map of the rabbit V_H chromosomal region. J. Immunol. 140, 1651–1659.

Dariavach, P., Lefranc, G., & Lefranc, M.-P. (1987). Human immunoglobulin $C_\lambda 6$ gene encodes the $Kern^+Oz^-$ λ chain and $\chi_\lambda 4$ and $C_\lambda 5$ are pseudogenes. Proc. Natl. Acad. Sci. USA 84, 9074–9078.

Decker, D.J., Boyle, N.E., Koziol, J.A., & Klinman, N.R. (1991). The expression of the Ig H chain repertoire in developing bone marrow B lineage cells. J. Immunol.146, 350–361.

de la Chapelle, A., Lenoir, G., Boué, J., Boué, A., Gallano, P., Huerre, C., Szajnert, M.-F., Jeanpierre, M., Lalouel, J.-M., & Kaplan, J.-C. (1983). Lambda Ig constant region genes are translocated to chromosome 8 in Burkitt's lymphoma with t(8;22). Nucleic Acids Res. 11, 1133–1142.

Delovitch, T.K., & Baglioni, C. (1973). Estimation of light-chain gene reiteration of mouse immunoglobulin by DNA-RNA hybridization. Proc. Natl. Acad. Sci. USA 70, 173–178.

Denzin, L.K., Gulliver, G.A., & Voss, E.W., Jr. (1993). Mutational analysis of active site contact residues in anti-fluorescein monoclonal antibody 4-4-20. Mol. Immunol. 30, 1331–1345.

Desiderio, S.V., Yancopoulos, G.D., Paskind, M., Thomas, E., Boss, M.A., Landau, N., Alt, F.W., & Baltimore, D. (1984). Insertion of *N* regions into heavy-chain genes is correlated with expression of terminal deoxytransferase in B cells. Nature 311, 752–755.

D'Eustachio, P., Bothwell, A.L.M., Takaro, T.K., Baltimore, D., & Ruddle, F.H. (1981). Chromosomal location of structural genes encoding murine immunoglobulin λ light chains. Genetics of murine λ light chains. J. Exp. Med. 153, 793–800.

D'Eustachio, P., Pravtcheva, D., Marcu, K., & Ruddle, F.H. (1980). Chromosomal location of the structural gene cluster encoding murine immunoglobulin heavy chains. J. Exp. Med. 151, 1545–1550.

D'Hoostelaere, L.A., Huppi, K., Mock, B., Mallett, C., & Potter, M. (1988). The IgκL chain allelic groups among the Igκ haplotypes and Igκ crossover populations suggest a gene order. J. Immunol. 141, 652–661.

Diamond, B., & Scharff, M.D. (1984). Somatic mutation of the T15 heavy chain gives rise to an antibody with autoantibody specificity. Proc. Natl. Acad. Sci. USA 81, 5841–5844.

Difilippantonio, M.J., McMahan, C.J., Eastman, Q.M., Spanopoulou, E., & Schatz, D.G. (1996). RAG1 mediates signal sequence recognition and recruitment of RAG2 in V(D)J recombination. Cell 87, 253–262.

Dildrop, R. (1984). A new classification of mouse V_H sequences. Immunol. Today 5, 85–86.

Dildrop, R., Gause, A., Müller, W., & Rajewsky, K. (1987). A new V gene expressed in lambda-2 light chains of the mouse. Eur. J. Immunol. 17, 731–734.

Dildrop, R., Krawinkel, U., Winter, E., & Rajewsky, K. (1985). V_H-gene expression in murine lipopolysaccharide blasts distributes over the nine known V_H-gene groups and may be random. Eur. J. Immunol. 15, 1154–1156.

Dray, S. & Nisonoff, A. (1963). Contribution of allelic genes A_b^4 and A_b^5 to formation of rabbit 7S γ-globulins. Proc. Soc. Exp. Biol. Med. 113, 20–26.

Dreyer, W.J. & Bennett, J.C. (1965). The molecular basis of antibody formation: A paradox. Proc. Natl. Acad. Sci. USA 54, 864–869.

Dreyfus, D.H. (1992). Evidence suggesting an evolutionary relationship between transposable elements and immune system recombination sequences. Mol. Immunol. 29, 807–810.

Dubiski, S. (1969). Immunochemistry and genetics of a "new" allotypic specificity A_e^{14} of rabbit γG immunoglobulins: recombination in somatic cells. J. Immunol. 103, 120–128.

Dugan, E.S., Bradshaw, R.A., Simms, E.S., & Eisen, H.N. (1973). Amino acid sequence of the light chain of a mouse myeloma protein (MOPC-315). Biochemistry 12, 5400–5416.

Dunnick, W., Hertz, G.Z., Scappino, L., & Gritzmacher, C. (1993). DNA sequences at immunoglobulin switch region recombination sites. Nucleic Acids Res. 21, 365–372.

Durdik, J., Gerstein, R.M., Rath, S., Robbins, P.F., Nisonoff, A., & Selsing, E. (1989). Isotype switching by a microinjected μ immunoglobulin heavy chain gene in transgenic mice. Proc. Natl. Acad. Sci. USA 86, 2346–2350.

Durdik, J., Moore, M.W., & Selsing, E. (1984). Novel κ light-chain gene rearrangements in mouse λ light chain-producing B lymphocytes. Nature 307, 749–752.

Early, P., Huang, H., Davis, M., Calame, K., & Hood, L. (1980a). An immunoglobulin heavy chain variable region gene is generated from three segments of DNA: V_H, D and J_H. Cell 19, 981–992.

Early, P., Rogers, J., Davis, M., Calame, K., Bond, M., Wall, R., & Hood, L. (1980b). Two mRNAs can be produced from a single immunoglobulin μ gene by alternative RNA processing pathways. Cell 20, 313–319.

Eastman, Q.M., Leu, T.M.J., & Schatz, D.G. (1996). Initiation of V(D)J recombination *in vitro* obeying the 12/23 rule. Nature 380, 85–88.

Ehlich, A., Schaal, S., Gu, H., Kitamura, D., Müller, W., & Rajewsky, K. (1993). Immunoglobulin heavy and light chain genes rearrange independently at early stages of B cell development. Cell 72, 695–704.

Eichmann, K. (1974). Idiotype suppression. I. Influence of the dose and of the effector functions of anti-idiotypic antibody on the production of an idiotype. Eur. J. Immunol. 4, 296–302.

Eichmann, K. & Kindt, T.J. (1971). The inheritance if individual antigenic specificities of rabbit antibodies to streptococcal carbohydrates. J. Exp. Med. 134, 532–552.

Ein, D. (1968). Nonallelic behavior of the Oz groups in human lambda immunoglobulin chains. Proc. Natl. Acad. Sci. USA 60, 982–985.

Ein, D. & Fahey, J.L. (1967). Two types of lambda polypeptide chains in human immunoglobulins. Science 156, 947–948.

Eisen, H.N. (1989). Affinity maturation: a retrospective view. In Molecular evolution on rugged landscapes: proteins, RNA and the immune system (Perelson, A.S. & Kauffman, A.A., Eds.), pp. 75–82. Addison Wesley, Redwood City.

Eisen, H.N., Simms, E.S., & Potter, M. (1968). Mouse myeloma proteins with antihapten antibody activity. The protein produced by plasma cell tumor MOPC-315. Biochemistry 7, 4126–4134.

Eisen, H.N., Sykulev, Y., & Tsomides, T.J. (1996). Antigen-specific T-cell receptors and their reactions with complexes formed by peptides with major histocompatibility complex proteins. Adv. Prot. Chem. 49, 1–56.

Elliott, B.W., Jr., Eisen, H.N., & Steiner, L.A. (1982). Unusual association of V, J, and C regions in a mouse immunoglobulin λ chain. Nature 299, 559–561.

Elliott, B.W., Jr., Eisen, H.N., & Steiner, L.A. (1984). Amino acid sequence diversity in mouse λ2 variable regions. J. Immunol. 133, 2757–2761.

Emorine, L., Kuehl, M., Weir, L., Leder, P., & Max, E.E. (1983). A conserved sequence in the immunoglobulin J_κ-C_κ intron: possible enhancer element. Nature 304, 447–449.

Falkner, F.G., & Zachau, H.G. (1984). Correct transcription of an immunoglobulin κ gene requires an upstream fragment containing conserved sequence elements. Nature 310, 71–74.

Fanger, M.W., Hart, D.A., Wells, J.V., & Nisonoff, A. (1970). Requirement for cross-linkage in the stimulation of transformation of rabbit peripheral lymphocytes by antiglobulin reagents. J. Immunol. 105, 1484–1492.

Farace, M.-G., Aellen, M.-F., Briand, P.-A., Faust, C.H., Vassalli, P., & Mach, B. (1976). No detectable reiteration of genes coding for mouse MOPC 41 immunoglobulin light-chain mRNA. Proc. Natl. Acad. Sci. USA 73, 727–731.

Faust, C.H., Diggelmann, H., & Mach, B. (1974). Estimation of the number of genes coding for the constant part of the mouse immunoglobulin kappa light chain. Proc. Natl. Acad. Sci. USA 71, 2491–2495.

Feddersen, R.M. & Van Ness, B.G. (1985). Double recombination of a single immunoglobulin κ-chain allele: implications for the mechanism of rearrangement. Proc. Natl. Acad. Sci. USA 82, 4793–4797.

Feddersen, R.M. & Van Ness, B.G. (1990). Corrective recombination of mouse immunoglobulin kappa alleles in Abelson murine leukemia virus-transformed pre-B cells. Mol. Cell. Biol. 10, 569–576.

Feeney, A.J., Lugo, G., & Escuro, G. (1997). Human cord blood κ repertoire. J. Immunol. 158, 3761–3768.

Feeney, A.J. & Riblet, R. (1993). D_{ST4}: a new, and probably the last, functional D_H gene in the BALB/c mouse. Immunogenetics 37, 217–221.

Feng, J.-A., Johnson, R.C., & Dickerson, R.E. (1994). Hin recombinase bound to DNA: The origin of specificity in major and minor groove interactions. Science 263, 348–355.

Fett, J.W. & Deutsch, H.F. (1975). A new λ-chain gene. Immunochemistry 12, 643–652.

Flanagan, J.G. & Rabbitts, T.H. (1982). Arrangement of human immunoglobulin heavy chain constant region genes implies evolutionary duplication of a segment containing γ, ε and α genes. Nature 300, 709–713.

Foote, J. & Eisen, H.N. (1995). Kinetic and affinity limits on antibodies produced during immune responses. Proc. Natl. Acad. Sci. USA 92, 1254–1256.

Foster, S.J., Brezinschek, H.-P., Brezinschek, R.I., & Lipsky, P.E. (1997). Molecular mechanisms and selective influences that shape the kappa gene repertoire of IgM^+ B cells. J. Clin. Invest. 99, 1614–1627.

Freitas, A.A., Andrade, L., Lembezat, M.-P., & Coutinho, A. (1990). Selection of V_H gene repertoires: differentiating B cells of adult bone marrow mimic fetal development. Int. Immunol. 2, 15–23.

Friedman, M.L., Tunyaplin, C., Zhai, S.K., & Knight, K.L. (1994). Neonatal V_H, D, and J_H gene usage in rabbit B lineage cells. J. Immunol. 152, 632–641.

Frippiat, J.-P., Williams, S.C., Tomlinson, I.M., Cook, G.P., Cherif, D., Le Paslier, D., Collins, J.E., Dunham, I., Winter, G., & Lefranc, M.-P. (1995). Organization of the human immunoglobulin lambda light-chain locus on chromosome 22q11.2. Hum. Mol. Genet. 4, 983–991.

Fu, S.M., Winchester, R.J., & Kunkel, H.G. (1975). Similar idiotypic specificity for the membrane IgD and IgM of human B lymphocytes. J. Immunol. 114, 250–252.

Fujieda, S., Lin, Y.Q., Saxon, A., & Zhang, K. (1996). Multiple types of chimeric germ-line Ig heavy chain transcripts in human B cells. Evidence for *trans*-splicing of human Ig RNA. J. Immunol. 157, 3450–3459.

Fulop, G.M. & Phillips, R.A. (1990). The *scid* mutation in mice causes a general defect in DNA repair. Nature 347, 479–482.

Fuschiotti, P., Harindranath, N., Mage, R.G., McCormack, W.T., Dhanarajan, P., & Roux, K.H. (1993). Recombination activating genes-1 and -2 of the rabbit: cloning and characterization of germline and expressed genes. Mol. Immunol. 30, 1021–1032.

Gallarda, J.L., Gleason, K.S., & Knight, K.L. (1985). Organization of rabbit immunoglobulins genes. I. Structure and multiplicity of germ-line VH genes. J. Immunol. 135, 4222–4228.

Galli, G., Guise, J., Tucker, P.W., & Nevins, J.R. (1988). Poly(A) site choice rather than splice site choice governs the regulated production of IgM heavy-chain RNAs. Proc. Natl. Acad. Sci. USA 85, 2439–2443.

Gay, D., Saunders, T., Camper, S., & Weigert, M. (1993). Receptor editing: an approach by autoreactive B cells to escape tolerance. J. Exp. Med. 177, 999–1008.

Gearhart, P.J. & Bogenhagen, D.F. (1983). Clusters of point mutations are found exclusively around rearranged antibody variable genes. Proc. Natl. Acad. Sci. USA 80, 3439–3443.

Gearhart, P.J., Johnson, N.D., Douglas, R., & Hood, L. (1981). IgG antibodies to phosphorylcholine exhibit more diversity than their IgM counterparts. Nature 291, 29–34.

Gearhart, P.J., Sigal, N.H., & Klinman, N.R. (1975). Production of antibodies of identical idiotype but diverse immunoglobulin classes by cells derived from a single stimulated B cell. Proc. Natl. Acad. Sci. USA 72, 1707–1711.

Gellert, M. (1997). Recent advances in understanding V(D)J recombination. Adv. Immunol. 64, 39–64.

George, J.B., Li, S., & Garrard, W.T. (1995). Yeast artificial chromosome contigs reveal that distal variable-region genes reside at least 3 megabases from the joining regions in the murine immunoglobulin κ locus. Proc. Natl. Acad. Sci. USA 92, 12421–12425.

Gershenfeld, H.K., Tsukamoto, A., Weissman, I.L., & Joho, R. (1981). Somatic diversification is required to generate the V_κ genes of MOPC 511 and MOPC 167 myeloma proteins. Proc. Natl. Acad. Sci. USA 78, 7674–7678.

Gerstein, R.M., Frankel, W.N., Hsieh, C.-L., Durdik, J.M., Rath, S., Coffin, J.M., Nisonoff, A., & Selsing, E. (1990). Isotype switching of an immunoglobulin heavy chain transgene occurs by DNA recombination between different chromosomes. Cell 63, 537–548.

Ghose, A.C. & Karush, F. (1973). The affinity and temporal variation of isoelectric fractions of rabbit anti-lactose antibody. Biochemistry 12, 2437–2443.

Gibson, D., Levanon, M., & Smithies, O. (1971). Heterogeneity of normal human immunoglobulin light chains. Nonallelic variation in the constant region of λ chains. Biochemistry 10, 3114–3122.

Gilfillan, S., Dierich, A., Lemeur, M., Benoist, C., & Mathis, D. (1993). Mice lacking TdT: mature animals with an immature lymphocyte repertoire. Science 261, 1175–1178.

Giusti, A.M., Chien, N.C., Zack, D.J., Shin, S.-U., & Scharff, M.D. (1987). Somatic diversification of S107 from an antiphosphorylcholine to an anti-DNA autoantibody is due to a single base change in its heavy chain variable region. Proc. Natl. Acad. Sci. USA 84, 2926–2930.

Giusti, A.M. & Manser, T. (1994). Somatic generation of hybrid antibody H chain genes in transgenic mice via interchromosomal gene conversion. J. Exp. Med. 179, 235–248.

Givol, D., Zakut, R., Effron, K., Rechavi, G., Ram, D., & Cohen, J.B. (1981). Diversity of germ-line immunoglobulin V_H genes. Nature 292, 426–430.

Goding, J.W. & Layton, J.E. (1976). Antigen-induced co-capping of IgM and IgD-like receptors on murine B cells. J. Exp. Med. 144, 852–857.

Goetzl, E.J. & Metzger, H. (1970). Affinity labeling of a mouse myeloma protein which binds nitrophenyl ligands. Sequence and position of a labeled tryptic peptide. Biochemistry 9, 3862–3871.

González-Fernández, A., Gupta, S.K., Pannell, R., Neuberger, M.S., & Milstein, C. (1994). Somatic mutation of immunoglobulin λ chains: a segment of the major intron hypermutates as much as the complementarity-determining regions. Proc. Natl. Acad. Sci. USA 91, 12614–12618.

Gorski, J., Rollini, P., & Mach, B. (1983). Somatic mutations of immunoglobulin variable genes are restricted to the rearranged V gene. Science 220, 1179–1181.

Gough, N.M. & Bernard, O. (1981). Sequences of the joining region genes for immunoglobulin heavy chains and their role in generation of antibody diversity. Proc. Natl. Acad. Sci. USA 78, 509–513.

Grawunder, U., Leu, T.M.J., Schatz, D.G., Werner, A., Rolink, A.G., Melchers, F., & Winkler, T.H. (1995). Down-regulation of *RAG1* and *RAG2* gene expression in PreB cells after functional immunoglobulin heavy chain rearrangement. Immunity 3, 601–608.

Green, N.M. (1969). Electron microscopy of the immunoglobulins. Adv. Immunol. 11, 1–30.

Greenhalgh, P., Olesen, C.E.M., & Steiner, L.A. (1993). Characterization and expression of recombination activating genes (RAG-1 and RAG-2) in *Xenopus laevis*. J. Immunol. 151, 3100–3110.

Greenhalgh, P. & Steiner, L.A. (1995). Recombination activating gene 1 (*Rag*1) in zebrafish and shark. Immunogenet. 41, 54–55.

Griffiths, G.M., Berek, C., Kaartinen, M., & Milstein, C. (1984). Somatic mutation and the maturation of immune response to 2-phenyl oxazolone. Nature 312, 271–275.

Gritzmacher, C.A. (1989). Molecular aspects of heavy-chain class switching. Crit. Rev. Immunol. 9, 173–200.

Grossberg, A.L., Roholt, O.A., & Pressman, D. (1976). Identification of heavy chain tyrosine 33 in the binding site of myeloma protein McPC 603 by paired label iodination. J. Immunol. 116, 1596–1600.

Gu, H., Tarlinton, D., Müller, W., Rajewsky, K., & Förster, I. (1991). Most peripheral B cells in mice are ligand selected. J. Exp. Med. 173, 1357–1371.

Han, S., Dillon, S.R., Zheng, B., Shimoda, M., Schlissel, M.S., & Kelsoe, G. (1997). V(D)J recombinase activity in a subset of germinal center B lymphocytes. Science 278, 301–305.

Han, S., Zheng, B., Schatz, D.G., Spanopoulou, E., & Kelsoe, G. (1996). Neoteny in lymphocytes: *Rag1* and *Rag2* expression in germinal center B cells. Science 274, 2094–2097.

Hanly, W.C. & Knight, K.L. (1975). Cellular distribution of rabbit IgA allotypes. J. Immunol. 115, 590–594.

Hansen, J.D. & Kaattari, S.L. (1995). The recombination activating gene 1 (*RAG1*) of rainbow trout (Oncorhynchus mykiss): cloning, expression and phylogenetic analysis. Immunogenetics 42, 188–195.

Harada, K. & Yamagishi, H. (1991). Lack of feedback inhibition of V_κ gene rearrangement by productively rearranged alleles. J. Exp. Med. 173, 409–415.

Harboe, M., Osterland, K.C., Mannik, M., & Kunkel, H.G. (1962). Genetic characters of human γ-globulins in myeloma proteins. J. Exp. Med. 116, 719–738.

Hardy, R.R. & Hayakawa, K. (1994). CD5 B cells, a fetal B cell lineage. Adv. Immunol. 55, 297–339.

Hart, D.A., Wang, A.-L., Pawlak, L.L., & Nisonoff, A. (1972). Suppression of idiotypic specificities in adult mice by administration of antiidiotypic antibody. J. Exp. Med. 135, 1293–1300.

Haughton, G., Arnold, L.W., Whitmore, A.C., & Clarke, S.H. (1993). B-1 cells are made, not born. Immunol. Today 14, 84–87.

Hayakawa, K., Hardy, R.R., Parks, D.R., & Herzenberg, L.A. (1983). The "Ly-1 B" cell subpopulation in normal, immunodefective, and autoimmune mice. J. Exp. Med. 157, 202–218.

Heinrich, G., Traunecker, A., & Tonegawa, S. (1984). Somatic mutation creates diversity in the major group of mouse immunoglobulin κ light chains. J. Exp. Med. 159, 417–435.

Held, W., Roland, J., & Raulet, D.H. (1995). Allelic exclusion of Ly49-family genes encoding class I MHC-specific receptors on NK cells. Nature 376, 355–358.

Hendrickson, E.A., Qin, X.-Q., Bump, E.A., Schatz, D.G., Oettinger, M., & Weaver, D.T. (1991). A link between double-strand break-related repair and V(D)J recombination: the *scid* mutation. Proc. Natl. Acad. Sci. USA 88, 4061–4065.

Herzenberg, L.A. (1964). A chromosome region for gamma$_{2a}$ and beta$_{2A}$ globulin H chain isoantigens in the mouse. Cold Spring Harbor Symp. Quant. Biol. 29, 455–462.

Hess, M., Hilschmann, N., Rivat, L., Rivat, C., & Ropartz, C. (1971). Isotypes in human immunoglobulin λ-chains. Nat. New Biol. 234, 58–61.

Hesse, J.E., Lieber, M.R., Gellert, M., & Mizuuchi, K. (1987). Extrachromosomal DNA substrates in pre-B cells undergo inversion or deletion at immunoglobulin V-(D)-J joining signals. Cell 49, 775–783.

Hesse, J.E., Lieber, M.R., Mizuuchi, K., & Gellert, M. (1989). V(D)J recombination: a functional definition of the joining signals. Genes Dev. 3, 1053–1061.

Hieter, P.A., Hollis, G.F., Korsmeyer, S.J., Waldmann, T.A., & Leder, P. (1981a). Clustered arrangement of immunoglobulin λ constant region genes in man. Nature 294, 536–540.

Hieter, P.A., Korsmeyer, S.J., Waldmann, T.A., & Leder, P. (1981b). Human immunoglobulin κ light-chain genes are deleted or rearranged in λ-producing B cells. Nature 290, 368–372.

Hieter, P.A., Maizel, J.V., Jr., & Leder, P. (1982). Evolution of human immunoglobulin κ J region genes. J. Biol. Chem. 257, 1516–1522.

Hieter, P.A., Max, E.E., Seidman, J.G., Maizel, J.V., Jr., & Leder, P. (1980). Cloned human and mouse kappa immunoglobulin constant and J region genes conserve homology in functional segments. Cell 22, 197–207.

Hikida, M., Mori, M., Takai, T., Tomochika, K., Hamatani, K., & Ohmori, H. (1996). Reexpression of RAG-1 and RAG-2 genes in activated mature mouse B cells. Science 274, 2092–2094.

Hinds-Frey, K.R., Nishikata, H., Litman, R.T., & Litman, G.W. (1993). Somatic variation precedes extensive diversification of germline sequences and combinatorial joining in the evolution of immunoglobulin heavy chain diversity. J. Exp. Med. 178, 815–824.

Hirama, T., Takeshita, S., Yoshida, Y., & Yamagishi, H. (1991). Structure of extrachromosomal circular DNAs generated by immunoglobulin light chain gene rearrangements. Immunol. Lett. 27, 19–23.

Höchtl, J., Müller, C.R., & Zachau, H.G. (1982). Recombined flanks of the variable and joining segments of immunoglobulin genes. Proc. Natl. Acad. Sci. USA 79, 1383–1387.

Höchtl, J. & Zachau, H.G. (1983). A novel type of aberrant recombination in immunoglobulin genes and its implications for *V-J* joining mechanism. Nature 302, 260–263.

Hofker, M.H., Walter, M.A., & Cox, D.W. (1989). Complete physical map of the human heavy chain constant region gene complex. Proc. Natl. Acad. Sci. USA 86, 5567–5571.

Honjo, T. & Kataoka, T. (1978). Organization of immunoglobulin heavy chain genes and allelic deletion model. Proc. Natl. Acad. Sci. USA 75, 2140–2144.

Honjo, T. & Matsuda, F. (1995). Immunoglobulin heavy chain loci of mouse and human. In: Immunoglobulin Genes, 2nd ed. (Honjo, T., Alt, F.W., eds.), pp. 145–171. Academic Press, London.

Honjo, T., Packman, S., Swan, D., & Leder, P. (1976). Quantitation of constant and variable region genes for mouse immunoglobulin λ chains. Biochemistry 15, 2780–2785.

Honjo, T., Packman, S., Swan, D., Nau, M., & Leder, P. (1974). Organization of immunoglobulin genes: reiteration frequency of the mouse κ chain constant region gene. Proc. Natl. Acad. Sci. USA 71, 3659–3663.

Hood, L., Gray, W.R., Sanders, B.G., & Dreyer, W.J. (1967). Light chain evolution. Cold Spring Harbor Symp. Quant. Biol. 32, 133–146.

Hood, L. & Talmage, D.W. (1970). Mechanism of antibody diversity: germ line basis for variability. Science 168, 325–334.

Hozumi, N. & Tonegawa, S. (1976). Evidence for somatic rearrangement of immunoglobulin genes coding for variable and constant regions. Proc. Natl. Acad. Sci. USA 73, 3628–3632.

Huang, H., Crews, S., & Hood, L. (1981). An immunoglobulin V_H pseudogene. J. Mol. Appl. Genet. 1, 93–101.

Huber, C., Klobeck, H.-G., & Zachau, H.G. (1992). Ongoing V_κ-J_κ recombination after formation of a productive V_κ-J_κ coding joint. Eur. J. Immunol. 22, 1561–1565.

Huetz, F., Carlsson, L., Tornberg, U.-C., & Holmberg, D. (1993). V-region directed selection in differentiating B lymphocytes. EMBO J. 12, 1819–1826.

Hurwitz, J.L., Coleclough, C., & Cebra, J.J. (1980). C_H gene rearrangements in IgM-bearing B cells and in the normal splenic DNA component of hybridomas making different isotypes of antibody. Cell 22, 349–359.

Ichihara, Y., Abe, M., Yasui, H., Matsuoka, H., & Kurosawa, Y. (1988a). At least five D_H genes of human immunoglobulin heavy chains are encoded in 9-kilobase DNA fragments. Eur. J. Immunol. 18, 649–652.

Ichihara, Y., Matsuoka, H., & Kurosawa, Y. (1988b). Organization of human immunoglobulin heavy chain diversity gene loci. EMBO J. 7, 4141–4150.

Iglesias, A., Kopf, M., Williams, G.S., Bühler, B., & Köhler, G. (1991). Molecular requirements for the μ-induced light chain gene rearrangement in pre-B cells. EMBO J. 10, 2147–2156.

Ignatovich, O., Tomlinson, I.M., Jones, P.T., & Winter, G. (1997). The creation of diversity in the human immunoglobulin V_λ repertoire. J. Mol. Biol. 268, 69–77.

Ishida, N., Ueda, S., Hayashida, H., Miyata, T., & Honjo, T. (1982). The nucleotide sequence of the mouse immunoglobulin epsilon gene: comparison with the human epsilon gene sequence. EMBO J. 1, 1117–1123.

Iwasato, T., Shimizu, A., Honjo, T., & Yamagishi, H. (1990). Circular DNA is excised by immunoglobulin class switch recombination. Cell 62, 143–149.

Jacob, F. (1982). The Possible and the Actual, U. of Washington Press, Seattle.

Jacob, J., Kelsoe, G., Rajewsky, K., & Weiss, U. (1991). Intraclonal generation of antibody mutants in germinal centres. Nature 354, 389–392.

Jaenisch, R. (1997). DNA methylation and imprinting: why bother? Trends Genet. 13, 323–329.

Jaton, J.-C. & Vassalli, P. (1980). A major structural difference between membrane-bound and secreted IgM of normal mouse spleen cells is located in the C-terminal region of their heavy chains. FEBS Lett. 116, 277–280.

Jeong, H.D. & Teale, J.M. (1988). Comparison of the fetal and adult functional B cell repertoires by analysis of V_H gene family expression. J. Exp. Med. 168, 589–603.

Jerne, N.K. (1955). The natural-selection theory of antibody formation. Proc. Natl. Acad. Sci. USA 41, 849–857.

Joho, R. & Weissman, I.L. (1980). V-J joining of immunoglobulin κ genes only occurs on one homologous chromosome. Nature 284, 179–181.

Kabat, E.A., Wu, T.T., Perry, H.M., Gottesman, K.S., & Foeller, C. (1991). Sequences of proteins of immunological interest. U.S. Department of Health and Human Services, Public Health Service, National Institutes of Health, Bethesda, 5th edition.

Kaartinen, M. & Makela, O. (1985). Reading of D genes in variable frames as a source of antibody diversity. Immunology Today 6, 324–327.

Kalinke, U., Bucher, E.M., Ernst, B., Oxenius, A., Roost, H.-P., Geley, S., Kofler, R., Zinkernagel, R.M., & Hengartner, H. (1996). The role of somatic mutation in the generation of the protective humoral immune response against vesicular stomatitis virus. Immunity 5, 639–652.

Karush, F. (1978). The affinity of antibody: range, variability, and the role of multivalence. In Comprehensive Immunology, Vol. 5, Immunoglobulins, pp. 85–116, (Litman, G.W., & Good, R.A., Eds.). Plenum Medical Book Company, New York.

Kataoka, T., Kawakami, T., Takahashi, N., & Honjo, T. (1980). Rearrangement of immunoglobulin γ1-chain gene and mechanism for heavy-chain class switch. Proc. Natl. Acad. Sci. USA 77, 919–923.

Katzenberg, D.R. & Birshtein, B.K. (1988). Sites of switch recombination in IgG2b- and IgG2a-producing hybridomas. J. Immunol. 140, 3219–3227.

Kawasaki, K., Minoshima, S., Nakato, E., Shibuya, K., Shintani, A., Schmeits, J.L., Wang, J., & Shimizu, N. (1997). One-megabase sequence analysis of the human immunoglobulin λ γene locus. Genome Res. 7, 250–261.

Kearney, J.F., & Vakil, M. (1986). Idiotype-directed interactions during ontogeny play a major role in the establishment of the adult B cell repertoire. Immunol. Revs. 94, 39–50.

Kehry, M., Ewald, S., Douglas, R., Sibley, C., Raschke, W., Fambrough, D., & Hood, L. (1980). The immunoglobulin μ chains of membrane-bound and secreted IgM molecules differ in their C-terminal segments. Cell 21, 393–406.

Kelley, D.E., Wiedemann, L.M., Pittet, A.-C., Strauss, S., Nelson, K.J., Davis, J., Van Ness, B., & Perry, R.P. (1985). Nonproductive kappa immunoglobulin genes: recombinational abnormalities and other lesions affecting transcription, RNA processing, turnover, and translation. Mol. Cell. Biol. 5, 1660–1675.

Kemp, D.J., Tyler, B., Bernard, O., Gough, N., Gerondakis, S., Adams, J.M., & Cory, S. (1981). Organization of genes and spacers within the mouse immunoglobulin V_H locus. J. Mol. Appl. Genet. 1, 245–261.

Kim, N., Kage, K., Matsuda, F., Lefranc, M.-P., & Storb, U. (1997). B lymphocytes of xeroderma pigmentosum or Cockayne syndrome patients with inherited defects in nucleotide excision repair are fully capable of somatic hypermutation of immunoglobulin genes. J. Exp. Med. 186, 413–419.

Kim, S., Davis, M., Sinn, E., Patten, P., & Hood, L. (1981). Antibody diversity: somatic hypermutation of rearranged V_H genes. Cell 27, 573–581.

Kinashi, T., Godal, T., Noma, Y., Ling, N.R., Yaoita, Y., & Honjo, T. (1987). Human neoplastic B cells express more than two isotypes of immunoglobulins without deletion of heavy-chain constant-region genes. Genes & Dev. 1, 465–470.

Kindt, T.J., Gris, C., Guenet, J.L., Bonhomme, F., & Cazenave, P.-A. (1985). Lambda light chain constant and variable gene complements in wild-derived inbred mouse strains. Eur. J. Immunol. 15, 535–540.

Kindt, T.J. & Mandy, W.J. (1972). Recombination of genes coding for constant and variable regions of immunoglobulin heavy chains. J. Immunol. 108, 1110–1113.

Kindt, T.J., Mandy, W.J., & Todd, C.W. (1970). Association of allotypic specificities of group a with allotypic specificities A11 and A12 in rabbit immunoglobulin. Biochemistry 9, 2028–2032.

Kipps, T.J. & Herzenberg, L.A. (1986). Homologous chromosome recombination generating immunoglobulin allotype and isotype switch variants. EMBO J. 5, 263–268.

Kirsch, I.R., Morton, C.C., Nakahara, K., & Leder, P. (1982). Human immunoglobulin heavy chain genes map to a region of translocations in malignant B lymphocytes. Science 216, 301–303.

Kirschbaum, T., Jaenichen, R., & Zachau, H.G. (1996). The mouse immunoglobulin κ locus contains about 140 variable gene segments. Eur. J. Immunol. 26, 1613–1620.

Kitamura, D. & Rajewsky, K. (1992). Targeted disruption of μ chain membrane exon causes loss of heavy-chain allelic exclusion. Nature 356, 154–156.

Kleinfeld, R., Hardy, R.R., Tarlinton, D., Dangl, J., Herzenberg, L.A., & Weigert, M. (1986) Recombination between an expressed immunoglobulin heavy-chain gene and a germline variable gene segment in a Ly 1^+ B-cell lymphoma. Nature 322, 843–846.

Klinman, N.R. (1996). The "clonal selection hypothesis" and current concepts of B cell tolerance. Immunity 5, 189–195.

Klinman, N.R. & Press, J.L. (1975). The B cell specificity repertoire: its relationship to definable subpopulations. Transplant. Rev. 24, 41–83.

Klobeck, H.-G., Combriato, G., & Zachau, H.G. (1987). N segment insertion and region-directed somatic hypermutation in a Kappa gene of a t(2;8) chromosomal translocation. Nucleic Acids Res. 15, 4877–4888.

Kluin, P.M., Kayano, H., Zani, V.J., Kluin-Nelemans, H.C., Tucker, P.W., Satterwhite, E., & Dyer, M.J.S. (1995). IgD class switching: identification of a novel recombination site in neoplastic and normal B cells. Eur. J. Immunol. 25, 3504–3508.

Knapp, M.R., Liu, C.-P., Newell, N., Ward, R.B., Tucker, P.W., Strober, S., & Blattner, F. (1982). Simultaneous expression of immunoglobulin μ and δ heavy chains by a cloned B-cell lymphoma: A single copy of the V_H gene is shared by two adjacent C_H genes. Proc. Natl. Acad. Sci. USA 79, 2996–3000.

Knight, K.L. & Becker, R.S. (1990). Molecular basis of the allelic inheritance of rabbit immunoglobulin VH allotypes: implications for the generation of antibody diversity. Cell 60, 963–970.

Knight, K.L., Kingzette, M., Crane, M.A., & Zhai, S.-K. (1995). Transchromosomally derived Ig heavy chains. J. Immunol. 155, 684–691.

Knight, K.L., Malek, T.R., & Hanly, W.C. (1974). Recombinant rabbit secretory immunoglobulin molecules: alpha chains with maternal (paternal) variable-region allotypes and paternal (maternal) constant-region allotypes. Proc. Natl. Acad. Sci. USA 71, 1169–1173.

Kodaira, M., Kinashi, T., Umemura, I., Matsuda, F., Noma, T., Ono, Y., & Honjo, T. (1986). Organization and evolution of variable region genes of the human immunoglobulin heavy chain. J. Mol. Biol. 190, 529–541.

Kofler, R., Geley, S., Kofler, H., & Helmberg, A. (1992). Mouse variable-region gene families: Compexity, polymorphism and use in non-autoimmune responses. Immunol. Revs. 128, 5–21.

Komaromy, M., Clayton, L., Rogers, J., Robertson, S., Kettman, J., & Wall, R. (1983). The structure of the mouse immunoglobulin in γ3 membrane gene segment. Nucleic Acids Res. 11, 6775–6785.

Komori, T., Okada, A., Stewart, V., & Alt, F.W. (1993). Lack of N regions in antigen receptor variable region genes of TdT-deficient lymphocytes. Science 261, 1171–1175.

Koshland, M.E. (1985). The coming of age of the immunoglobulin J chain. Annu. Rev. Immunol. 3, 425–453.

Kraal, G., Weissman, I.L., & Butcher, E.C. (1982). Germinal centre B cells: antigen specificity and changes in heavy chain class expression. Nature 298, 377–379.

Kraj, P., Rao, S.P., Glas, A.M., Hardy, R.R., Milner, E.C.B., & Silberstein, L.E. (1997). The human heavy chain Ig V region gene repertoire is biased at all stages of B cell ontogeny, including early pre-B cells. J. Immunol. 158, 5824–5832.

Kronenberg, M., Siu, G., Hood, L.E., & Shastri, N. (1986). The molecular genetics of the T-cell antigen receptor and T-cell antigen recognition. Annu. Rev. Immunol. 4, 529–591.

Kubagawa, H., Cooper, M.D., Carroll, A.J., & Burrows, P.D. (1989). Light-chain gene expression before heavy-chain gene rearrangement in pre-B cells transformed by Epstein-Barr virus. Proc. Natl. Acad. Sci. USA 86, 2356–2360.

Kuettner, M.G., Wang, A.-L., & Nisonoff, A. (1972). Quantitative investigations of idiotypic antibodies. VI. Idiotypic specificity as a potential genetic marker for the variable regions of mouse immunoglobulin polypeptide chains. J. Exp. Med.135, 579–595.

Kunkel, H.G., Smith, W.K., Joslin, F.G., Natvig, J.B., & Litwin, S.D. (1969). Genetic marker of the γA2 subgroup of γA immunoglobulins. Nature 223, 1247–1248.

Küppers, R., Zhao, M., Hansmann, M.-L., & Rajewsky, K. (1993). Tracing B cell development in human germinal centres by molecular analysis of single cells picked from histological sections. EMBO J. 12, 4955–4967.

Kurosawa, Y. & Tonegawa, S. (1982). Organization, structure, and assembly of immunoglobulin heavy chain diversity DNA segments. J. Exp. Med. 155, 201–218.

Kurosawa, Y., von Boehmer, H., Haas, W., Sakano, H., Trauneker, A., & Tonegawa, S. (1981). Identification of D segments of immunoglobulin heavy-chain genes and their rearrangement in T lymphocytes. Nature 290, 565–570.

Kwan, S.-P., Max, E.E, Seidman, J.G., Leder, P., & Scharff, M.D. (1981). Two kappa immunoglobulin genes are expressed in the myeloma S107. Cell 26, 57–66.

Lafaille, J.J., DeCloux, A., Bonneville, M., Takagaki, Y., & Tonegawa, S. (1989). Junctional sequences of T cell receptor γδ genes: implications for γδ T cell lineages and for a novel intermediate of V-(D)-J joining. Cell 59, 859–870.

Lai, M.M.C. (1992). RNA recombination in animal and plant viruses. Microbiological Reviews 56, 61–79.

Landucci Tosi, S., Mage, R.G., & Dubiski, S. (1970). Distribution of allotypic specifities A1, A2, A14, and A15 among immunoglobulin G molecules. J. Immunol. 104, 641–647.

Landucci Tosi, S. & Tosi, R.M. (1973). Recombinant IgG molecules in rabbits doubly heterozygous for group *a* and group *e* allotypic specificities. Immunochemistry 10, 65–71.

Langman, R.E. & Cohn, M. (1987). The E-T (elephant-tadpole) paradox necessitates the concept of a unit of B-cell function: the protecton. Mol. Immunol. 24, 675–697.

Lawler, A.M., Lin, P.S., & Gearhart, P.J. (1987). Adult B-cell repertoire is biased toward two heavy-chain variable-region genes that rearrange frequently in fetal pre-B cells. Proc. Natl. Acad. Sci. USA 84, 2454–2458.

Lebecque, S.G. & Gearhart, P.J. (1990). Boundaries of somatic mutation in rearranged immunoglobulin genes: 5′ boundary is near the promoter, and 3′ boundary is ~ 1 kb from V(D)J gene. J. Exp. Med. 172, 1717–1727.

Lederberg, J. (1959). Genes and antibodies: do antigens bear instructions for antibody specificity or do they select cells lines that arise by mutation? Science 129, 1649–1653.

Lee, S.K., Bridges, S.L., Jr., Koopman, W.J., & Schroeder, H.W., Jr. (1992). The immunoglobulin kappa light chain repertoire expressed in the synovium of a patient with rheumatoid arthritis. Arthritis Rheum. 35, 905–913.

Lefranc, M.-P., Lefranc, G., & Rabbitts, T.H. (1982). Inherited deletion of immunoglobulin heavy chain constant region genes in normal human individuals. Nature 300, 760–762.

Lenhard-Schuller, R., Hohn, B., Brack, C., Hirama, M., & Tonegawa, S. (1978). DNA clones containing mouse immunoglobulin κ chain genes isolated by *in vitro* packaging into phage λ coats. Proc. Natl. Acad. Sci. USA 75, 4709–4713.

Lennox, E.S. & Cohn, M. (1967). Immunoglobulins. Annu. Rev. Biochem. 36, 365–406.

Leu, T.M.J., Eastman, Q.M., & Schatz, D.G. (1997). Coding joint formation in a cell-free V(D)J recombination system. Immunity 7, 303–314.

Levitt, D. & Cooper, M.D. (1980). Mouse pre-B cells synthesize and secrete μ heavy chains but not light chains. Cell 19, 617–625.

Levy, S., Campbell, M.J., & Levy, R. (1989). Functional immunoglobulin light chain genes are replaced by ongoing rearrangements of germline V_κ genes to downstream J_κ segments in a murine B cell line. J. Exp. Med. 170, 1–13.

Lewis, S.M. (1994a). The mechanism of V(D)J joining: lessons from molecular, immunological, and comparative analyses. Adv. Immunol. 56, 27–150.

Lewis, S.M. (1994b). P nucleotide insertions and the resolution of hairpin DNA structures in mammalian cells. Proc. Natl. Acad. Sci. USA 91, 1332–1336.

Lewis, S., Gifford, A., & Baltimore, D. (1984). Joining of V_κ to J_κ gene segments in a retroviral vector introduced into lymphoid cells. Nature 308, 425–428.

Lewis, S., Rosenberg, N., Alt, F., & Baltimore, D. (1982). Continuing kappa-gene rearrangement in a cell line transformed by Abelson murine leukemia virus. Cell 30, 807–816.

Lewis, S.M. & Wu, G.E. (1997). The origins of V(D)J recombination. Cell 88, 159–162.

Li, Y.-S., Hayakawa, K., & Hardy, R.R. (1993). The regulated expression of B lineage associated genes during B cell differentiation in bone marrow and fetal liver. J. Exp. Med. 178, 951–960.

Lieber, M.R. (1991). Site-specific recombination in the immune system. FASEB J. 5, 2934–2944.

Lieber, M.R., Hesse, J.E., Lewis, S., Bosma, G.C., Rosenberg, N., Mizuuchi, K., Bosma, M.J., & Gellert, M. (1988). The defect in murine severe combined immunodeficiency: joining of signal sequences but not coding segments in V(D)J recombination. Cell 55, 7–16.

Lieberman, R. & Potter, M. (1966). Close linkage in genes controlling γA and γG heavy chain structure in BALB/c mice. J. Mol. Biol. 18, 516–528.

Liu. C.-P., Tucker, P.W., Mushinski, J.F., & Blattner, F.R. (1980). Mapping of heavy chain genes for mouse immunoglobulins M and D. Science 209, 1348–1353.

Liu, Y.-J. & Banchereau, J. (1996). The paths and molecular controls of peripheral B-cell development. The Immunologist 4, 55–66.

Liu, Y.J., Malisan, F., de Bouteiller, O., Guret, C., Lebecque, S., Banchereau, J., Mills, F.C., Max, E.E., & Martinez-Valdez, H. (1996). Within germinal centers, isotype switching of immunoglobulin genes occurs after the onset of somatic mutation. Immunity 4, 241–250.

Livant, D., Blatt, C., & Hood, L. (1986). One heavy chain variable region gene segment subfamily in the BALB/c mouse contains 500–1000 or more members. Cell 47, 461–470.

Löffert, D., Ehlich, A., Müller, W., & Rajewsky, K. (1996). Surrogate light chain expression is required to establish immunoglobulin heavy chain allelic exclusion during early B cell development. Immunity 4, 133–144.

Loken, M.R., Shah, V.O., Dattilio, K.L., & Civin, C.I. (1987). Flow cytometric analysis of human bone marrow. II. Normal B lymphocyte development. Blood 70, 1316–1324.

Loor, F., Forni, L., & Pernis, B. (1972). The dynamic state of the lymphocyte membrane. Factors affecting the distribution and turnover of surface immunoglobulins. Eur. J. Immunol. 2, 203–212.

Luning Prak, E. & Weigert, M. (1995). Light chain replacement: a new model for antibody gene rearrangement. J. Exp. Med. 182, 541–548.

Ma, A., Fisher, P., Dildrop, R., Oltz, E., Rathbun, G., Achacoso, P., Stall, A., & Alt, F.W. (1992). Surface IgM mediated regulation of *RAG* gene expression in Eμ-N-*myc* B cell lines. EMBO J. 11, 2727–2734.

Mach, B., Faust, C., & Vassalli, P. (1973). Purification of 14S messenger RNA of immunoglobulin light chain that codes for a possible light chain precursor. Proc. Natl. Acad. Sci. USA 70, 451–455.

MacLennan, I.C.M. (1994). Germinal centers. Annu. Rev. Immunol. 12, 117–139.

MacLennan, I.C.M. & Gray, D. (1986). Antigen-driven selection of virgin and memory B cells. Immunol. Rev. 91, 61–85.

Maeda, T., Sugiyama, H., Tani, Y., Miyake, S., Oka, Y., Ogawa, H., Komori, T., Soma, T., & Kishimoto, S. (1987). Start of μ-chain production by the further two-step rearrangements of immunoglobulin heavy chain genes on one chromosome from a DJ_H/DJ_H configuration in an Abelson virus-transformed cell line: evidence of secondary DJ_H complex formation. J. Immunol. 138, 2305–2310.

Mage, R.G., Young-Cooper, G.O., & Alexander, C. (1971). Genetic control of variable and constant regions of immunoglobulin heavy chains. Nature New Biology 230, 63–64.

Mainville, C.A., Sheehan, K.M., Klaman, L.D., Giorgetti, C.A., Press, J.L., & Brodeur, P.H. (1996). Deletional mapping of fifteen mouse V_H gene families reveals a common organization for three *Igh* haplotypes. J. Immunol. 156, 1038–1046.

Maki, R., Roeder, W., Traunecker, A., Sidman, C., Wabl, M., Raschke, W., & Tonegawa, S. (1981). The role of DNA rearrangement and alternative RNA processing in the expression of immunoglobulin delta genes. Cell 24, 353–365.

Malcolm, S., Barton, P., Murphy, C., Ferguson-Smith, M.A., Bentley, D.L., & Rabbits, T.H. (1982). Localization of human immunoglobulin κ light chain variable region genes to the short arm of chromosome 2 by *in situ* hybridization. Proc. Natl. Acad. Sci. USA 79, 4957–4961.

Malissen, M., Trucy, J., Jouvin-Marche, E., Cazenave, P.-A., Scollay, R., & Malissen, B. (1992). Regulation of TCR α and β gene allelic exclusion during T-cell development. Immunology Today 13, 315–322.

Malynn, B.A., Yancopoulos, G.D., Barth, J.E., Bona, C.A., & Alt, F.W. (1990). Biased expression of J_H-proximal V_H genes occurs in the newly generated repertoire of neonatal and adult mice. J. Exp. Med. 171, 843–859.

Mandy, W.J. & Todd, C.W. (1970). Rabbit immunoglobulin allotype A12: a new agglutinating specificity. Biochem. Genetics 4, 59–71.

Manser, T. & Gefter, M.L. (1986). The molecular evolution of the immune response: idiotype-specific suppression indicates that B cells express germ-line-encoded V genes prior to antigenic stimulation. Eur. J. Immunol. 16, 1439–1444.

Manser, T., Parhami-Seren, B., Margolies, M.N., & Gefter, M.L. (1987). Somatically mutated forms of a major anti-*p*-azophenylarsonate antibody variable region with drastically reduced affinity for *p*-azophenylarsonate. By-products of an antigen-driven immune response? J. Exp. Med. 166, 1456–1463.

Manz, J., Denis, K., Witte, O., Brinster, R., & Storb, U. (1988). Feedback inhibition of immunoglobulin gene rearrangement by membrane μ, but not by secreted μ heavy chains. J. Exp. Med. 168, 1363–1381.

Marchalonis, J.J. (1969). An enzymic method for the trace iodination of immunoglobulins and other proteins. Biochem. J. 113, 299–305.

Marchalonis, J.J., Cone, R.E., & Santer, V. (1971). Enzymic iodination. A probe for accessible surface proteins of normal and neoplastic lymphocytes. Biochem. J. 124, 921–927.

Mårtensson, L. (1961). Gm characters of M-components. Acta Medica Scand. 170, suppl 367, 87–93.

Martin, T., Blaison, G., Levallois, H., & Pasquali, J.L. (1992). Molecular analysis of the VκIII-Jκ junctional diversity of polyclonal rheumatoid factors during rheumatoid arthritis frequently reveals N addition. Eur. J. Immunol. 22, 1773–1779.

Matsuda, F. & Honjo, T. (1996). Organization of the human immunoglobulin heavy-chain locus. Adv. Immunol. 62, 1–29.

Matsuda, F., Shin, E.K., Nagaoka, H., Matsumura, R., Haino, M., Fukita, Y., Taka-ishi, S., Imai, T., Riley, J.H., Anand, R., Soeda, E., & Honjo, T. (1993). Structure and physical map of 64 variable segments in the 3′ 0.8-megabase region of the human immunoglobulin heavy-chain locus. Nature Genet. 3, 88–94.

Matsuoka, M., Yoshida, K., Maeda, T., Usuda, S., & Sakano, H. (1990). Switch circular DNA formed in cytokine-treated mouse splenocytes: evidence for intramolecular DNA deletion in immunoglobulin class switching. Cell 62, 135–142.

Matthyssens, G. & Tonegawa, S. (1978). V and C parts of immunoglobulin κ-chain genes are separate in myeloma. Nature 273, 763–765.

Max, E.E., Maizel, J.V., Jr., & Leder, P. (1981). The nucleotide sequence of a 5.5-kilobase DNA segment containing the mouse κ immunoglobulin J and C region genes. J. Biol. Chem. 256, 5116–5120.

Max, E.E., Seidman, J.G., & Leder, P. (1979). Sequences of five potential recombination sites encoded close to an immunoglobulin κ constant region gene. Proc. Natl. Acad. Sci. USA 76, 3450–3454.

Max, E.E., Seidman, J.G., Miller, H., & Leder, P. (1980). Variation in the crossover point of kappa immunoglobulin gene recombination: evidence from a cryptic gene. Cell 21, 793–799.

Maxam, A.M. & Gilbert, W. (1977). A new method for sequencing DNA. Proc. Natl. Acad. Sci. USA 74, 560–564.

McBlane, J.F., van Gent, D.C., Ramsden, D.A., Romeo, C., Cuomo, C.A., Gellert, M., & Oettinger, M.A. (1995). Cleavage at a V(D)J recombination signal requires only RAG1 and RAG2 proteins and occurs in two steps. Cell 83, 387–395.

McCormack, W.T., Tjoelker, L.W., Carlson, L.M., Petryniak, B., Barth, C.F., Humphries, E.H., & Thompson, C.B. (1989). Chicken Ig_L gene rearrangement involves deletion of a circular episome and addition of single nonrandom nucleotides to both coding segments. Cell 56, 785–791.

McCormack, W.T., Tjoelker, L.W., & Thompson, C.B. (1991). Avian B-cell development: generation of an immunoglobulin repertoire by gene conversion. Annu. Rev. Immunol. 9, 219–241.

McCune, J.M., Lingappa, V.R., Fu, S.M., Blobel, G., & Kunkel, H.G. (1980). Membrane-bound and secreted immunoglobulins. I. Two distinct translation products of human μ-chain, with identical N-termini and different C-termini. J. Exp. Med. 152, 463–468.

McIntire, K.R. & Rouse, A.M. (1970). Mouse immunoglobulin light chains: alteration of κ:λ ratio. Fed. Proc. 29,704 (abstr).

McKean, D., Huppi, K., Bell, M., Staudt, L., Gerhard, W., & Weigert, M. (1984). Generation of antibody diversity in the immune response of BALB/c mice to influenza virus hemagglutinin. Proc. Natl. Acad. Sci. USA 81, 3180–3184.

Meek, K.D., Hasemann, C.A., & Capra, J.D. (1989). Novel rearrangements at the immunoglobulin D locus. Inversions and fusions add to IgH somatic diversity. J. Exp. Med. 170, 39–57.

Meek, K., Rathbun, G., Reininger, L., Jaton, J.-C., Kofler, R., Tucker, P.W., & Capra, J.D. (1990). Organization of the murine immunoglobulin VH complex: placement of two new VH families (VH10 and VH11) and analysis of VH family clustering and interdigitation. Mol. Immunol. 27, 1073–1081.

Meier, J.T. & Lewis, S.M. (1993). P Nucleotides in V(D)J recombination: a fine-structure analysis. Mol. Cell. Biol. 13, 1078–1092.

Melcher, U. & Uhr, J.W. (1976). Cell surface immunoglobulin. XVI. Polypeptide chain structure of mouse IgM and IgD-like molecule. J. Immunol. 116, 409–415.

Melcher, U. & Uhr, J.W. (1977). Density differences between membrane and secreted immunoglobulins of murine splenocytes. Biochemistry 16, 145–152.

Melcher, U., Vitetta, E.S., McWilliams, M., Lamm, M.E., Phillips-Quagliata, J.M., & Uhr, J.W. (1974). Cell surface immunoglobulin. X. Identification of an IgD-like molecule on the surface of murine splenocytes. J. Exp. Med. 140, 1427–1431.

Melchers, F., Karasuyama, H., Haasner, D., Bauer, S., Kudo, A., Sakaguchi, N., Jameson, B., & Rolink, A. (1993). The surrogate light chain in B-cell development. Immunology Today 14, 60–68.

Melli, M., Whitfield, C., Rao, K.V., Richardson, M., & Bishop, J.O. (1971). DNA-RNA hybridization in vast DNA excess. Nature New Biol. 231, 8–12.

Metzger, H. (1970). Structure and function of γM macroglobulins. Adv. Immunol. 12, 57–116.

Meyer, K.B. & Neuberger, M.S. (1989). The immunoglobulin k locus contains a second, stronger B-cell-specific enhancer which is located downstream of the constant region. EMBO J. 8, 1959–1964.

Miller, J., Bothwell, A., & Storb, U. (1981). Physical linkage of the constant region genes for immunoglobulins λ_I and λ_{III}. Proc. Natl. Acad. Sci. USA 78, 3829–3833.

Miller, J., Ogden, S., McMullen, M., Andres, H., & Storb, U. (1988). The order and orientation of mouse λ-genes explain λ-rearrangement patterns. J. Immunol. 141, 2497–2502.

Miller, J., Selsing, E., & Storb, U. (1982). Structural alterations in J regions of mouse immunoglobulin λ genes are associated with differential gene expression. Nature 295, 428–430.

Milstein, C., Brownlee, G.G., Harrison, T.M., & Mathews, M.B. (1972). A possible precursor of immunoglobulin light chains. Nature New Biol. 239, 117–120.

Milstein, C.P., Deverson, E.V., & Rabbitts, T.H. (1984). The sequence of the human μ-δ intron reveals possible vestigial switch segments. Nucleic Acids Res. 12, 6523–6535.

Milstein, C., Even, J., Jarvis, J.M., Gonzalez-Fernandez, A., & Gherardi, E. (1992). Non-random features of the repertoire expressed by the members of one V_κ gene family and of the V-J recombination. Eur. J. Immunol. 22, 1627–1634.

Milstein, C. & Neuberger, M.S. (1996). Maturation of the immune response. Adv. Prot. Chem. 49, 451–485.

Möller, G. (1961). Demonstration of mouse isoantigens at the cellular level by the fluorescent antibody technique. J. Exp. Med. 114, 415–434.

Mombaerts, P., Iacomini, J., Johnson, R.S., Herrup, K., Tonegawa, S., & Papaioannou, V.E. (1992). RAG1-deficient mice have no mature B and T lymphocytes. Cell 68, 869–877.

Moore, K.W., Rogers, J., Hunkapiller, T., Early, P., Nottenburg, C., Weissman, I., Bazin, H., Wall, R., & Hood, L.E. (1981). Expression of IgD may use both DNA rearrangement and RNA splicing mechanisms. Proc. Natl. Acad. Sci. USA 78, 1800–1804.

Motoyama, N., Miwa, T., Suzuki, Y., Okada, H., & Azuma, T. (1994). Comparison of somatic mutation frequency among immunoglobulin genes. (1994). J. Exp. Med. 179, 395–403.

Motoyama, N., Okada, H., & Azuma, T. (1991). Somatic mutation in constant regions of λ1 light chains. Proc. Natl. Acad. Sci. USA 88, 7933–7937.

Nahm, M.H., Kroese, F.G.M., & Hoffman, J.W. (1992). The evolution of immune memory and germinal centers. Immunology Today 13, 438–441.

Nemazee, D. (1995). B lymphocyte tolerance in the mouse. In: Immunoglobulin Genes, 2nd ed. (Honjo, T., Alt, F.W., Eds.), pp. 365–378. Academic Press, London.

Newell, N., Richards, J.E., Tucker, P.W., & Blattner, F.R. (1980). J genes for heavy chain immunoglobulins of mouse. Science 209, 1128–1132.

Newman, M.A., Mainhart, C.R., Mallett, C.P., Lavoie, T.B., & Smith-Gill, S.J. (1992). Patterns of antibody specificity during the BALB/c immune response to hen eggwhite lysozyme. J. Immunol. 149, 3260–3272.

Nieuwenhuis, P. & Opstelten, D. (1984). Functional anatomy of germinal centers. Amer. J. Anatomy 170, 421–435.

Niewold, T.A., Murphy, C.L., Weiss, D.T., & Solomon, A. (1996). Characterization of a light chain product of the human JC λ7 gene complex. J. Immunol. 157, 4474–4477.

Nitschke, L., Kosco, M.H., Köhler, G., & Lamers, M.C. (1993). Immunoglobulin D-deficient mice can mount normal immune responses to thymus-independent and -dependent antigens. Proc. Natl. Acad. Sci. USA 90, 1887–1891.

Nolan-Willard, M., Berton, M.T., & Tucker, P. (1992). Coexpression of μ and γ1 heavy chains can occur by a discontinuous transcription mechanism from the same unrearranged chromosome. Proc. Natl. Acad. Sci. USA 89, 1234–1238.

Nossal, G.J.V. (1983). Cellular mechanisms of immunologic tolerance. Annu. Rev. Immunol. 1, 33–62.

Nossal, G.J.V., Szenberg, A., Ada, G.L., & Austin, C.M. (1964). Single cell studies on 19S antibody production. J. Exp. Med. 119, 485–502.

Nottenburg, C. & Weissman, I.L. (1981). *C*μ gene rearrangement of mouse immunoglobulin genes in normal B cells occurs on both the expressed and nonexpressed chromosomes. Proc. Natl. Acad. Sci. USA 78, 484–488.

Nussenzweig, M.C., Shaw, A.C., Sinn, E., Danner, D.B., Holmes, K.L., Morse, H.C. III, & Leder, P. (1987). Allelic exclusion in transgenic mice that express the membrane form of immunoglobulin μ. Science 236, 816–819.

Obata, M., Kataoka, T., Nakai, S., Yamagishi, H., Takahashi, N., Yamawaki-Kataoka, Y., Nikaido, T., Shimizu, A., & Honjo, T. (1981). Structure of a rearranged γ1 chain gene and its implication to immunoglobulin class-switch mechanism. Proc. Natl. Acad. Sci. USA 78, 2437–2441.

O'Brien, R.L., Brinster, R.L., & Storb, U. (1987). Somatic hypermutation of an immunoglobulin transgene in κ transgenic mice. Nature 326, 405–409.

Oettinger, M.A., Schatz, D.G., Gorka, C., & Baltimore, D. (1990). RAG-1 and RAG-2, adjacent genes that synergistically activate V(D)J recombination. Science 248, 1517–1523.

Okada, A. & Alt, F.W. (1995). The variable region gene assembly mechanism. In: Immunoglobulin genes, 2nd edition. (Honjo, T., Alt, F.W., Eds.), pp. 205–234. Academic Press, London.

Oudin, J. (1960). Allotypy of rabbit serum proteins. I. Immunochemical analysis leading to the individualization of seven main allotypes. J. Exp. Med. 112, 107–124.

Owen, M.J. & Venkitaraman, A.R. (1996). Signalling in lymphocyte development. Curr. Op. Immunol. 8, 191–198.

Papavasiliou, F., Casellas, R., Suh, H., Qin, X.-F., Besmer, E., Pelanda, R., Nemazee, D., Rajewsky, K., & Nussenzweig, M.C. (1997). V(D)J recombination in mature B cells: a mechanism for altering antibody responses. Science 278, 298–301.

Parkhouse, R.M.E., Lifter, J., & Choi, Y.S. (1980). Chemical characterisation of the Fab and Fc fragments from surface immunoglobulin. Nature 284, 280–281.

Parslow, T.G., Blair, D.L., Murphy, W.J., & Granner, D.K. (1984). Structure of the 5′ ends of immunoglobulin genes: a novel conserved sequence. Proc. Natl. Acad. Sci. USA 81, 2650–2654.

Pawlak, L.L., Mushinski, E.B., Nisonoff, A., & Potter, M. (1973). Evidence for the linkage of the IGC_H locus to a gene controlling the idiotypic specificity of anti-*p*-azophenylarsonate antibodies in strain A mice. J. Exp. Med. 137, 22–31.

Pech, M., Höchtl, J., Schnell, H., & Zachau, H.G. (1981). Differences between germ-line and rearranged immunoglobulin V_κ coding sequences suggest a localized mutation mechanism. Nature 291, 668–670.

Pech, M. & Zachau, H.G. (1984). Immunoglobulin genes of different subgroups are interdigitated within the V_κ locus. Nucleic Acids Res. 12, 9229–9236.

Pelanda, R., Schwers, S., Sonoda, E., Torres, R.M., Nemazee, D., & Rajewsky, K. (1997). Immunity 7, 765–775.

Peng, C., Davis, F.M., Sun, L.K., Liou, R.S., Kim, Y.-W., & Chang, T.W. (1992). A new isoform of human membrane-bound IgE. J. Immunol. 148, 129–136.

Perlmutter, A.P. & Gilbert, W. (1984). Antibodies of the secondary response can be expressed without switch recombination in normal mouse B cells. Proc. Natl. Acad. Sci. USA 81, 7189–7193.

Perlmutter, R.M., Kearney, J.F., Chang, S.P., & Hood, L.E. (1985). Developmentally controlled expression of immunoglobulin V_H genes. Science 227, 1597–1601.

Pernis, B., Brouet, J.C., & Seligmann, M. (1974). IgD and IgM on the membrane of lymphoid cells in macroglobulinemia. Evidence for identity of membrane IgD and IgM antibody activity in a case with anti-IgG receptors. Eur. J. Immunol. 4, 776–778.

Pernis, B., Chiappino, G., Kelus, A.S., & Gell, P.G.H. (1965). Cellular localization of immunoglobulins with different allotypic specificities in rabbit lymphoid tissues. J. Exp. Med. 122, 853–876.

Pernis, B., Forni, L., & Amante, L. (1970). Immunoglobulin spots on the surface of rabbit lymphocytes. J. Exp. Med. 132, 1001–1018.

Pernis, B., Forni, L., Dubiski, S., Kelus, A.S., Mandy, W.J., & Todd, C.W. (1973). Heavy chain variable and constant region allotypes in single rabbit plasma cells. Immunochemistry 10, 281–285.

Perry, R.P., Kelley, D.E., Coleclough, C., Seidman, J.G., Leder, P., Tonegawa, S., Matthyssens, G., & Weigert, M. (1980). Transcription of mouse κ chain genes: implications for allelic exclusion. Proc. Natl. Acad. Sci. USA 77, 1937–1941.

Peters, A. & Storb, U. (1996). Somatic hypermutation of immunoglobulin genes is linked to transcription initiation. Immunity 4, 57–65.

Peterson, M.L. & Perry, R.P. (1986). Regulated production of μ_m and μ_s mRNA requires linkage of the poly(A) addition sites and is dependent on the length of the μ_s-μ_m intron. Proc. Natl. Acad. Sci. USA 83, 8883–8887.

Peterson, M.L. & Perry, R.P. (1989). The regulated production of μ_m and μ_s mRNA is dependent on the relative efficiencies of μ_s poly(A) site usage and the Cμ4-to-M1 splice. Mol. Cell. Biol. 9, 726–738.

Phillips, D.R. & Morrison, M. (1970). The arrangement of proteins in the human erythrocyte membrane. Biochem. Biophys. Res. Commun. 40, 284–289.

Picard, D. & Schaffner, W. (1984). A lymphocyte-specific enhancer in the mouse immunoglobulin κ gene. Nature 307, 80–82.

Queen, C. & Stafford, J. (1984). Fine mapping of an immunoglobulin gene activator. Mol. Cell. Biol. 4, 1042–1049.

Rabbitts, T.H. (1974). Hybridization characteristics of enzymatically synthesised DNA complementary to mouse immunoglobulin messenger RNA. FEBS Letters 42, 323–326.

Rabbitts, T.H. (1978). Evidence for splicing of interrupted immunoglobulin variable and constant region sequences in nuclear RNA. Nature 275, 291–296.

Rabbitts, T.H. & Forster, A. (1978). Evidence for noncontiguous variable and constant region genes in both germ line and myeloma DNA. Cell 13, 319–327.

Rabbitts, T.H., Forster, A., Dunnick, W., & Bentley, D.L. (1980). The role of gene deletion in the immunoglobulin heavy chain switch. Nature 283, 351–356.

Rabellino, E., Colon, S., Grey, H.M., & Unanue, E.R. (1971). Immunoglobulins on the surface of lymphocytes. I. Distribution and quantitation. J. Exp. Med. 133, 156–167.

Rada, C., González-Fernández, A., Jarvis, J.M., & Milstein, C. (1994). The 5′ boundary of somatic hypermutation in a V_κ gene is in the leader intron. Eur. J. Immunol. 24, 1453–1457.
Rada, C., Gupta, S.K., Gherardi, E., & Milstein, C. (1991). Mutation and selection during the secondary response to 2-phenyloxazolone. Proc. Natl. Acad. Sci. USA 88, 5508–5512.
Radic, M.Z., Erikson, J., Litwin, S., & Weigert, M. (1993). B lymphocytes may escape tolerance by revising their antigen receptors. J. Exp. Med. 177, 1165–1173.
Raff, M.C., Feldmann, M., & de Petris, S. (1973). Monospecificity of bone marrow-derived lymphocytes. J. Exp. Med. 137, 1024–1030.
Raff, M.C., Sternberg, M., & Taylor, R.B. (1970). Immunoglobulin determinants on the surface of mouse lymphoid cells. Nature (London) 225, 553–554.
Raman, C., Spieker-Polet, H., Yam, P.-C., & Knight, K.L. (1994). Preferential V_H gene usage in rabbit Ig-secreting heterohybridomas. J. Immunol. 152, 3935–3945.
Ramsden, D.A. & Gellert, M. (1995). Formation and resolution of double-strand break intermediates in *V(D)J* rearrangement. Genes and Dev. 9, 2409–2420.
Ramsden, D.A., Paull, T.T., & Gellert, M. (1997). Cell-free *V(D)J* recombination. Nature 388, 488–491.
Rathbun, G.A., Capra, J.D., & Tucker, P.W. (1987). Organization of the murine immunoglobulin V_H complex in the inbreds strains. EMBO J. 6, 2931–2937.
Ravetch, J.V., Siebenlist, U., Korsmeyer, S., Waldmann, T., & Leder, P. (1981). Structure of the human immunoglobulin μ locus: characterization of embryonic and rearranged J and D genes. Cell 27, 583–591.
Reidl, L.S., Kinoshita, C.M., & Steiner, L.A. (1992). Wild mice express an Ig Vλ gene that differs from any Vλ in BALB/c but resembles a human Vλ subgroup. J. Immunol. 149, 471–480.
Reilly, E.B., Blomberg, B., Imanishi-Kari, T., Tonegawa, S., & Eisen, H.N. (1984). Restricted association of *V* and *J-C* gene segments for mouse λ light chains. Proc. Natl. Acad. Sci. USA 81, 2484–2488.
Reth, M. (1992). Antigen receptors on B lymphocytes. Annu. Rev. Immunol. 10, 97–121.
Reth, M.G., Ammirati, P., Jackson, S., & Alt, F.W. (1985). Regulated progression of a cultured pre-B-cell line to the B–cell stage. Nature 317, 353–355.
Reth, M., Gehrmann, P., Petrac, E., & Wiese, P. (1986b). A novel V_H to V_HDJ_H joining mechanism in heavy-chain-negative (null) pre-B cells results in heavy-chain production. Nature 322, 840–842.
Reth, M., Hombach, J., Wienands, J., Campbell, K.S., Chien, N., Justement, L.B., & Cambier, J.C. (1991). The B-cell antigen receptor complex. Immunology Today 12, 196–201.
Reth, M.G., Jackson, S., & Alt, F.W. (1986a). V_HDJ_H formation and DJ_H replacement during pre-B differentiation: non-random usage of gene segments. EMBO J. 5, 2131–2138.
Reth, M., Petrac, E., Wiese, P., Lobel, L., & Alt, F.W. 1987. Activation of Vκ gene rearrangement in pre-B cells follows the expression of membrane-bound immunoglobulin heavy chains. EMBO J. 6, 3299–3305.
Reynaud, C.-A., Bertocci, B., Dahan, A., & Weill, J.-C. (1994). Formation of the chicken B-cell repertoire: ontogenesis, regulation of Ig gene rearrangement, and diversification by gene conversion. Adv. Immunol. 57, 353–378.
Reynaud, C.-A., Garcia, C., Hein, W.R., & Weill, J.-C. (1995). Hypermutation generating the sheep immunoglobulin repertoire is an antigen-independent process. Cell 80, 115–125.
Reynaud, C.-A., Mackay, C.R., Müller, R.G., & Weill, J.-C. (1991). Somatic generation of diversity in a mammalian primary lymphoid organ: the sheep ileal Peyer's patches. Cell 64, 995–1005.
Ritchie, K.A., Brinster, R.L., & Storb, U. (1984). Allelic exclusion and control of endogenous immunoglobulin gene rearrangement in κ transgenic mice. Nature 312, 517–520.
Rodwell, J.D. & Karush, F. (1980). Restriction in IgM expression-I. The V_H regions of equine anti-lactose antibodies. Mol. Immunol. 17, 1553–1561.
Roes, J., Hüppi, K., Rajewsky, K., & Sablitzky, F. (1989). V gene rearrangement is required to fully activate the hypermutation mechanism in B cells. J. Immunol. 142, 1022–1026.

Roes, J. & Rajewsky, K. (1993). Immunoglobulin D (IgD)-deficient mice reveal an auxiliary receptor function for IgD in antigen-mediated recruitment of B cells. J. Exp. Med. 177, 45–55.

Rogers, J., Choi, E., Souza, L., Carter, C., Word, C., Kuehl, M., Eisenberg, D., & Wall, R. (1981). Gene segments encoding transmembrane carboxyl termini of immunoglobulin γ chains. Cell 26, 19–27.

Rogers, J., Early, P., Carter, C., Calame, K., Bond, M., Hood, L., & Wall, R. (1980). Two mRNAs with different 3′ ends encode membrane-bound and secreted forms of immunoglobulin μ chain. Cell 20, 303–312.

Rogers, J. & Wall, R. (1984). Immunoglobulin RNA rearrangements in B lymphocyte differentiation. Adv. Immunol. 35, 39–59.

Rogerson, B.J. (1994). Mapping the upstream boundary of somatic mutations in rearranged immunoglobulin transgenes and endogenous genes. Mol. Immunol. 31, 83–98.

Rogozin, I.B. & Kolchanov, N.A. (1992). Somatic hypermutagenesis in immunoglobulin genes. II. Influence of neighboring base sequences on mutagenesis. Biochim. Biophys. Acta 1171, 11–18.

Roholt, O.A., Friedenson, B., Radzimski, G., & Pressman, D. (1973). A light chain tyrosyl sequence in the antibody site of a rabbit anti-*p*-azobenzoate antibody. J. Immunol. 111, 1367–1375.

Roost, H.-P., Bachmann, M.F., Haag, A., Kalinke, U., Pliska, V., Hengartner, H., & Zinkernagel, R.M. (1995). Early high-affinity neutralizing anti-viral IgG responses without further overall improvements of affinity. Proc. Natl. Acad. Sci. USA 92, 1257–1261.

Rose, S.M., Kuehl, W.M., & Smith, G.P. (1977). Cloned MPC 11 myeloma cells express two kappa genes: a gene for a complete light chain and a gene for a constant region polypeptide. Cell 12, 453–462.

Roth, D.B., Menetski, J.P., Nakajima, P.B., Bosma, M.J., & Gellert, M. (1992b). V(D)J recombination: broken DNA molecules with covalently sealed (hairpin) coding ends in scid mouse thymocytes. Cell 70, 983–991.

Roth, D.B., Nakajima, P.B., Menetski, J.P., Bosma, M.J., & Gellert, M. (1992a). V(D)J recombination in mouse thymocytes: double-strand breaks near T cell receptor δ rearrangement signals. Cell 69, 41–53.

Roth, D.B., Zhu, C., & Gellert, M. (1993). Characterization of broken DNA molecules associated with V(D)J recombination. Proc. Natl. Acad. Sci. USA 90, 10788–10792.

Rowe, D.S. & Fahey, J.L. (1965a). A new class of human immunoglobulins. I. A unique myeloma protein. J. Exp. Med. 121, 171–184.

Rowe, D.S. & Fahey, J.L. (1965b). A new class of human immunoglobulins. II. Normal serum IgD. J. Exp. Med. 121, 185–199.

Rowe, D.S., Hug, K., Faulk, W.P., McCormick, J.N., & Gerber, H. (1973a). IgD on the surface of peripheral blood lymphocytes of the human newborn. Nature New Biol. 242, 155–157.

Rowe, D.S., Hug, K., Forni, L., & Pernis, B. (1973b). Immunoglobulin D as a lymphocyte receptor. J. Exp. Med. 138, 965–972.

Royal, A., Garapin, A., Cami, B., Perrin, F., Mandel, J.L., LeMeur, M., Brégégègre, F., Gannon, F., LePennec, J.P., Chambon, P., & Kourilsky, P. (1979). The ovalbumin gene region: common features in the organisation of three genes expressed in chicken oviduct under hormonal control. Nature 279, 125–132.

Rudikoff, S., Giusti, A.M., Cook, W.D., & Scharff, M.D. (1982). Single amino acid substitution altering antigen-binding specificity. Proc. Natl. Acad. Sci. USA 79, 1979–1983.

Rudikoff, S., Pawlita, M., Pumphrey, J., & Heller, M. (1984). Somatic diversification of immunoglobulins. Proc. Natl. Acad. Sci. USA 81, 2162–2166.

Sablitzky, F., Wildner, G., & Rajewsky, K. (1985). Somatic mutation and clonal expansion of B cells in an antigen-driven immune response. EMBO J. 4, 345–350.

Sakano, H., Hüppi, K., Heinrich, G., & Tonegawa, S. (1979a). Sequences at the somatic recombination site of immunoglobulin light-chain genes. Nature 280, 288–294.

Sakano, H., Kurosawa, Y., Weigert, M., & Tonegawa, S. (1981). Identification and nucleotide sequence of a diversity DNA segment (D) of immunoglobulin heavy-chain genes. Nature 290, 562–565.

Sakano, H., Maki, R., Kurosawa, Y., Roeder, W., & Tonegawa, S. (1980). Two types of somatic recombination are necessary for the generation of complete immunoglobulin heavy-chain genes. Nature 286, 676–683.

Sakano, H., Rogers, J.H., Hüppi, K., Brack, C., Traunecker, A., Maki, R., Wall, R., & Tonegawa, S. (1979b). Domains and the hinge region of an immunoglobulin heavy chain are encoded in separate DNA segments. Nature 277, 627–633.

Salsano, F., Froland, S.S., Natvig, J.B., & Michaelsen, T.E. (1974). Same idiotype of B-lymphocyte membrane IgD and IgM. Formal evidence for monoclonality of chronic lymphocytic leukemia cells. Scand. J. Immunol. 3, 841–846.

Sanchez, P., Marche, P.N., Le Guern, C., & Cazenave, P.-A. (1987). Structure of a third murine immunoglobulin λ light chain variable region that is expressed in laboratory mice. Proc. Natl. Acad. Sci. USA 84, 9185–9188.

Sanchez, P., Nadel, B., & Cazenave, P.-A. (1991). V_λ-J_λ rearrangements are restricted within a V-J-C recombination unit in the mouse. Eur. J. Immunol. 21, 907–911.

Sanz, I. (1991). Multiple mechanisms participate in the generation of diversity of human H chain CDR3 regions. J. Immunol. 147, 1720–1729.

Schäble, K.F. & Zachau, H.G. (1993). The variable genes of the human immunoglobulin κ locus. Biol. Chem. Hoppe-Seyler 374, 1001–1022.

Scharff, M.D., Poltoratsky, V., & Green, N.S. (1997). The promotion of V region hypermutation. J. Exp. Med. 185, 185–188.

Schatz, D.G. (1997). V(D)J recombination moves *in vitro*. Semin. Immunol. 9, 149–159.

Schatz, D.G. & Baltimore, D. (1988). Stable expression of immunoglobulin gene V(D)J recombinase activity by gene transfer into 3T3 fibroblasts. Cell 53, 107–115.

Schatz, D.G., Oettinger, M.A., & Baltimore, D. (1989). The V(D)J recombination activating gene, *RAG1*. Cell 59, 1035–1048.

Schechter, I. (1973). Biologically and chemically pure mRNA coding for a mouse immunoglobulin L-chain prepared with the aid of antibodies and immobilized oligothymidine. Proc. Natl. Acad. Sci. USA 70, 2256–2260.

Schilling, J., Clevinger, B., Davie, J.M., & Hood, L. (1980). Amino acid sequence of homogeneous antibodies to dextran and DNA rearrangements in heavy chain V-region gene segments. Nature 283, 35–40.

Schlissel, M.S. & Baltimore, D. (1989). Activation of immunoglobulin kappa gene rearrangement correlates with induction of germline kappa gene transcription. Cell 58, 1001–1007.

Schlissel, M., Constantinescu, A., Morrow, T., Baxter, M., & Peng, A. (1993). Double-strand signal sequence breaks in *V(D)J* recombination are blunt, 5′-phosphorylated, RAG-dependent, and cell cycle regulated. Genes and Dev. 7, 2520–2532.

Schroeder, H.W., Jr., Walter, M.A., Hofker, M.H., Ebens, A., Willems van Dijk, K., Liao, L.C., Cox, D.W., Milner, E.C.B., & Perlmutter, R.M. (1988). Physical linkage of a human immunoglobulin heavy chain variable region gene segment to diversity and joining region elements. Proc. Natl. Acad. Sci. USA 85, 8196–8200.

Schroeder, H.W., Jr., & Wang, J.Y. (1990). Preferential utilization of conserved immunoglobulin heavy chain variable gene segments during fetal life. Proc. Natl. Acad. Sci. USA 87, 6146–6150.

Schulenburg, E.P., Simms, E.S., Lynch, R.G., Bradshaw, R.A., & Eisen, H.N. (1971). Amino acid sequence of the light chain from a mouse myeloma protein with anti-hapten activity: evidence for a third type of light chain. Proc. Natl. Acad. Sci. USA 68, 2623–2626.

Schulze, D.H. & Kelsoe, G. (1987). Genotypic analysis of B cell colonies by in situ hybridization. Stoichiometric expression of three V_H families in adult C57BL/6 and BALB/c mice. J. Exp. Med. 166, 163–172.

Schupp, I.W., Schlake, T., Kirschbaum, T., Zachau, H.G., & Boehm, T. (1997). A yeast artificial chromosome contig spanning the mouse immunoglobulin kappa light chain locus. Immunogenetics 45, 180–187.

Schwager, J., Bürckert, N., Courtet, M., & Du Pasquier, L. (1989). Genetic basis of the antibody repertoire in *Xenopus*: analyis of the V_H diversity. EMBO J. 8, 2989–3001.

Schwartz, D.C. & Cantor, C.R. (1984). Separation of yeast chromosome-sized DNA by pulsed field gradient gel electrophoresis. Cell 37, 67–75.

Schwarz, K., Gauss, G.H., Ludwig, L., Pannicke, U., Li, Z., Lindner, D., Friedrich, W., Seger, R.A., Hansen-Hagge, T.E., Desiderio, S., Lieber, M.R., & Bartram, C.R. (1996). RAG mutations in human B cell-negative SCID. Science 274, 97–99.

Scott, C.L. & Potter, M. (1984). Polymorphism of Cλ genes and units of duplication in the genus *Mus*. J. Immunol. 132, 2630–2637.

Seidman, J.G. & Leder, P. (1978). The arrangement and rearrangement of antibody genes. Nature 276, 790–795.

Seidman, J.G. & Leder, P. (1980). A mutant immunoglobulin light chain is formed by aberrant DNA- and RNA-splicing events. Nature 286, 779–783.

Seidman, J.G., Leder, A., Edgell, M.H., Polsky, F., Tilghman, S.M., Tiemeier, D.C., & Leder, P. (1978). Multiple related immunoglobulin variable-region genes identified by cloning and sequence analyis. Proc. Natl. Acad. Sci. USA 75, 3881–3885.

Seidman, J.G., Nau, M.M., Norman, B., Kwan, S.-P., Scharff, M., & Leder, P. (1980). Immunoglobulin V/J recombination is accompanied by deletion of joining site and variable region segments. Proc. Natl. Acad. Sci. USA 77, 6022–6026.

Sell, S. & Gell, P.G.H. (1965). Studies on rabbit lymphocytes in vitro. I. Stimulation of blast transformation with an antiallotype serum. J. Exp. Med. 122, 423–439.

Selsing, E., Miller, J., Wilson, R., & Storb, U. (1982). Evolution of mouse immunoglobulin λ genes. Proc. Natl. Acad. Sci. USA 79, 4681–4685.

Selsing, E. & Storb, U. (1981a). Mapping of immunoglobulin variable region genes: relationhip to the 'deletion' model of immunoglobulin gene rearrangement. Nucleic Acids Res. 9, 5725–5735.

Selsing, E. & Storb, U. (1981b). Somatic mutation of immunoglobulin light-chain variable-region genes. Cell 25, 47–58.

Selsing, E., Voss, J., & Storb, U. (1984). Immunoglobulin gene "remnant" DNA—implications for antibody gene recombination. Nucleic Acids Res. 12, 4229–4246.

Shapiro, A.M., Schlissel, M.S., Baltimore, D., & De Franco, A.L. (1993). Stimulation of κ light-chain gene rearrangement by the immunoglobulin μ heavy chain in a pre-B-cell line. Mol. Cell. Biol. 13, 5679–5690.

Shapiro, M.A. & Weigert, M. (1987). How immunoglobulin V_κ genes rearrange. J. Immunol. 139, 3834–3839.

Sharon, J. (1990). Structural correlates of high antibody affinity: three engineered amino acid substitutions can increase the affinity of an anti-*p*-azophenylarsonate antibody 200-fold. Proc. Natl. Acad. Sci. USA 87, 4814–4817.

Sharp, P.A. (1987). *Trans* splicing: variation on a familiar theme? Cell 50, 147–148.

Sharpe, M.J., Milstein, C., Jarvis, J.M., & Neuberger, M.S. (1991). Somatic hypermutation of immunoglobulin κ may depend on sequences 3′ of C_κ and occurs on passenger transgenes. EMBO J. 10, 2139–2145.

Sher, A. & Cohn, M. (1972). Inheritance of an idiotype associated with the immune response of inbred mice to phosphorylcholine. Eur. J. Immunol. 2, 319–326.

Shimizu, A., Nussenzweig, M.C., Han, H., Sanchez, M., & Honjo, T. (1991). *Trans*-splicing as a possible molecular mechanism for the multiple isotype expression of the immunoglobulin gene. J. Exp. Med. 173, 1385–1393.

Shimizu, A., Nussenzweig, M.C., Mizuta, T.-R., Leder, P., & Honjo, T. (1989). Immunoglobulin double-isotype expression by trans-mRNA in a human immunoglobulin transgenic mouse. Proc. Natl. Acad. Sci. USA 86, 8020–8023.

Shimizu, A., Takahashi, N., Yaoita, Y., & Honjo, T. (1982). Organization of the constant-region gene family of the mouse immunoglobulin heavy chain. Cell 28, 499–506.

Shin, E.K., Matsuda, F., Nagaoka, H., Fukita, Y., Imai, T., Yokoyama, K., Soeda, E., & Honjo, T. (1991). Physical map of the 3′ region of the human immunoglobulin heavy chain locus: clustering of autoantibody-related variable segments in one haplotype. EMBO J. 10, 3641–3645.

Shinkai, Y., Rathbun, G., Lam, K.-P., Oltz, E.M., Stewart, V., Mendelsohn, M., Charron, J., Datta, M., Young, F., Stall, A.M., & Alt, F.W. (1992). RAG-2-deficient mice lack mature lymphocytes owing to inability to initiate V(D)J rearrangement. Cell 68, 855–867.

Sibley, C.H., Ewald, S.J., Kehry, M.R., Douglas, R.H., Raschke, W.C., & Hood, L.E. (1980). Characterization of multiple immunoglobulin μ-chains synthesized by two clones of a B cell lymphoma. J. Immunol. 125, 2097–2105.

Siden, E., Alt, F.W., Shinefeld, L., Sato, V., & Baltimore, D. (1981). Synthesis of immunoglobulin μ chain gene products precedes synthesis of light chains during B-lymphocyte development. Proc. Natl. Acad. Sci. USA 78, 1823–1827.

Siebenlist, U., Ravetch, J.V., Korsmeyer, S., Waldmann, T., & Leder, P. (1981). Human immunoglobulin D segments encoded in tandem multigenic families. Nature 294, 631–635.

Siekevitz, M., Huang, S.Y., & Gefter, M.L. (1983). The genetic basis of antibody production: a single heavy chain variable region gene encodes all molecules bearing the dominant anti-arsonate idiotype in the strain A mouse. Eur. J. Immunol. 13, 123–132.

Siminovitch, K.A., Bakhshi, A., Goldman, P., & Korsmeyer, S.J. (1985). A uniform deleting element mediates the loss of κ genes in human B cells. Nature 316, 260–262.

Simon, M., Zieg, J., Silverman, M., Mandel, G., & Doolittle, R. (1980). Phase variation: evolution of a controlling element. Science 209, 1370–1374.

Singer, P.A., Singer, H.H., & Williamson, A.R. (1980). Different species of messenger RNA encode receptor and secretory IgM μ chains differing at their carboxy termini. Nature 285, 294–300.

Smider, V. & Chu, G. (1997). The end-joining reaction in V(D)J recombination. Semin. Immunol. 9, 189–197.

Smith, G.P. (1976). The significance of hybridization kinetic experiments for theories of antibody diversity. Cold Spring Harbor Symp. Quant. Biol. 41, 863–875.

Sonoda, E., Pewzner-Jung, Y., Schwers, S., Taki, S., Jung, S., Eilat, D., & Rajewsky, K. (1997). B cell development under the condition of allelic inclusion. Immunity 6, 225–233.

Southern, E.M. (1975). Detection of specific sequences among DNA fragments separated by gel electrophoresis. J. Mol. Biol. 98, 503–517.

Spanopoulou, E., Zaitseva, F., Wang, F.-H., Santagata, S., Baltimore, D., & Panayotou, G. (1996). The homeodomain region of Rag-1 reveals the parallel mechanisms of bacterial and V(D)J recombination. Cell 87, 263–276.

Stavnezer, J. (1996). Antibody class switching. Adv. Immunol. 61, 79–146.

Stavnezer, J. & Huang, R.C.C. (1971). Synthesis of a mouse immunoglobulin light chain in a rabbit reticulocyte cell-free system. Nature New Biol. 230, 172–176.

Stavnezer, J., Huang, R.C.C., Stavnezer, E., & Bishop, J.M. (1974). Isolation of messenger RNA for an immunoglobulin kappa chain and enumeration of the genes for the constant region of kappa chain in the mouse. J. Mol. Biol. 88, 43–63.

Steinmetz, M., Altenburger, W., & Zachau, H.G. (1980). A rearranged DNA sequence possibly related to the translocation of immunoglobulin gene segments. Nucleic Acids Res. 8, 1709–1720.

Steinmetz, M., Zachau, H.G., & Mach, B. (1979). Cloning of immunoglobulin kappa light chain genes from mouse liver and myeloma MOPC 173. Nucleic Acids Res. 6, 3213–3229.

Stern, C. & McConnell, I. (1976). Immunoglobulins M and D as antigen-binding receptors on the same cell, with shared specificity. Eur. J. Immunol. 6, 225–227.

Sterzl, J. & Silverstein, A.M. (1967). Developmental aspects of immunity. Adv. Immunol. 6, 337–459.

Stiernholm, N.B.J., Verkoczy, L.K., & Berinstein, N.L. (1995). Rearrangement and expression of the human ψCλ6 gene segment results in a surface Ig receptor with a truncated light chain constant region. J. Immunol. 154, 4583–4591.

Stollar, B.D. (1991). Autoantibodies and autoantigens: a conserved system that may shape a primary immunoglobulin pool. Mol. Immunol. 28, 1399–1412.

Storb, U. (1972). Quantitation of immunoglobulin genes by nucleic acid hybridization with RNA from myeloma and spleen microsomes. J. Immunol. 108, 755–764.

Storb, U. (1995). Ig gene expression and regulation in Ig transgenic mice. In: Immunoglobulin genes, 2nd edition. (Honjo, T., Alt, F.W., Eds.), pp. 345–363. Academic Press, London.

Storb, U. (1996). The molecular basis of somatic hypermutation of immunoglobulin genes. Curr. Op. Immunol. 8, 206–214.

Storb, U., Haasch, D., Arp, B., Sanchez, P., Cazenave, P.–A., & Miller, J. (1989). Physical linkage of mouse λ genes by pulsed-field gel electrophoresis suggests that the rearrangement process favors proximate target sequences. Mol. Cell. Biol. 9, 711–718.

Swan, D., Aviv, H., & Leder, P. (1972). Purification and properties of biologically active messenger RNA for a myeloma light chain. Proc. Natl. Acad. Sci. USA 69, 1967–1971.

Swan, D., D'Eustachio, P., Leinwand, L., Seidman, J., Keithley, D., & Ruddle, F.H. (1979). Chromosomal assignment of the mouse κ light chain genes. Proc. Natl. Acad. Sci. USA 76, 2735–2739.

Takahashi, N., Ueda, S., Obata, M., Nikaido, T., Nakai, S., & Honjo, T. (1982). Structure of human immunoglobulin gamma genes: implications for evolution of a gene family. Cell 29, 671–679.

Takeda, S., Zou, Y.-R., Bluethmann, H., Kitamura, D., Muller, U., & Rajewsky, K. (1993). Deletion of the immunoglobulin κ chain intron enhancer abolishes κ chain gene rearrangement in *cis* but not λ chain gene rearrangement in *trans*. EMBO J. 12, 2329–2336.

Taki, S., Meiering, M., & Rajewsky, R. (1993). Targeted insertion of a variable gene locus into the immunoglobulin heavy chain locus. Science 262, 1268–1271.

Taki, S., Schwenk, F., & Rajewsky, K. (1995). Rearrangement of upstream D_H and V_H to a rearranged immunoglobulin variable region gene inserted into the DQ52-J_H region of the immunoglobulin heavy chain locus. Eur. J. Immunol. 25, 1888–1896.

Talmage, D.W. (1957). Allergy and immunology. Annu. Rev. Med. 8, 239–256.

Taub, R.A., Hollis, G.F., Hieter, P.A., Korsmeyer, S., Waldmann, T.A., & Leder, P. (1983). Variable amplification of immunoglobulin λ light-chain genes in human populations. Nature 304, 172–174.

Taylor, R.B., Duffus, W.P.H., Raff, M.C., & de Petris, S. (1971). Redistribution and pinocystosis of lymphocyte surface immunoglobulin molecules induced by anti-immunoglobulin antibody. Nature New Biol. 233, 225–229.

Thomas, M., White, R.L., & Davis, R.W. (1976). Hybridization of RNA to double-stranded DNA: formation of R-loops. Proc. Natl. Acad. Sci. USA 73, 2294–2298.

Thompson, C.B. (1995). New insights into V(D)J recombination and its role in the evolution of the immune system. Immunity 3, 531–539.

Thorbecke, G.J. (1959). Some histological and functional aspects of lymphoid tissue in germfree animals. I. Morphological studies. Ann. N.Y. Acad. Sci. 78, 237–246.

Thorbecke, G.J., Amin, A.R., & Tsiagbe, V.K. (1994). Biology of germinal centers in lymphoid tissue. FASEB J. 8, 832–840.

Tiegs, S.L., Russell, D.M., & Nemazee, D. (1993). Receptor editing in self-reactive bone marrow B cells. J. Exp. Med. 177, 1009–1020.

Toda, M., Hirama, T., Takeshita, S., & Yamagishi, H. (1989). Excision products of immunoglobulin gene rearrangements. Immunol. Lett. 21, 311–316.

Todd, C.W. (1963). Allotypy in rabbit 19S protein. Biochem. Biophys. Res. Commun. 11, 170–175.

Todd, C.W. & Inman, F.P. (1967). Comparison of the allotypic combining sites on H-chains of rabbit IgG and IgM. Immunochemistry 4, 407–417.

Tomlinson, I.M., Cox, J.P.L., Gherardi, E., Lesk, A.M., & Chothia, C. (1995). The structural repertoire of the human V_κ domain. EMBO J. 14, 4628–4638.

Tomlinson, I.M., Walter, G., Jones, P.T., Dear, P.H., Sonnhammer, E.L.L., & Winter, G. (1996). The imprint of somatic hypermutation on the repertore of human germline V genes. J. Mol. Biol. 256, 813–817.

Tonegawa, S. (1976). Reiteration frequency of immunoglobulin light chain genes: further evidence for somatic generation of antibody diversity. Proc. Natl. Acad. Sci. USA 73, 203–207.

Tonegawa, S. (1983). Somatic generation of antibody diversity. Nature 302, 575–581.

Tonegawa, S. & Baldi, I. (1973). Electrophoretically homogeneous myeloma light chain mRNA and its translation in vitro. Biochem. Biophys. Res. Commun. 51, 81–87.

Tonegawa, S., Brack, C., Hozumi, N., & Pirrotta, V. (1977a). Organization of immunoglobulin genes. Cold Spring Harb. Symp. Quant. Biol. 42, 921–931.

Tonegawa, S., Brack, C., Hozumi, N., & Schuller, R. (1977c). Cloning of an immunoglobulin variable region gene from mouse embryo. Proc. Natl. Acad. Sci. USA 74, 3518–3522.

Tonegawa, S., Hozumi, N., Brack, C., & Schuller, R. (1977b). Arrangement and rearrangement of immunoglobulin genes. In: ICN-UCLA Symposia on the Immune System: Genetics and Regulation, Vol. VI (Sercarz, E.E., Herzenberg, L.A., Fox, C.F., Eds.), pp. 43–55, Academic Press Inc., New York.

Tonegawa, S., Hozumi, N., Matthyssens, G., & Schuller, R. (1976). Somatic changes in the content and context of immunoglobulin genes. Cold Spring Harbor Symp. Quant. Biol. 41, 877–889.

Tonegawa, S., Maxam, A.M., Tizard, R., Bernard, O., & Gilbert, W. (1978). Sequence of a mouse germ-line gene for a variable region of an immunoglobulin light chain. Proc. Natl. Acad. Sci. USA 75, 1485–1489.

Tsurushita, N., Avdalovic, N.M., & Korn, L.J. (1987). Regulation of differential processing of mouse immunoglobulin μ heavy-chain mRNA. Nucleic Acids Res. 15, 4603–4615.

Tucker, P.W., Slightom, J.L., & Blattner, F.R. (1981). Mouse IgA heavy chain gene sequence: implications for evolution of immunoglobulin hinge exons. Proc. Natl. Acad. Sci USA 78, 7684–7688.

Tumas-Brundage, K. & Manser, T. (1997). The transcriptional promoter regulates hypermutation of the antibody heavy chain locus. J. Exp. Med. 185, 239–250.

Turka, L.A., Schatz, D.G., Oettinger, M.A., Chun, J.J.M., Gorka, C., Lee, K., McCormack, W.T., & Thompson, C.B. (1991). Thymocyte expression of RAG-1 and RAG-2: termination by T cell receptor cross-linking. Science 253, 778–781.

Tyler, B.M., Cowman, A.F., Gerondakis, S.D., Adams, J.M., & Bernard, O. (1982). mRNA for surface immunoglobulin γ chains encodes a highly conserved transmembrane sequence and a 28-residue intracellular domain. Proc. Natl. Acad. Sci. USA 79, 2008–2012.

Udey, J.A. & Blomberg, B. (1987). Human λ light chain locus: organization and DNA sequences of three genomic *J* regions. Immunogenetics 25, 63–70.

Uhr, J.W. & Finkelstein, M.S. (1963). Antibody formation. IV. Formation of rapidly and slowly sedimenting antibodies and immunological memory to bacteriophage ϕX 174. J. Exp. Med. 117, 457–477.

Umar, A. & Gearhart, P.J. (1995). Reciprocal homologous recombination in or near antibody VDJ genes. Eur. J. Immunol. 25, 2392–2400.

Usuda, S., Takemori, T., Matsuoka, M., Shirasawa, T., Yoshida, K., Mori, A., Ishizaka, K., & Sakano, H. (1992). Immunoglobulin V gene replacement is caused by the intramolecular DNA deletion mechanism. EMBO J. 11, 611–618.

Valbuena, O., Marcu, K.B., Weigert, M., & Perry, R.P. (1978). Multiplicity of germline genes specifying a group of related mouse κ chains with implications for the generation of immunoglobulin diversity. Nature 276, 780–784.

van Boxel, J.A., Paul, W.E., Terry, W.D., & Green, I. (1972). IgD-bearing human lymphocytes. J. Immunol. 109, 648–651.

van Gent, D.C., McBlane, J.F., Ramsden, D.A., Sadofsky, M.J., Hesse, J.E., & Gellert, M. (1995). Initiation of V(D)J recombination in a cell-free system. Cell 81, 925–934.

van Gent, D.C., Mizuuchi, K., & Gellert, M. (1996b). Similarities between initiation of V(D)J recombination and retroviral integration. Science 271, 1592–1594.

van Gent, D.C., Ramsden, D.A., & Gellert, M. (1996a). The RAG1 and RAG2 proteins establish the 12/23 rule in V(D)J recombination. Cell 85, 107–113.

van Loghem, E., Natvig, J.B., & Matsumoto, H. (1970). Genetic markers of immunoglobulins in Japanese families. Inheritance of associated markers belonging to one IgA and three IgG subclasses. Ann. Hum. Genet. Lond. 33, 351–359.

Van Ness, B.G., Coleclough, C., Perry, R.P., & Weigert, M. (1982). DNA between variable and joining gene segments of immunoglobulin κ light chain is frequently retained in cells that rearrange the κ locus. Proc. Natl. Acad. Sci. USA 79, 262–266.

Vasicek, T.J. & Leder, P. (1990). Structure and expression of the human immunoglobulin λ genes. J. Exp. Med. 172, 609–620.

Vassalli, P., Tedghi, R., Lisowska-Bernstein, B., Tartakoff, A., & Jaton, J.-C. (1979). Evidence for hydrophobic region within heavy chains of mouse B lymphocyte membrane-bound IgM. Proc. Natl. Acad. Sci. USA 76, 5515–5519.

Victor, K.D. & Capra, J.D. (1994). An apparently common mechanism of generating antibody diversity: length variation of the V_L-J_L junction. Mol. Immunol. 31, 39–46.

Victor, K.D., Vu, K., & Feeney, A.J. (1994). Limited junctional diversity in κ light chains. Junctional sequences from $CD43^+B220^+$ early B cell progenitors resemble those from peripheral B cells. J. Immunol. 152, 3467–3475.

Vitetta, E.S., Baur, S., & Uhr, J.W. (1971). Cell surface immunoglobulin. II. Isolation and characterization of immunoglobulin from mouse splenic lymphocytes. J. Exp. Med. 134, 242–264.

Vitetta, E.S. & Uhr, J.W. (1975). Immunoglobulin-receptors revisited. Science 189, 964–969.

von Schwedler, U., Jäck, H.-M., & Wabl, M. (1990). Circular DNA is a product of the immunoglobulin class switch rearrangement. Nature 345, 452–456.

Wabl, M.R., Johnson, J.P., Haas, I.G., Tenkhoff, M., Meo, T., & Inan, R. (1980). Simultaneous expression of mouse immunoglobulins M and D is determined by the same homolog of chromosome 12. Proc. Natl. Acad. Sci. USA 77, 6793–6796.

Wabl, M., Meyer, J., Beck-Engeser, G., Tenkhoff, M., & Burrows, P.D. (1985). Critical test of a sister chromatid exchange model for the immunoglobulin heavy-chain class switch. Nature 313, 687–689.

Wabl, M. & Steinberg, C. (1982). A theory of allelic and isotypic exclusion for immunoglobulin genes. Proc. Natl. Acad. Sci. USA 79, 6976–6978.

Wagner, S.D., Elvin, J.G., Norris, P., McGregor, J.M., & Neuberger, M.S. (1996). Somatic hypermutation of Ig genes in patients with xeroderma pigmentosum (XP-D). Int. Immunol. 8, 701–705.

Wagner, S.D., Milstein, C., & Neuberger, M.S. (1995). Codon bias targets mutation. Nature 376, 732.

Walfield, A.M., Storb, U., Selsing, E., & Zentgraf, H. (1980). Comparison of different rearranged immunoglobulin kappa genes of a myeloma by electromicroscopy and restriction mapping of cloned DNA: implications for "allelic exclusion". Nucleic Acids Res. 8, 4689–4707.

Walter, G., Tomlinson, I.M., Cook, G.P., Winter, G., Rabbitts, T.H., & Dear, P.H. (1993). HAPPY mapping of a YAC reveals alternative haplotypes in the human immunoglobulin V_H locus. Nucleic Acids Res. 21, 4524–4529.

Wang, A.-C., Wang, I.Y., & Fudenberg, H.H. (1977). Immunoglobulin structure and genetics. Identity between variable regions of a μ and a γ2 chain. J. Biol. Chem. 252, 7192–7199.

Wang, A.-C., Wang, I.Y.F., McCormick, J.N., & Fudenberg, H.H. (1969). The identity of light chains of monoclonal IgG and monoclonal IgM in one patient. Immunochemistry 6, 451–459.

Wang, A.C., Wilson, S.K., Hopper, J.E., Fudenberg, H.H., & Nisonoff, A. (1970). Evidence for control of synthesis of the variable regions of the heavy chains of immunoglobulins G and M by the same gene. Proc. Natl. Acad. Sci. USA 66, 337–343.

Warner, N.L. (1974). Membrane immunoglobulins and antigen receptors on B and T lymphocytes. Adv. Immunol. 19, 67–216.

Warner, N.L., Byrt, P., & Ada, G.L. (1970). Blocking of the lymphocyte antigen receptor site with anti-immunoglobulin sera *in vitro*. Nature 226, 942–943.

Weber, J.S., Berry, J., Manser, T., & Claflin, J.L. (1991). Position of the rearranged V_κ and its 5′ flanking sequences determines the location of somatic mutations in the J_κ locus. J. Immunol. 146, 3652–3655.

Webster, R.G. & Laver, W.G. (1975). Antigenic variation of influenza viruses. In The Influenza Viruses and Influenza (Kilbourne, E.D., Ed.), pp. 269–314, Academic Press, New York.

Weichhold, G.M., Klobeck, H.-G., Ohnheiser, R., Combriato, G., & Zachau, H.G. (1990). Megabase inversions in the human genome as physiological events. Nature 347, 90–92.

Weigert, M.G., Cesari, I.M., Yonkovich, S.J., & Cohn, M. (1970). Variability in the lambda light chain sequences of mouse antibody. Nature 228, 1045–1047.

Weigert, M., Gatmaitan, L., Loh, E., Schilling, J., & Hood, L. (1978). Rearrangement of genetic information may produce immunoglobulin diversity. Nature 276, 785–790.

Weigert, M., Perry, R., Kelley, D., Hunkapiller, T., Schilling, J., & Hood, L. (1980). The joining of V and J gene segments creates antibody diversity. Nature 283, 497–499.

Weigert, M. & Riblet, R. (1976). Genetic control of antibody variable regions. Cold Spring Harb. Symp. Quant. Biol. 41, 837–846.

Weiler, E. (1965). Differential activity of allelic γ-globulin genes in antibody-producing cells. Proc. Natl. Acad. Sci. USA 54, 1765–1772.

Weis-Garcia, F., Besmer, E., Sawchuk, D.J., Yu, W., Hu, Y., Cassard, S., Nussenzweig, M.C., & Cortes, P. (1997). V(D)J recombination: in vitro coding joint formation.

Weiss, S. & Wu, G.E. (1987). Somatic point mutations in unrearranged immunoglobulin gene segments encoding the variable region of λ light chains. EMBO J. 6, 927–932.

Whitehurst, C., Henney, H.R., Max, E.E., Schroeder, H.W., Jr., Stüber, F., Siminovitch, K.A., & Garrard, W.T. (1992). Nucleotide sequence of the intron of the germline human κ immunoglobulin gene connecting the J and C regions reveals a matrix association region (MAR) next to the enhancer. Nucleic Acids Res. 20, 4929–4930.

Willems van Dijk, K., Milner, L.A., Sasso, E.H., & Milner, E.C.B. (1992). Chromosomal organization of the heavy chain variable region gene segments comprising the human fetal antibody repertoire. Proc. Natl. Acad. Sci. USA 89, 10430–10434.

Willett, C.E., Cherry, J.J., & Steiner, L.A. (1997). Characterization and expression of the recombination activating genes (*rag1* and *rag2*) of zebrafish. Immunogenetics 45, 394–404.

Williams, P.B., Kubo, R.T., & Grey, H.M. (1978). μ-Chains from a nonsecretor B cell line differ from secreted μ-chains at the C-terminal end. J. Immunol. 121, 2435–2439.

Williams, S.C., Frippiat, J.-P., Tomlinson, I.M., Ignatovich, O., Lefranc, M.-P., & Winter, G. (1996). Sequence and evolution of the human germline V_λ repertoire. J. Mol. Biol. 264, 220–232.

Wilson, M., Hsu, E., Marcuz, A., Courtet, M., Du Pasquier, L., & Steinberg, C. (1992). What limits affinity maturation in *Xenopus*—the rate of somatic mutation or the ability to select mutants? EMBO J. 11, 4337–4347.

Wilson, R., Miller, J., & Storb, U. (1979). Rearrangement of immunoglobulin genes. Biochemistry 18, 5013–5021.

Winchester, R.J., Fu, S.M., Hoffman, T., & Kunkel, H.G. (1975). IgG on lymphocyte surfaces: technical problems and the significance of a third cell population. J. Immunol. 114, 1210–1212.

Wood, C. & Tonegawa, S. (1983). Diversity and joining segments of mouse immunoglobulin heavy chain genes are closely linked and in the same orientation: implications for the joining mechanism. Proc. Natl. Acad. Sci. USA 80, 3030–3034.

Woodruff, M.F.A., Reid, B., & James, K. (1967). Effect of antilymphocytic antibody and antibody fragments on human lymphocytes *in vitro*. Nature 215, 591–594.

Word, C.J., Mushinski, J.F., & Tucker, P.W. (1983). The murine immunoglobulin α gene expresses multiple transcripts from a unique membrane exon. EMBO J. 2, 887–898.

Wortis, H.H. (1992). Surface markers, heavy chain sequences and B cell lineages. Intern. Rev. Immunol. 8, 235–246.

Wu, G.E., Govindji, N., Hozumi, N., & Murialdo, H. (1982). Nucleotide sequence of a chromosomal rearranged λ_2 immunoglobulin gene of mouse. Nucleic Acids Res. 10, 3831–3843.

Wu, G.E. & Paige, C.J. (1986). V_H gene family utilization in colonies derived from B and pre-B cells detected by the RNA colony blot assay. EMBO J. 5, 3475–3481.

Wu, T.T. & Kabat, E.A. (1970). An analysis of the sequences of the variable regions of Bence Jones proteins and myeloma light chains and their implications for antibody complementarity. J. Exp. Med. 132, 211–250.

Wu, T.T., Kabat, E.A., & Bilofsky, H. (1975). Similarities among hypervariable segments of immunoglobulin chains. Proc. Natl. Acad. Sci. USA 72, 5107–5110.

Wysocki, L., Manser, T., & Gefter, M.L. (1986). Somatic evolution of variable region structures during an immune response. Proc. Natl. Acad. Sci. USA 83, 1847–1851.

Yaoita, Y. & Honjo, T. (1980). Deletion of immunoglobulin heavy chain genes from expressed allelic chromosome. Nature 286, 850–853.

Yaoita, Y., Kumagai, Y., Okumura, K., & Honjo, T. (1982). Expression of lymphocyte surface IgE does not require switch recombination. Nature 297, 697–699.

Yamawaki-Kataoka, Y., Nakai, S., Miyata, T., & Honjo, T. (1982). Nucleotide sequences of gene segments encoding membrane domains of immunoglobulin γ chains. Proc. Natl. Acad. Sci. USA 79, 2623–2627.

Yancopoulos, G.D., Desiderio, S.V., Paskind, M., Kearney, J.F., Baltimore, D., & Alt, F.W. (1984). Preferential utilization of the most J_H-proximal V_H gene segments in pre-B-cell lines. Nature 311, 727–733.

Yancopoulos, G.D., Malynn, B.A., & Alt, F.W. (1988). Developmentally regulated and strain-specific expression of murine V_H gene families. J. Exp. Med. 168, 417–435.

Yasui, H., Akahori, Y., Hirano, M., Yamada, K., & Kurosawa, Y. (1989). Class switch from μ to δ is mediated by homologous recombination between σ_μ and Σ_μ sequences in human immunoglobulin gene loci. Eur. J. Immunol. 19, 1399–1403.

Yèlamos, J., Klix, N., Goyenechea, B., Lozano, F., Chui, Y.L., González Fernández, A., Pannell, R., Neuberger, M.S., & Milstein, C. (1995). Targeting of non-Ig sequences in place of the V segment by somatic hypermutation. Nature 376, 225–229.

Yu, L., Peng, C., Starnes, S.M., Liou, R.S., & Chang, T.W. (1990). Two isoforms of membrane-bound α Ig resulting from alternative mRNA splicing in the membrane segment. J. Immunol. 145, 3932–3936.

Yuan, D. & Vitetta, E.S. (1978). Cell surface immunoglobulin. XXI. Appearance of IgD on murine lymphocytes during differentiation. J. Immunol. 120, 353–356.

Zachau, H.G. (1995). The human immunoglobulin κ genes. In: Immunoglobulin Genes, 2nd ed. (Honjo, T. & Alt, F.W., Eds.), pp. 173–191. Academic Press Ltd, London.

Zachau, H.G. (1996). The human immunoglobulin κ genes. The Immunologist 4, 49–54.

Zeelon, E.P., Bothwell, A.L.M., Kantor, F., & Schechter, I. (1981). An experimental approach to enumerate the genes coding for immunoglobulin variable-regions. Nucleic Acids Res. 9, 3809–3820.

Zhang, K., Saxon, A., & Max, E.E. (1992). Two unusual forms of human immunoglobulin E encoded by alternative RNA splicing of ε heavy chain membrane exons. J. Exp. Med. 176, 233–243.

Zocher, I., Röschenthaler, F., Kirschbaum, T., Schäble, K.F., Hörlein, R., Fleischmann, B., Kofler, R., Geley, S., Hameister, H., & Zachau, H.G. (1995). Clustered and interspersed gene families in the mouse immunoglobulin κ locus. Eur. J. Immunol. 25, 3326–3331.

Zou, Y.-R., Takeda, S., & Rajewsky (1993). Gene targeting in the Igκ locus: efficient generation of λ chain-expressing B cells, independent of gene rearrangements in Igκ. EMBO J. 12, 811–820.

Chapter 2

FROM TRANSPLANT TO TRANSCRIPT

Cheryll Tickle

INTRODUCTION

The first Nobel prize for embryology was awarded in 1935 to Hans Spemann[1] for his discovery of the "organizer"; the second, 60 years later in 1995, to Christiane Nusslein-Volhard,[3] Eric Weischaus[2] and Edward Lewis[4] for the discovery of the genetic basis of development. These two prizes illustrate very nicely the transition from "transplant" to "transcript." Spemann's discoveries hinged on the results of grafting experiments in amphibian embryos and the recent recipients of the award are *Drosophila* geneticists.

Foundations of Modern Biochemistry,
Volume 4, pages 97–120.

ISBN: 0-7623-0351-4

These prizes are landmarks in the astonishing history of unravelling the most fundamental problem in biology; that of embryonic development. Consider, for example, that, in just 21 days, the seemingly lifeless contents of a hen's egg are transformed into a baby chick. The sheer speed of development together with the dramatic change in form, probably fueled one early school of thought that the embryo was preformed (in the sperm!) and only needed the provision of food from the mother to allow its appearance. This idea however only pushed the question further back to "How is the embryo preformed in the sperm?" In addition, this type of theory could not stimulate much interest in the embryo itself. It was after the discovery that most organisms were made up of many cells and experimentalists began to disassemble and reassemble organisms, that some of the fundamental questions about developmental mechanisms began to emerge.

In a series of experiments around the turn of the century Wilson (1907) made the remarkable observation that individual cells of a dissociated marine sponge could reaggregate into a new sponge. This finding immediately raised important issues. Did the individual cells rearrange themselves to take up their appropriate positions within the new sponge, or did the cells come together and then adopt the character appropriate to their position? Wilson attempted to answer these questions by aggregating just two types of cells. Because no other cell types appeared in the reconstituted sponges, he concluded that cells maintained their original identity and reassembled to take up their proper positions.

The same problem is also posed by multicellular embryos. How do cells become different and arranged in their proper positions? At that time, it was not possible to pull apart the cells of vertebrate embryos because disaggregation treatments had not been devised. (Similar experiments to those of Wilson on amphibian embryos were not carried out until the 1950s). It was possible, however, to separate individual cells at very early stages when a fertilized egg had only divided a few times, and rearrange them. These early experiments indicated that position within the embryo was paramount in determining how a cell should behave.

It was around this time that Spemann perfected ways of cutting out and grafting small pieces of amphibian embryos and the stage was then set for his classical experiments (Spemann, 1938). With his student, Mangold, he showed that when the blastopore lip from one embryo was transplanted ectopically to a second embryo, this then induced a complete second axis to develop. Cells in the embryo, which were in the position where belly normally formed, now changed their fate to become nervous tissue and other tissues such as muscle. These experiments demonstrated that interactions between cells govern the positioning of tissues within the embryo and the blastopore lip became known as the "organizer."

It must have been very frustrating that so little progress was made in immediately discovering the chemical nature of the "organizer." As documented in Hamburger (1988), this was not for want of trying but the factor or factors remained elusive and many agents were found to induce neural tissues (but not a complete second axis). These agents included fixed and boiled tissues.

An important advance in the search for the organizer was the development of the animal cap assay for mesoderm induction by Nieuwkoop (1969). This allowed the dissection of the steps that lead to primary axis formation and definition of the three step model; signaling between endoderm and ectoderm to generate mesoderm and the organizer followed by mesodermal patterning and then generation of head to tail axis (Smith et al., 1985; Slack, 1991; Figure 1). The first step—induction of mesoderm—could be assayed in culture. When the animal pole, the pigmented upper region, of an amphibian blastula, is cut out and explanted on its own, it will give rise to epidermis. But when the vegetal (lower) part of the blastula is cocultured with the animal cap, then mesoderm (the layer that forms muscle, blood, etc., in the

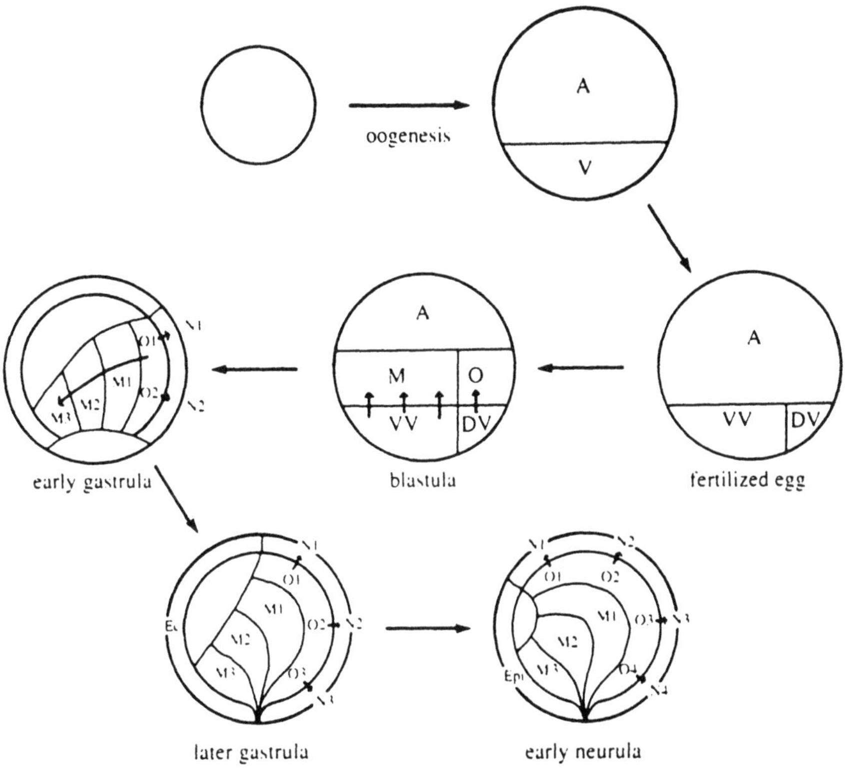

Figure 1. Diagram summarizing the events of mesoderm induction and mesoderm patterning in frog embryos. Signaling (arrows) by ventral vegetal region (VV) of the blastula induces mesoderm (M) and signaling (arrow) by dorsal vegetal (DV) region induces organizer (O). A is the animal region that has not been induced to form mesoderm and will go on to form ectoderm. A signal produced by the organizer then patterns the mesoderm. This model is known as the three signal model (after Slack, 1991).

embryo) is produced. The first leap in molecular analysis towards "transcript" was finding a factor that specified mesoderm in animal caps. Jim Smith used the animal cap assay to screen culture media conditioned by Xenopus cell lines for mesoderm inducing activity and identified an active cell line (Smith et al., 1990). Then it was a question of purifying the active component. This turned out to be a member of the transforming growth factor β (TGFβ) growth factor family named activin. Ironically, it now seems that activin is probably not the signal that operates in the embryo and another related growth factor encoded by the gene called *Vg1* is currently the best candidate for the mesoderm-inducing factor (Kessler and Melton, 1995). Another member of the TGFβ superfamily, a Bone Morphogenetic Factor (BMP4) specifies ventral mesoderm. The action of BMP4 is antagonized by chordin, noggin and follistatin which are produced by the organizer and all three of these molecules have organizer activity in that they can induce secondary axes though their activity varies (Lemaire and Kodjabachian, 1996; Thomsen, 1997). Nevertheless, with the discovery of activin, the jump from transplant to transcript had been achieved.

THE LIMB

Development

Nowhere has the remarkable progress from transplant to transcript been more striking than in the development of the vertebrate limb (Table 1). This has been accomplished over the last 50 years or so with the adoption of chick embryos as experimental material. When a window is cut into an eggshell, the embryo can be seen lying on its left side and the developing right limbs are clearly visible. Manipulations can be performed while the embryo remains within the egg, the egg then re-sealed and the effects on limb structure monitored several days later. Most

Table 1. From Transplant to Transcript in the Developing Limb

1948	Discovery of apical ridge
1968	Discovery of polarizing region
1974	Discovery of ectoderm signaling
1973–1976	Models for signaling
1982	Retinoic acid found to mimic polarizing region signaling
1989	Homeobox genes expressed in vertebrate limb buds
1993	Sonic hedgehog as polarizing region signal
1993, 1994	Fibroblast growth factors as apical ridge signals
1995	*Wnt7a* as dorsal ectoderm signal
1995	FGFs initiate limb development
1996	Human *synpolydactyly* shown to involve mutation in *Hox* gene
1997	Fringe and induction of the apical ridge

work has been carried out on the wing bud which, although near the heart, is very accessible.

Limbs of chick embryos can first be made out as slight bulges from the body wall at about 3 days after the start of incubation. These bulges soon become distinct buds and, at this stage, consist of an apparently homogeneous population of mesenchyme cells encased in ectoderm. The mesenchyme cells will give rise to most of the limb structures for example, the connective tissues, including skeleton, and musculature. The patterning process—that is the specification of the position in which main parts of the chick limb will arise—takes around 3 days; cells of the early skeleton then differentiate into cartilage and secrete extracellular matrix that can be stained with alcian green. The complete skeleton of a chick limb can be visualized in cartilage at about 3 days after the end of the patterning process. In the wing, where it joins the body proximally, is the humerus articulating with the shoulder girdle. This is followed by two elements, radius and ulna, and at the tip of the limb, there are 3 digits, digit 2,3 and 4. Digit 2 is at the anterior (nearest the head) while digit 4 is at the posterior (nearest the tail) and digit 3 is in the middle (Figure 2). By the time the skeleton has been laid down separate muscles have also formed with the tendons that connect the muscles to the skeleton. Flexors are formed ventrally and extensors dorsally.

Development of limbs poses the same fundamental problems as establishing the main body axis. How do structures arise in their correct positions? For the chick wing, these can be couched in specific terms such as "How is it that digits form at the tip?" and "What ensures that digit 2 arises at the anterior and digit 4 at the posterior?"

In order to begin to address these questions, it was necessary to understand which cells give rise to the different limb parts. Fate maps of the limb bud were con-

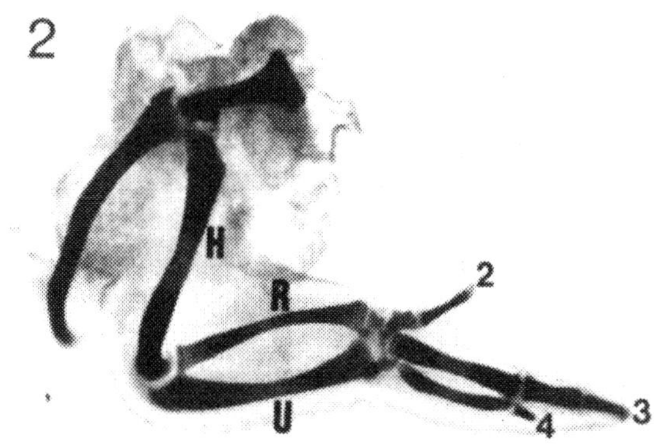

Figure 2. Whole mount of a chicken wing showing the skeletal pattern. H = humerus, R = radius, U = ulna, digits 2, 3 and 4.

structed. These early fate maps (Saunders, 1948) were made by marking cells with carbon particles and showed, for example, that the anterior of the early wing bud gives rise to proximal structures and contributes to the shoulder girdle while digits originate from posterior parts of the bud. There was always the worry that carbon marks could become displaced from cells with which they were initially associated. Nevertheless subsequent generations of fate maps have shown that, by and large, many of the original observations were accurate. Similar conclusions to those above have been reached more recently by grafting "labeled" cells into the limb bud (Bowen et al., 1989) and by marking cells with lipophilic dyes that are only transferred to their progeny and not to neighboring cells (Vargesson et al., 1997).

Of major importance in discovering the origin of many parts of the avian embryo has been the use of grafts of quail tissue and the construction of chick/quail chimaeras as pioneered by Le Douarin (1973). Quail cells can be distinguished histologically from chick cells and recently specific antibodies have become available. Although some of the most spectacular lineage tracing has been performed with migratory cells of the neural crest (Le Douarin, 1982), chick/quail chimaeras have also revealed the fate of some of the different regions of the limb bud (Bowen et al., 1989) and shown that the myogenic cells of limb muscles originate in the lateral parts of the somites (Ordahl and Le Douarin, 1992).

Fate maps only reveal contributions of cells in different regions but do not show when decisions to form a particular structure have been taken. Once cells have taken this decision, which seals their fate irreversibly, they are said to be determined and the traditional test for determination is transplantation. If cells are determined, they will develop autonomously when transplanted into a different region of an embryo; if they are not, they will change fate and develop according to their new site (Slack, 1991).

In a classical transplantation experiment of this type in chick embryos, Saunders and colleagues (Saunders et al., 1959; reviewed in Wolpert,[5] 1978) transplanted tissue from the base of an early limb bud to the tip. In order to follow the contribution of transplanted cells, Saunders grafted tissue from leg bud into wing bud. The dramatic result was that toes developed at the wing tip. Thus proximal leg bud tissue, although it will form femur if left in its original position, can be reprogrammed when placed at the tip of the wing bud to give distal leg structures—toes!

The results of the transplantation experiment just outlined, suggested the important principle that the fate of limb bud cells is determined by their position. This harks back to issues raised by the sponge experiments at the beginning of the century. By this time, in the 1960s, disaggregating embryos into single cells had become routine and considerable interest was generated by the experiments of Steinberg on positioning of cells in reaggregates (Steinberg, 1964, 1970). When cells of different types were mixed together in aggregates, they sorted out, so that, for example, heart cells stuck to heart cells, liver to liver and so on. Furthermore, within a mixed aggregate of two cell types, cells of one type adopted a particular position; heart cells sorted to the inside of mixed reaggregates with liver (Steinberg,

1964). From his examination of many different combinations of cell types, Steinberg deduced a hierarchy of adhesiveness and proposed that quantitative differences in adhesiveness could account for "sorting out." More recently molecules that mediate cell adhesion have been identified and precise patterns of cell associations have been reproduced in mixed cell reaggregation systems with cells manipulated to express different cell adhesion molecules and with cells expressing different amounts of the same cell adhesion molecule (Takeichi, 1991; Steinberg and Takeichi, 1994).

Does "sorting out" play any role in the development of limb patterning? What happens if limb bud mesenchyme is disaggregated into single cells, the cells mixed together, then reaggregated and stuffed into a limb ectodermal jacket? Such reaggregates do not form well-patterned limbs but, rather surprisingly, some recognizable digits—sometimes branched—develop (Zwilling, 1964). These experiments still present the old puzzle. Did chick limb bud cells change fate according to their position or did they re–assort to form these digit-like structures? Recently this puzzle has been solved for patterning of the slug of the slime mold *Dictyostelium,* and here, both mechanisms—positional specification and sorting out—occur (Early et al., 1995).

Signaling Regions of the Limb

The first important signaling region to be discovered in the chick limb was the apical ectodermal ridge (reviewed Saunders, 1977). This thickened epithelium at the tip of the limb bud can be seen as a transparent rim. Work by Saunders in the 1940s (Saunders, 1948; later Summerbell, 1974) showed that when this rim is removed, the limb that develops is truncated. The time at which the apical ridge is removed determines the level of truncation; when the ridge is removed early, only proximal structures (such as humerus) are formed, whereas when the ridge is removed later on, more of the limb develops and it is the tip of the limb that is missing. The interpretation of these findings was, at the time, rather controversial and opinion was divided between those who thought that the apical ridge provided a special signal that governed the generation of limb structures and those who thought that removal of the ridge simply led to death of mesenchyme cells. The issue was resolved by transplantation experiments carried out by Zwilling (1956) who showed that when an apical ridge is grafted to the dorsal surface of a limb bud, a new outgrowth perpendicular to the normal limb is induced which can go on to form a limb tipped with digits. This result confirmed the view that the apical ridge has special properties that are necessary for laying down the proper sequence of structures along the long axis (proximo–distal axis) of the limb.

It was soon recognized that signaling between apical ridge and mesenchyme in the developing limb is reciprocal. When tissue from the flank, for example, is grafted at the tip of the limb bud, the apical ridge flattens. Therefore, it was suggested that limb mesenchyme produces an apical ectodermal ridge maintenance factor. The apical ridge over the posterior part of the limb bud is more pronounced than over the anterior and transplantation experiments show that the height of the ridge is dictated by the mesenchyme. When ectoderm is separated from mesenchyme, rotated through 180° and then recombined with the mesenchyme, the region of ridge that was originally anterior now lies over posterior mesenchyme and becomes thickened and the posterior ridge now associated with anterior mesenchyme becomes flatter. From these observations it was deduced that the apical ridge maintenance factor is present at higher levels in posterior mesenchyme than in anterior mesenchyme.

At first sight, it seems surprising that another 20 years or so passed before the cells providing the signals that control pattern along antero–posterior and dorso–ventral axes were identified. The reason for this was probably because there was nothing particularly remarkable about the appearance of these signaling cells. Indeed, if it had not been for the fact that the polarizing region of the chick wing bud colocalizes with a region of programmed cell death, it may have remained undiscovered for much longer. It was during the course of a series of transplantations to investigate how programmed cell death is controlled in the chick wing buds, that Saunders and Gasseling (1968) apparently first saw the dramatic effect of grafting a small piece of posterior tissue to the anterior of the limb bud. Instead of the usual 3 digits (234) an additional 3 digits developed in mirror image symmetry with the normal set giving the pattern 432234 (Figure 3). The region of the limb bud with this special duplicating property was called the zone of polarizing activity (ZPA) or polarizing region and this induction of additional digits provides an assay for polarizing activity.

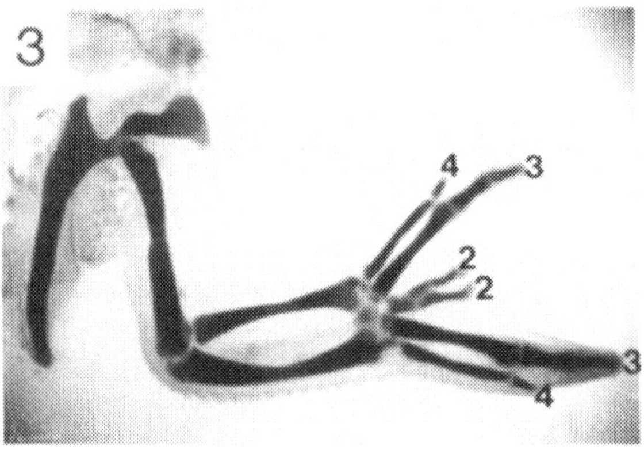

Figure 3. Whole mount of a chicken wing showing the skeletal pattern following implantation of a bead soaked in retinoic acid to the anterior margin. The digit pattern is duplicated 4 3 2 2 3 4.

A few years later, recombination experiments between limb ectoderm and mesoderm showed that control of dorso-ventral patterning is exerted by nonridge ectoderm (reviewed Saunders, 1977). In these experiments dorsal and ventral tissues are apposed, so that, for example, dorsal ectoderm is placed over ventral mesoderm. In the resulting limb, the dorso-ventral pattern of structures laid down after the operation conformed to the polarity of the ectoderm rather than that of the mesoderm. In addition, if an apical ridge was grafted to the dorsal surface or to the ventral surface, either a bidorsal or a biventral outgrowth, respectively, was induced.

For some time, it was an article of faith that work on chick embryos was relevant to understanding the development of the limbs of other vertebrates including man. Therefore it was an important step to show that limb buds of other vertebrates had the same signaling regions as chick limb buds. Posterior tissue from limb buds of mammals—mice, hamsters and even human—was found to induce digit duplications when implanted at the anterior margin of a chick wing bud showing that the polarizing signal is conserved (reviewed Saunders, 1977). It should be noted that the additional digits induced by these grafts of tissue from various different animals are chick digits because the signal is "interpreted" by chick wing cells (see later). Recombination experiments also showed that a mouse apical ridge would support development of a chick wing. Now that some of the molecules produced by signaling regions have been identified and found to be the same in different vertebrate limbs (see later), this "cross-talk" between cells of different species is not that surprising.

Even more striking than the conservation of signaling in limbs of different vertebrates, was the finding of similarities between limb signaling regions and signaling regions that mediate patterning in other parts of vertebrate embryos. Thus, it was found that when Hensen's node (the signaling region in the early chick embryo that controls axis formation) is grafted to the chick wing, additional digits are specified and a number of other signaling regions, such as notochord and the floor plate of the neural tube can also induce digit duplications. Again, the reasons for this have become clear now that the molecular basis of signaling is known.

Breeding stocks of a small number of chicken mutants exist and the usefulness of these mutants in providing a different slant on developmental mechanisms has been recognized for some time. Far from being just curiosities, mutant embryos have been exploited in grafting and tissue recombination experiments with normal embryos. Recombination experiments, for example, between mesenchyme of the polydactylous *duplicate* mutant and ectoderm of normal embryos, bolstered the idea of an apical ridge maintenance factor. The abnormally widespread production of this factor in the mutant was suggested to contribute to the broad limb bud and formation of additional digits. In contrast, polarizing activity in this mutant and in other polydactylous mutants was found to be normal. Another set of striking mutants comprises limb deficiencies, *limbless* and *wingless*. In *limbless*, limb buds are initiated but no apical ridge develops. Again recombination experiments between normal and mutant tissues established that in *limbless*, it is the ectoderm that is defective. Perhaps rather surprising, given that there are at least a hundred known mouse limb mutants, interspecific grafting to pinpoint the affected tissues has not been exploited very widely (reviewed by Tickle, 1996).

MODELS

The mechanisms that lead to patterning of the limb have been the subject of intense speculation. How is it that the correct arrangement of structures is generated? Wolpert suggested that the essence of this general problem is the same as that of forming the French flag (Wolpert, 1969; see also Wolpert, 1996). How could an array of cells give rise to the blue, white and red stripes of the French flag? The solution is that cells are informed of their position and once they know this, they can produce the appropriate color. The importance of this concept, which has been called positional information, is that it provides a framework for thinking about pattern formation. According to the ideas of positional information, pattern formation is a two-step process. In the first step, cells are informed of their position and, in the second, interpret this information in accordance with their previous history to undergo appropriate cytodifferentiation.

One way in which cell position could be specified is by a morphogen gradient. Early in this century, Child, for example, invoked metabolic gradients to account for regional differences in cell behavior in many systems (reviewed Child, 1915). The term morphogen was coined by Turing (1952) and refers to a diffusible factor that controls cell behavior. The clearest example of the graded distribution of a known molecule specifying cell position is in the *Drosophila* egg. This is the gradient of bicoid protein which specifies the head to tail axis and which was discovered by Nusslein-Volhard and co-workers (Driever and Nusslein-Volhard, 1988). She, together with Weischaus (see for example, Nusselein-Volhard and Wieschaus, 1980), was responsible for discovering, in addition, 15 of the genes in the three classes of segmentation genes that act sequentially to create a series of discrete body segments. The first class of these genes to be expressed are the *gap* genes which are activated in response to particular threshold levels of bicoid protein; activation of *gap* genes then leads on to activation of pair-rule genes and hence, in turn, to activation of segment-polarity genes (reviewed Wolpert, 1991).

A gradient model has been suggested to account for signaling by the polarizing region and control of antero-posterior pattern in the developing vertebrate limb. According to this model, the polarizing region secretes a diffusible morphogen that becomes distributed in a graded fashion. A series of predetermined threshold values would then specify the position in which different structures, for example, each of the different digits, form (Tickle et al., 1975). In this model, signaling is dose dependent and long range. In a contrasting type of model based on that put forward by French et al., (1976), it was proposed that short range interactions could operate and lead to intercalation. In this model, cells in the vertebrate limb bud possess a positional value that represents their position along the antero-posterior axis. Transplantation of posterior cells (the polarizing region) to the anterior margin juxtaposes cells with different positional values and it is proposed that this disparity stimulates intercalation of the missing positional values thus generating a duplicate set of digits (Iten, 1982). These different models stimulated many new experiments and this debate was prominent at the end of the 1970s. Compelling evidence against an intercalation model was the demonstration that cells could be reprogrammed by a polarizing signal through a barrier of intervening tissue (Honig, 1981). Thus cells that have never been adjacent to the polarizing region can be induced to change fate. However this does not mean that the polarizing region signal must act long range; another possibility is that a polarizing signal could be relayed by the cells in the barrier. With the identification of the signaling molecules, these models can now be directly tested (see later). The process that patterns structures along the proximo-distal axis of the developing limb appears to be determined by a timing mechanism. As the limb bud grows out, a zone of undifferentiated mesenchyme cells, known as the progress zone, is maintained at the distal tip of the limb by signaling of the apical ectodermal ridge (Summerbell et al., 1973). The signal from the apical ectodermal ridge does not appear to alter with time nor is there any evidence that signals from proximal tissue govern which structure forms next. Thus, it appears that it is the length of time that cells spend in the progress zone that determines whether they form distal or proximal structures. Cells that leave the zone early form proximal structures while cells that leave later form more distal ones.

A striking feature of the limb skeleton is its repeated pattern; a single element (e.g., humerus) followed by two elements (e.g., radius/ulna) followed by three-five elements (digits). Various reaction diffusion models have been suggested in which morphogen(s) become distributed in waves (Turing, 1952). Such a wave-like morphogen distribution could pre-figure the individual skeletal elements and account for periodicity of limb pattern. Another way of looking at periodicity would be to consider it in terms of a spacing pattern involving lateral inhibition. Lateral inhibition has been widely applied to spacing of skin appendages—feathers and hairs—and to cell arrangements in the nervous system.

POLARIZING SIGNALS

The first defined molecule that was discovered that could repolarize the chick wing was a vitamin A derivative, retinoic acid (Tickle et al., 1982). Retinoic acid was applied to chick limbs because it had been reported to inhibit cell–cell communication via gap junctions in cultured cells and the possibility that gap junctions might be involved in signaling in embryos was attractive. It was completely unexpected that application of retinoic acid would mimic signaling by the polarizing region and lead to mirror image duplicated wings. Subsequent work using antibodies to block gap junctions has suggested that such cell–cell communication may contribute to patterning in early amphibian embryos and in chick limbs but the precise details are still unknown.

The discovery of the signaling ability of retinoic acid gave an important stimulus to the field, particularly because at the same time, Malcolm Maden extended some observations first made in India by Niazi and Saxena on the effects of retinoic acid on adult regenerating amphibian limbs (reviewed Summerbell and Maden, 1990). Unlike the embryonic chick wing, amphibian limb regenerates treated with retinoic acid are duplicated along the proximo–distal axis. This violates a cardinal rule of regeneration that had long been recognized: that only those parts that are missing are replaced. More recently, work in India by Mohanty-Heijmadi and her colleagues (1992), this time on tail regeneration in the exotically named Marbled balloon frog, has led to the discovery of a still more bizarre effect of retinoic acid. When amputated tail stumps are treated with retinoic acid, up to nine legs can be produced.

The demonstration that retinoic acid could be extracted from chick limb buds raised the exciting possibility that retinoic acid was the polarizing morphogen. In an heroic experiment, Gregor Eichele and Christina Thaller dissected over 300 chick limb buds and showed that retinoic acid is enriched in the posterior part of the bud where the polarizing region is located (Thaller and Eichele, 1987). At more or less the same time, Chambon in France and Evans in the United States identified retinoic acid receptors which are related to the steroid hormone family of receptors (reviewed Ragsdale and Brockes, 1991; Hofmann and Eichele, 1994). Transcripts of genes encoding these receptors were found to be expressed in developing limb buds. Now, for the first time, a potential signaling molecule had been identified together with the molecular machinery that could mediate the response.

Retinoic acid has many features that makes it an attractive signal. It is a small molecule and readily diffusible in the wing bud; its lipophilic nature restricts its diffusion to cell membranes, thus avoiding problems of dilution in the vascular supply. Retinoic acid can be generated from retinol and has a short half-life. Because it had dose-dependent effects and there is evidence that it could act long range, it was a good candidate for the polarizing region morphogen. But it soon became apparent that it was more complicated (reviewed by Brockes, 1991). Susan Bryant and colleagues and Sumihare Noji and his group independently showed that mesenchymal tissue taken from next to a retinoic bead could act as a polarizing region and induce additional digits when grafted into the anterior margin of a wing bud. Noji also showed that retinoic acid beads activated expression of transcripts of RARβ (one of the retinoic acid receptors) in anterior mesenchyme whereas polarizing region grafts did not. The first retinoic acid receptor "knockouts" had no limb defects (reviewed Kastner et al., 1995). Thus, the idea that a retinoic acid gradient directly specified antero–posterior position in the limb bud was beginning to appear too simplistic but there were no other candidate molecules for the polarizing region morphogen. This was all to change dramatically at the end of 1993 when the trio of Philip Ingham, Andrew McMahon and Cliff Tabin reported the identification of a gene family in vertebrates—zebra fish (Krauss et al., 1993), mice (Echelard et al., 1993), and chickens (Riddle et al., 1993)—related to the signaling molecule in *Drosophila* encoded by the gene *hedgehog*.

Cliff Tabin and his group showed that a vertebrate *hedgehog* gene dubbed *sonic hedgehog* (with reference to a hedgehog in a cartoon) is expressed at the posterior margin of chick wing buds and furthermore, when cells expressing the *sonic hedgehog* gene are grafted to the anterior margin of a wing bud, additional digits develop in mirror-image symmetrical patterns (Riddle et al., 1993). They directly compared the distribution of *Shh* transcripts with distribution of polarizing activity determined by Honig and Summerbell (1985). The map of polarizing activity was made by systematically assaying the ability of tissue in different regions of the bud to induce additional digits when transplanted to the anterior margin of a chick wing bud. The impressive congruence of these two maps—one demonstrated by transplant, the other by transcript—was key to the argument that *Shh* encodes the limb polarizing signal. It was also shown that retinoic acid can activate *sonic hedgehog* expression in anterior mesenchyme and this could explain the duplicating effect of retinoic acid. Thus the focus in the search for the morphogen shifted from retinoic acid to *sonic hedgehog*. Very recently, it has been shown by several different groups, that interference with retinoic acid signaling at early prelimb stages in chick embryos prevents limb buds developing, thus providing new evidence that retinoic acid is required to start off limb development. This is also consistent with the phenotype of the *Shh* knockout mouse, in which proximal limb parts develop in the absence of functional *Shh* but distal structures do not form (Chiang et al., 1996).

The precise mechanism by which *sonic hedgehog* acts to control antero-posterior pattern in the limb is still not clear. The molecule encoded by the *sonic hedgehog* gene is processed in a very interesting way so that it is associated with the plasma membrane of the cell that produces it (Porter et al., 1996). Such membrane-tethering would not be predicted for a molecule that acts as a long range signal. In *Drosophila*, *hedgehog* appears to act in some situations as a long range signal but, in some of these cases, for example, in the insect wing, it acts indirectly by inducing a second signaling molecule, encoded by the *decapentaplegic* (*dpp*) gene (reviewed Johnson and Tabin, 1995). Rather unexpectedly, this signaling cascade is conserved in vertebrates; a bone morphogenetic protein, BMP2, is the product of the vertebrate homolog of *dpp* and *Bmp-2* transcripts are expressed in the polarizing region and can be induced by *Shh* in chick wing bud. BMP2 alone cannot polarize the limb bud but it is possible that several BMPs could act together as a long range signal to control vertebrate digit patterning or that *Shh* and BMP cooperate in some, at present, unknown way.

THE INSECT–VERTEBRATE CONNECTION

The discovery of the vertebrate *Shh* gene depended on the idea that signaling mechanisms are conserved in vertebrate and insect embryos. Based on anatomy and descriptive embryology, this possibility seems, at first sight, very remote. Despite obvious differences in body form and the way in which insects and vertebrates develop, fundamental molecular homologies have emerged and *Drosophila* genetics has led the way in identifying genes that regulate development. It was for this pioneering work in uncovering such genes that the 1995 Nobel prize was awarded (see above). A recent culmination of the insect-vertebrate connection in limb development has been the discovery of a common molecular signaling system that defines the margin or tip and thus controls production of growth and patterning signals (Rodriguez-Esteban et al., 1997; Laufer et al., 1997; reviewed Gaunt, 1997). Since many developmentally important genes are even more widely conserved, studies on other creatures such as nematodes are also proving to be very germane to vertebrate embryos (see for example, Hill and Sternberg, 1992).

In *Drosophila*, the important initial breakthrough in the molecular analysis of development was the identification of patterning genes. These genes act after the segmentation genes and govern development of different body segments. Mutants that affect such "master" genes had been recognized in the last century and were known as homeotic mutants. When homeotic genes are mutated, this results in patterning defects such that a particular part of the body is replaced by an another inappropriate part. Edward Lewis, one of the recent Nobel laureates, was instrumental in the analysis of the organization of one of the two homeotic gene clusters in *Drosophila*—the one which contains genes that specify the different thoracic and abdominal segments (Lewis, 1978). A major contribution was his demonstration of structural colinearity—that is that the order of the genes in the cluster is the same as the order in which the genes are expressed along the body axis.

With the advent of new molecular techniques, it became possible to identify those genes which have such profound influences on the body plan. As these genes became characterized, it became apparent that they contained a conserved 180bp sequence that encodes a DNA binding domain in the protein. This sequence became known as the homeobox and its identification was of prime importance in moving into the new molecular era (reviewed Slack, 1984). It was a bold step to embark upon a search for similar genes in vertebrates because, although homeotic mutations are well known in *Drosophila*, e.g., *Antennapaedia* in which the antenna is replaced by a leg, no such extraordinary mutations had ever been described in vertebrates. In addition, homeotic genes are involved in conveying segment identity and, while the insect body plan is clearly segmented, segmentation in vertebrates is more cryptic. In 1984, McGinnis and colleagues (and independently Muller and co-workers) were the first to discover vertebrate genes containing homeoboxes. The surprising conservation of developmental genes between vertebrates and insects was revealed for the first time. The making of such connections has been one of the most intriguing and satisfying aspects of the "transcript" age.

Once the first vertebrate homeobox genes had been found, very rapid progress was made in identifying such genes in mice and in humans. At first, the relationships between the increasing numbers of new mouse genes were bewildering but, largely due to the work of Duboule and others (see for example, Duboule and Dollé, 1989; reviewed Krumlauf, 1994), this confusion was resolved within a couple of years. Then everything became satisfyingly neat; in vertebrates there are four clusters of genes (now called *Hox* genes) which are homologous to an ancestral single gene cluster in insects (in *Drosophila* this cluster is split into two complexes) (reviewed Akam, 1989). More extraordinarily, the way in which the clusters operate in relationship to the development of the body plan turns out to be similar and in vertebrates as well as in insects, genes at the 3′ end of each cluster are expressed anteriorly at the head end and genes at the 5′ end posteriorly at the tail end of the embryo (De Robertis et al., 1990). In addition the expression of genes in each cluster shows temporal colinearity with 3′ genes being expressed before 5′ genes. In 1992, a logical nomenclature for the *Hox* gene clusters was adopted and a chart was produced that plots both these new names and previous names (Scott, 1992). This valuable "aide–memoire" for reading current and past literature is pinned to every developmental biologist's wall (Figure 4).

HOX GENES IN THE LIMB

Studies of expression patterns of the first *Hox* genes to be discovered revealed that these were expressed not only along the main body axis but also in limbs. As more genes became identified, 5′ members of two *Hox* clusters (the *Hoxa* and *Hoxd* clusters) were found to be expressed in a series of overlapping domains such that mesenchyme cells in different regions of the limb bud express different combinations of these genes (Izpisúa-Belmonte and Duboule, 1992). The rules of structural and temporal colinearity are obeyed and the most 5′ genes are switched on last in cells at the distal posterior tip of the bud. At that time, evidence was accumulating that *Hox* gene expression is related to axial pattern (reviewed by McGinnis and Krumlauf, 1992) and therefore an attractive possibility was that, in the limb, *Hox* gene expression encodes position too. In experiments in which anterior mesenchyme was respecified either by a polarizing region graft or by retinoic acid, posterior *Hoxd* genes were activated anteriorly. These mirror image patterns of gene expression presaged the mirror image patterns of digits. This result was very exciting because it was the first molecular demonstration of an alteration in cell specification. The overlapping domains of the 5′ *Hoxd* genes expressed in developing vertebrate limbs generate five different *Hox* gene "codes." This led some to wonder whether it was just a coincidence that vertebrate limbs have five digits! (Tabin, 1992).

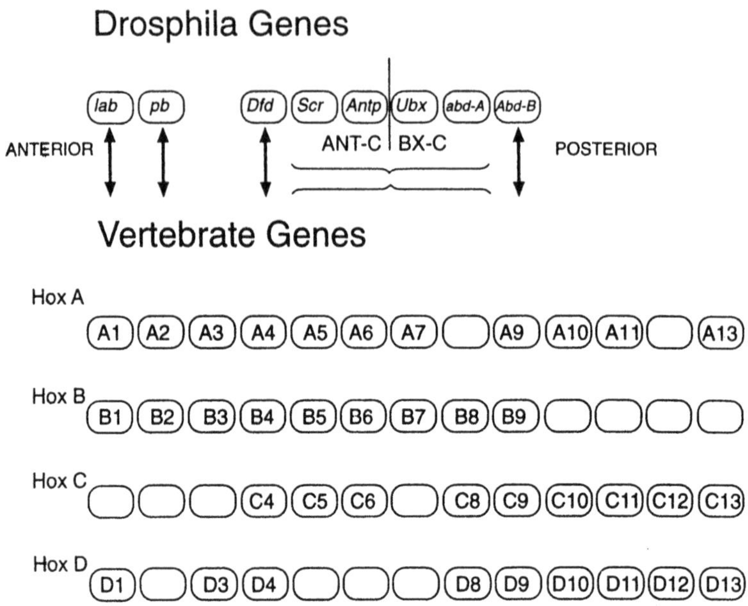

Figure 4. Simplified chart showing the new names for the vertebrate *Hox* genes and the general relationship between genes in the four different clusters and between *Hox* genes and the *Drosophila* genes (after Scott, 1992).

Although there is a good correlation between patterns of gene expression and patterns of limb structures that develop, there was no direct evidence that the pattern of *Hox* gene expression in a particular position of the bud determines which structure will develop there. In order to test this hypothesis, it was necessary to alter the pattern of gene expression and show that the digit pattern altered in a predictable fashion. At about this time an important new tool for transgenesis, a replication competent chicken retrovirus, became available (reviewed by Tickle, 1992). This technique is proving to be very powerful in manipulating gene expression in chick embryos; virus can be injected into particular regions such as the limb bud at any chosen time. When a *Hox* gene, that is normally expressed at the posterior of the bud, was expressed throughout the entire developing leg bud, the most anterior digit (big toe) resembled a more posterior digit in a third of the cases; and in the wing, an extra digit 2 was produced. This result is consistent with the idea that the character of a structure is indeed dependent on the combination of *Hox* genes expressed by these cells. However, functional inactivation of the most 5′ gene in the *Hoxd* cluster, *Hoxd13*, which is expressed most posteriorly and distally, produced a surprise. The expectation was that posterior digits would be affected in these "knockout" mice; but instead there were abnormalities in several digits, growth was delayed, and even some rudimentary extra digits were sometimes seen.

This general affect on digit morphology may be connected to later phases of *Hoxd* gene expression in which *Hoxd13* (and the other *Hoxd* genes) are expressed throughout the developing hand plate. As more *Hox* genes have been knocked out and even double knockouts constructed, it has become apparent that there are interactions both between genes in the same cluster and between genes in different clusters. Some sophisticated genetic tinkering has already started to dissect how these gene clusters operate (Van der Hoeven et al., 1996) and this will be necessary in order to interpret how different limb phenotypes are generated. Although it now seems clear that *Hox* gene expression does govern cell behavior, such as growth, in broad regions of the developing limb, for example *Hoxa11/Hoxd11* in the region that will give rise to forearm and *Hoxd13/Hoxa13* in presumptive digits, the way in which this then produces different structures is still a mystery.

If *Hox* genes are of fundamental importance in axial development it seemed likely that mutations in these genes would be early lethals and thus it was not expected that *Hox* gene mutations would be responsible for limb defects in humans. Indeed, up until a few years ago, there were no such inherited conditions known in vertebrates that involved *Hox* genes. Very recently, however, a mutation in a homeobox gene was shown to be responsible for the limb defects in the *hypodactyly* mouse mutant and even more excitingly, two human conditions with limb abnormalities, synpolydactyly and hand–foot–genital syndrome, are now thought to be based on mutations in *Hox* genes. These conditions are semi-dominant and seem best interpreted as being due to production of a mutant protein that fails not only to activate normal downstream targets but also blocks functioning of the products of other *Hox* genes (reviewed Scott, 1997).

GROWTH FACTORS IN THE LIMB

The first "growth factor" to be uncovered was nerve growth factor but it was clear from early cell culture studies that cells required factors for proliferation and survival and these could be supplied for animal cells by adding tissue extracts and/or serum to the culture medium (coconut milk for plant cells, see Osborne, Chapter 6). The history of growth factors is linked to tumor biology and it was in the context of cell transformation, recognized by changes in growth patterns of cells in culture, that some growth factors were defined, hence names such as Transforming Growth Factor α and Transforming Growth Factor β. In some cases, growth factor is a misnomer. For example, TGFβ inhibits proliferation of many cell types. It is only recently, with the ability to identify genes, that members of growth factor families have become recognized and it has turned out that many members of these families are expressed in embryos.

Fibroblast growth factors are expressed in the apical ridge of vertebrate limb buds and mediate bud outgrowth. Two fibroblast growth factors (acidic FGF and basic

FGF) had been extracted from bovine tissues but other members of this family soon began to be identified because of their involvement in tumors. As a result of this flurry of activity at least 10 different FGFs are now known (reviewed Mason, 1994). In addition to basic FGF, FGF2, at least three other members of the family, *Fgf4*,*Fgf8*, and *Fgf10*, are expressed in developing limbs. The expression pattern of *Fgf4* in the apical ectodermal ridge was first described in mouse embryos. Some preliminary functional analysis carried out in short term cultures of mouse limbs showed that FGF4 could maintain bud outgrowth. But it was experiments on chick embryos which showed dramatically that FGF4 (and FGF2) could substitute for the apical ridge and maintain outgrowth and patterning of the limbs. Subsequently another FGF, *Fgf8*, was shown to be expressed in apical ridges of both chick and mouse limb buds, in addition to *Fgf2* and *Fgf4*. FGF8 can also maintain outgrowth of the chick limb. The identification of signals such as FGFs in limbs of different vertebrates vindicates the conclusions of the transplantation experiments described earlier which first suggested that signaling was conserved. Many of the same signals also crop up at different times and places during embryonic development (for example *Shh* in Hensen's node and in notochord) and this explains why signaling regions can be interchanged (see above).

The *Wnt* gene family also contains signaling molecules that operate in vertebrate limbs. The name of this gene family encapsulates the two routes by which these molecules were identified, one route was again derived from *Drosophila*, this time from the *wingless* mutant; the second from a gene integration site that invokes mammary tumors in mice (Nusse and Varmus, 1992). In limbs of both chick and mouse embryos, transcripts of *Wnt7a* are expressed in dorsal but not ventral ectoderm and there is good evidence that signaling through *Wnt7a* is necessary for the development of dorsal structures. This conclusion stems from the limb phenotype in a strain of mice in which *Wnt7a* is functionally inactivated; the paws of these mice have a double ventral phenotype.

Members of all major families of signaling molecules appear to be expressed in developing limbs, often both in early limb and then again as tissues are formed. It should be possible to begin to dissect out these various roles by employing not only mouse transgenic approaches but also retroviral experiments and experimental manipulations in chick embryos. In the insect wing it has been possible to identify precisely the roles of dynamic patterns of gene expression by focusing on an analysis at particular times by using, for example, temperature sensitive techniques (Brook et al., 1996). It will be important to develop similar strategies in vertebrates to uncover precise roles of gene expression at different times during development.

While it is of considerable interest to identify "transcripts" that mediate patterning, the bigger challenge is to find out how molecules work together to produce limb structures. An important observation was the mutual maintenance of *Shh* expression in the polarizing region and *Fgf4* expression in the ridge because this is a molecular reflection of the mutual interactions between mesenchyme and ectoderm that were recognized long ago (reviewed Maden, 1994). This mutual maintenance of *Shh* and *Fgf4* expression together with the unexpected help of the dorsal ectodermal signal, *Wnt7a*, in maintaining *Shh* expression, serves to coordinate patterning along all three axes of the limb during development (reviewed Tickle, 1995).

Many of the "old" chicken mutants studied in the transplant era have now been revisited in the new transcript age. Although the affected genes are still unknown, expression patterns of signaling molecules and their downstream effectors such as *Hox* genes have proved informative. In the polydactylous *talpid* mutants, for example, *Shh* expression is posteriorly restricted as is high polarizing activity but *Bmps* are uniformly expressed throughout the bud tip. This pattern of *Bmp* expression mirrors the distribution of the putative apical ridge maintenance factor. In the *limbless* mutant, in which a limb bud forms but apical ridge formation and bud outgrowth does not occur, no *Shh* or *Fgf4* expression can be detected. In some remarkable recent work, it has been shown that limb bud outgrowth in the mutant can be rescued by applying FGF (Ros et al., 1996; Greishammer et al., 1996). But there is an unexpected twist. In the absence of the ridge, *Wnt7a* which is normally restricted to dorsal limb ectoderm is now expressed throughout mutant limb ectoderm both dorsally and ventrally. Thus when the mutant limbs are "rescued," they are bidorsal.

LIMB INITIATION

Limb patterning is important with respect to how the precise arrangement of structures is generated within the limb but the pattern of limb initiation is also central to the body plan (reviewed Cohn and Tickle, 1996). What ensures that limbs develop in the proper positions and that the type of limb is appropriate to that position?

A few years ago, it was shown that development of an extra limb could be induced from the flank region of a chick embryo by implanting a bead soaked in a fibroblast growth factor (Cohn et al., 1995). This experiment was stimulated by the initial observations of John Heath and colleagues that bud-like structures develop from the flank of chimeric mice which contained cells constitutively expressing FGF4—another example of the value of integrating information from different approaches in different vertebrates. Rather remarkably, this same finding had in essence been reported by Balinsky in the 1930s (reviewed Balinsky, 1965). Balinsky had transplanted nasal placode into the flank of an amphibian embryo and shown that an ectopic limb was produced. Nasal placodes in mice and chicks are now known to express FGFs. Therefore it seems likely that Balinsky was in fact applying FGF to the flank!

What makes wing and leg different? Transplantation experiments in chick embryos showed that cells very early on at particular axial levels have a property that can be described as "wingness" or "legness" and, in the classical experiment by Saunders (see earlier), grafts of presumptive thigh placed at the tip of the wing form toes. Very recently genes that may be candidates for conveying these particular properties have been identified (reviewed Papaioannou, 1997). This holds out the promise of fleshing out what was a very abstract concept. A recent successful model is the identification of an evolutionarily conserved gene for "eyeness" (Halter et al., 1995) which is conserved in animals running from *Drosophila* to man and is even involved in those spectacular eyes of molluscs.

CONNECTIONS TO HUMAN CONDITIONS

One of the most satisfying recent developments has been the making of connections between experimental embryology and clinical genetics. This could only be accomplished once the molecules involved in embryonic development were known and genes responsible for human conditions identified. Now that there is rapid progress on both fronts, the connections between the two are being revealed at an increasing pace. It will be important to uncover all components of the signaling pathways in embryos (Wolpert, 1997) because alterations in any of these could lead to abnormal development.

For some time the only gene known to be responsible for a human limb defect was *Gli3*. This human condition is Greig cephalosyndactyly and the same gene also appears to be involved in a mouse mutant *extra toes*. Embarrassingly, the role of this gene in limb development was at that time unknown. Now it has emerged that *Gli3* is the vertebrate homolog of the insect gene known as *cubitus interruptus* which is involved in intracellular transduction of *Shh* signaling. Recent work has established the expression of this gene and other related vertebrate *Gli* genes in the limb and given some indications as to their role (Marigo et al., 1997).

If one considers the genes now known to be expressed in developing limbs, there is a very impressive list of human conditions that are presently known to be associated with mutations in these genes. Mutations have been identified in genes not only encoding transcription factors and molecules associated with signal transduction (see above) but also in genes encoding growth factor receptors (Wilkie et al., 1994). This list is likely to grow rapidly. It is rather extraordinary that these human connections have been made as a result of work on frog transplants and on fly transcripts.

ACKNOWLEDGMENTS

I am grateful to Leslie Dale for comments, Anne Crawley for preparing the figures and to Konstandina Kostakopoulou for her very valuable help in putting the manuscript together.

REFERENCES

Akam, M. (1989). Hox and HOM: homologous gene clusters in insects and vertebrates. Cell 57, 347–349.

Balinsky, B.I. (1965). An Introduction to Embryology, 2nd. ed., pp. 673. W.B. Saunders Company: Philadelphia and London.

Bowen, J., Hinchliffe, J.R., Horder, T.J., & Reeve, A.M.F. (1989). The fate map of the chick forelimb-bud and its bearing on hypothesized developmental control mechanisms. Anat. Embryol. 179, 269–283.

Brockes, J.P. (1991). We may not have a morphogen. Nature (London) 350, 15.

Brook, W.J., Diaz-Benjumea, F.J., & Cohen, S.M. (1996). Organizing spatial pattern in limb development. Ann. Rev. Cell Dev. Biol. 12, 161–180.

Chiang, C., Litingtung, Y., Lee, E., Young, K.E., Corden, J.L., Westphal, H., & Beachy, P.A. (1996). Cyclopia and defective axial patterning in mice lacking Sonic hedgehog gene function. Nature (London), 383, 407–413.

Child, C.M. (1915). Senescence and Rejuvenescence, pp. 481. The University of Chicago Press, Chicago, Il.

Cohn, M., Izpisúa-Belmonte, J.C., Abud, H., Heath, J., & Tickle, C. (1995). Fibroblast growth factors induce additional limb development from the flank of chick embryos. Cell 80, 739–746.

Cohn, M. & Tickle, C. (1996). Limbs: a model for pattern formation within the vertebrate body plan. Trends Genet. 12(7), 253–257.

De Robertis, E.M., Oliver, G., & Wright, C.V.E. (1990). Homeobox genes and the vertebrate body plan. Sci. Am. 263(1), 26–33.

Driever, W. & Nusslein-Volhard, C. (1988). A gradient of *bicoid* protein in *Drosophila* embryos. Cell 54, 83–93.

Duboule, D. & Dollé, P. (1989). The structural and functional organization of the murine HOX gene family resembles that of *Drosophila* homeotic genes. EMBO J. 8, 1497–1505.

Early, A., Abe, T., & Williams, J. (1995). Evidence for positional differentiation of pre-stalk cells and for a morphogenetic gradient in *Dictystelium*. Cell 83, 91–99.

Echelard, Y., Epstein, D.J., St Jacques, B., Shen, L., Mohler, J., McMahon, J.A., & McMahon, A.P. (1993). Sonic hedgehog, a member of a family of putative signalling molecules, is implicated in the regulation of CNS polarity. Cell 75, 1417–1430.

French, V., Bryant, P., & Bryant, S.V. (1976). Pattern regulation in epimorphic fields. Science 193(4257), 969–981.

Gaunt, S.J. (1997). Chick limbs, fly wings and homology at the fringe. Nature (London) 386, 324–325.

Greishammer, U., Minowada, G., Pisenti, J.M., Abbott, U.K., & Martin, G.R. (1996). The chick *limbless* mutation causes abnormalities in limb bud dorso-ventral patterning: implications for the mechanism of apical ridge formation. Development 122, 3851–3861.

Halter, G., Callaerts, P., & Gehring, W.J. (1995). New perspectives on eye evolution. Curr. Opinion. Genet. Dev. 5, 602–609.

Hamburger, V. (1988). The heritage of experimental embryology: Hans Spemann and the organizer, pp. 196. Oxford University Press, New York and Oxford.

Hill, R.J. & Sternberg, P.W. (1992). The gene *lin-3* encodes an inductive signal for vulval development in *C. elegans*. Nature (London) 358, 470–476.

Hofmann, C. & Eichele, G. (1994). Retinoids in development. In: The Retinoids: Biology, Chemistry and Medicine (Sporn, M.B., Roberts A.B., & Goodman, D.S., Eds.), pp. 387–441. Raven Press, New York.

Honig, L.S. (1981). Positional signal transmission in the developing chick limb. Nature (London) 291, 72–73.

Honig, L.S. & Summerbell, D. (1985). Maps of strength of positional signalling activity in the developing chick wing bud. J. Embryol. Exp. Morphol. 87, 163–174.

Iten, L. (1982). Pattern specification and pattern regulation in the embryonic chick limb bud. Amer. Zool. 22, 117–129.

Izpisúa-Belmonte, J-C. & Duboule, D. (1992). Homeobox genes and pattern formation in the vertebrate limb. Dev. Biol. 152, 26–36.

Johnson, R.L. & Tabin, C. (1995). The long and short of hedgehog signalling. Cell 81, 313–316.

Kastner, P., Mark, M., & Chambon, P. (1995). Non-steroid nuclear receptors: what are genetic studies telling us about their role in real life? Cell 83, 859–869.

Kessler, D.E. & Melton, D.A. (1995). Induction of dorsal mesoderm by soluble, mature Vg1 protein. Development 121, 2155–2164.

Krauss, S., Concordet, J.P., & Ingham, P.W. (1993). A functionally conserved homolog of the Drosophila segment polarity gene hh is expressed in tissues with polarising activity in zebrafish embryos. Cell 75, 1431–1444.

Krumlauf, R. (1994) *Hox* genes in vertebrate development. Cell 78, 191–201.

Laufer, E., Dahn, R., Orozco, O.E., Yeo, C-Y., Pisenti, J., Henriques, D., Abbott, U.K., Fallon, J.F., & Tabin, C. (1997). Expression of Radical fringe in limb-bud ectoderm regulates apical ectodermal ridge formation. Nature (London) 386, 366–373.

Le Douarin, N.M. (1973). A cell biological labelling technique and its use in experimental embryology. Devl Biol. 30, 217–222.

Le Douarin, N.M. (1982). The Neural Crest. Developmental and Cell Biology, Series 12 (Barlow, P.W. Green, P.B., & Wylie, C.C., Eds.), pp. 259. Cambridge University Press, Cambridge.

LeMaire, P. & Kodjabachian, L. (1996). The vertebrate organizer: structure and molecules. Trends Genetics 12, 525–531.

Lewis, E.B. (1978). A gene complex controlling segmentation in Drosophila. Nature (London) 276, 565–570.

Maden, M. (1994). The limb bud—part two. Nature (London) 371, 560–561.

Marigo, V., Johnson, R.L., Vortkamp, A., & Tabin, C. (1996). Sonic hedgehog differentially regulates expression of GLI and GLI3 during limb development. Dev. Biol. 180(1), 273–283.

Mason, I. (1994) The ins and outs of fibroblast growth factors. Cell 78, 547–552.

McGinnis, W., Garber, R.L., Wirz, J., Kuriowa, A., & Gehring, W.J. (1984). A conserved DNA sequence in homeotic genes of the *Drosophila Antennapedia* and *bithorax* complexes. Nature (London) 308, 428–433.

McGinnis, W. & Krumlauf, R. (1992). Homeobox genes and axial patterning. Cell 68(2), 283–302.

Mohanty-Hejmadi, P., Dutta, S.K., & Mahapatra, P. (1992). Limbs generated at site of tail amputation in marbled balloon frog after vitamin A treatment. Nature (London) 355, 352–353.

Muller, M.M., Carrasco, A.E., & De Robertis, E.M. (1984). A homeobox-containing gene expressed during oogenesis in Xenopus. Cell 39, 157–162.

Nieuwkoop, P.D. (1969). The formation of the mesoderm in urodelean Amphibians. I Induction by the endoderm. Wilhelm Roux Archiv 162, 341–373.

Nusse, N. & Varmus, I. (1992). *Wnt* genes. Cell 69, 1073–1087.

Nusslein-Volhard, C. & Wieschaus, E. (1980). Mutations affecting segment number and polarity in Drosophila. Nature (London) 287, 795–801.

Ordahl, C.P. & Le Douarin, N. (1992). Two myogenic lineages within the developing somite. Development 114(2), 339–353.

Papaioannou, V.E. (1997). T-box family reunion. Trends Genet. 13(6), 212–213.

Porter, J.A., Ekker, S.C., Pard, W.J., Von-Kessler, D.P., Young, K.E., Chen, C.H., Ma, Y., Wood, A.S., Cotter, R.J., Koonin, E.V., & Beachey, P.A. (1996). Hedgehog patterning activity: a role of a lipophilic modification mediated by the carboxy-terminal autoprocessing domain. Cell 86, 21–34.

Ragsdale, C.W. & Brockes, J.P. (1991). Retinoids and their targets in vertebrate development. Curr. Opin. Cell Biol. 928–934.

Riddle, R., Johnson, R., Laufer, E., & Tabin, C. (1993). Sonic Hedgehog mediates the polarising activity of the ZPA. Cell 75, 1401–1411.

Rodriguez-Esteban, C., Schwabe, J.W.R., De La Pena, J., Foys, B., Eshelman, B., & Izpisúa-Belmonte, J.C. (1997). Radical fringe positions the apical ectodermal ridge at the dorsoventral boundary of the vertebrate limb. Nature (London) 386, 360–365.

Ros, M.A., Lopez-Martinez, A., Simandl, B.K., Rodriguez, C., Izpisúa-Belmonte, J.C., Dahn, R., & Fallon, J.F. (1996). The limb field mesoderm determines initial limb bud anteroposterior asymmetry and budding independent of *sonic hedgehog* or apical ridge expressions. Development 122, 2319–2330.

Saunders, J.W. (1948). The proximo-distal sequence of origin the parts of the chick wing and the role of the ectoderm. J. Exp. Zool. 108, 363–403.

Saunders, J.W. Jr. (1977). The experimental analysis of the chick limb bud development. In: Vertebrate Limb and Somite Morphogenesis (Ede, D.A., Hinchliffe, J.R., & Balls, M., Eds.), pp. 1–24. Cambridge Univ. Press, Cambridge.

Saunders, J.W., Gasseling, M.T., & Cairns, J.M. (1959). Differentiation of prospective thigh mesoderm grafted beneath the apical ectodermal ridge of the wing in the chick embryo. Dev Biol. 1, 281–301.

Saunders, J.W. Jr. & Gasseling, M.T. (1968). Ectodermal mesenchymal interactions in the origin of limb symmetry. In: Epithelial-Mesenchymal Interactions (Fleischmajer, R. & Billingham, R., Eds.), pp. 78–97. Williams and Wilkins, Baltimore.

Scott, M.P. (1992). Vertebrate homeobox gene nomenclature. Cell 71, 551–553.

Scott, M.P. (1997). Hox genes, Arms and the Man. Nature Genetics 15, 117–118.

Slack, J.M.W. (1984). A Rosetta stone for pattern formation in animals. Nature (London) 310, 364–365.

Slack, J.M.W. (1991). From Egg to Embryo. Developmental and Cell Biology, Series 26 (Barlow, P.W., Bray, D., Green, P.B., & Slack, J.M.W., Eds.). Cambridge Univ. Press, Cambridge.

Smith, J.C., Dale, L., & Slack, J.M.W. (1985). Cell lineage labels and region specific markers in the analysis of inductive interactions. J. Embryol. exp. Morph. 89 suppl. 317–331.

Smith, J.C., Price, B.M., Van-Nimmen, K., & Huylebroeck, D. (1990). Identification of a potent Xenopus mesoderm-inducing factor as a homologue of activin A. Nature (London) 345, 729–731.

Spemann, H. (1938). Embryonic Development and Induction. Yale University Press, New Haven, CT. (Reprinted, New York: Garland, 1988.)

Steinberg, M.S. (1964). The problem of adhesive selectivity in cellular interactions. In Cellular Membranes in Development, (Locke, M., Ed.), pp. 321–366. Academic Press, New York.

Steinberg, M.S. (1970). Does differential adhesion govern self-assembly processes in histogenesis? Equilibrium configurations and the emergence of a hierarchy among populations of embryonic cells. J. Exp. Zool. 173, 395–434.

Steinberg, M.S. & Takeichi, M. (1994). Experimental modification of cell sorting, tissue spreading and specific spatial patterning by quantitative differences in cadherin expression. Proc. Nat. Acad. Sci. USA, 91, 206–209.

Summerbell, D. (1974). A quantitative analysis of the effect of excision of the AER from the chick limb bud. J. Embryol. Exp. Morph. 32, 651–660.

Summerbell, D., Lewis, J.H., & Wolpert, L. (1973). Positional information in chick limb morphogenesis. Nature (London) 244, 492–496.

Summerbell, D. & Maden, M. (1990). Retinoic acid, a developmental signalling molecule. Trends Neurosci. 13, 142–147.

Tabin, C.J. (1992). Why we have (only) five fingers per hand: hox genes and the evolution of paired limbs. Development 116(2), 289–296.

Takeichi, M. (1991). Cadherin cell adhesion receptors as a morphogenetic regulator. Science 251, 1451–1455.

Thaller, G. & Eichele, G. (1987). Identification and spatial distribution of retinoids in the developing chick limb bud. Nature (London) 327, 625–628.

Thomsen, G.H. (1997). Anatagonism in and around the organizer; BMP inhibitors in vertebrate body patterning. Trends. Genet. 13, 209–211.

Tickle, C. (1992). Vertebrate development. A tool for transgenesis. Nature (London) 358, 188–189.

Tickle, C. (1995). Vertebrate limb development. Curr. Opin. Genet. Dev. 5, 478–484.
Tickle, C. (1996). Genetics and limb development. Developmental Genetics 19(1), 1–8.
Tickle, C., Alberts, B., Wolpert, L., & Lee, J. (1982). Local application of retinoic acid to the limb bud mimics the action of the polarizing region. Nature (London) 296, 564–566.
Tickle, C., Summerbell, D., & Wolpert, L. (1975). Positional signalling and specification of digits in chick limb morphogenesis. Nature (London) 254, 199–202.
Turing, A.M. (1952). The chemical nature of morphogenesis. Phil. Trans. Roy. Soc. Lond. 237, 37–72.
Van der Hoeven, F., Zakany, J., & Duboule, D. (1996). Gene transpositions in the HoxD complex reveal a hierarchy of regulatory controls. Cell 85, 1025–1035.
Vargesson, N., Clarke, J.D.W., Vincent, K., Coles, C., Wolpert, L., & Tickle, C. (1997). Cell fate in the chick limb bud and relationship to gene expression. Development, 124, 1909–1918.
Wilkie, A.O.M., Morriss-Kay, G.M., Jones, E.Y., & Heath, J.K. (1995). Functions of fibroblast growth factors and their receptors. Curr. Biol. 5, 500–507.
Wilson, E.V. (1907). On some phenomena of coalescence and regeneration in sponges. J. Exp. Zool. 5, 245–258.
Wolpert, L. (1969). Positional information and the spatial pattern of cellular differentiation. J. Theor. Biol. 25, 1–47.
Wolpert, L. (1978). Pattern formation in biological development. Scientific American 239(4), 154–164.
Wolpert, L. (1991). The Triumph of the Embryo. Oxford Univ. Press, Oxford.
Wolpert, L. (1996). One hundred years of positional information. Trends Genet. 12(9) 359–364.
Wolpert, L. (1997). Genomes. Genes and Function 1, 3–4.
Zwilling, E. (1956). Interaction between limb bud ectoderm and mesoderm in the chick embryo. II. Experimental limb duplication. J. Exp. Zol. 132, 173–187.
Zwilling, E. (1964). Development of fragmented and dissociated limb bud mesoderm. Dev. Biol. 9, 20–37.

Chapter 3

THE DISCOVERY OF REPAIR IN DNA

Brian Cox

INTRODUCTION

Disciplines in biochemistry all tend to move through an epistemological progression, starting with the identification of a process, recognition of the substrates and products involved (not necessarily in that order), and leading finally to the purification of the relevant enzymes whose kinetics and structures can then be studied more closely. In more recent years, the interests of biochemists have been enlarged by a return to the living organism in order to track the components of a particular process in space and time, exploring its integration with the whole living system through the controls that act on it.

Foundations of Modern Biochemistry,
Volume 4, pages 121–149.

ISBN: 0-7623-0351-4

Plate 1. Participants at the NAS-NRC Conference on Radiation Microbiology at the University of Chicago, October 1965

NAS - NRC Conference - Chicago - 1965

Key to Photo

1. E. P. Geiduschek
2. D. M. Freifelder
3. R. B. Setlow
4. C. S. Rupert
5. R. F. Hill
6. E. M. Witkin
7. W. Szybalski
8. J. K. Setlow
9. L. Grossman
10. A. Loveless
11. S. Wolff
12. W. Harm
13. R. F. Kimball
14. R. B. Uretz
15. R. P. Boyce
16. H. E. Johns
17. R. E. Zirkle
18. R. K. Mortimer
19. L. D. Hamilton
20. P. C. Hanawalt
21. J. E. Till
22. M. Delbrück
23. F. L. Haas
24. H. I. Adler
25. D. Billen
26. D. R. Krieg
27. H. E. Kubitschek
28. R. A. Deering
29. E. C. Pollard
30. A. Rörsch
31. E. Freese
32. B. S. Strauss
33. H. Marcovich
34. K. A. Stacey
35. P. Howard-Flanders
36. H. S. Kaplan
37. R. H. Haynes
38. K. C. Smith

Absent from picture: K. C. Atwood, V. P. Bond, R. S. Caldecott, J. W. Drake, A. Hollaender.

PROCESS

Experimentalists had been damaging DNA long before the ability of the cell to repair the damage was discovered. Motivations were various. Roentgen's discovery of X-rays in 1896 was followed by their use as a plaything, often by doctors who never tired of visualizing the bones of their hands (and were less amused later to discover the peculiar changes brought about in the skin). Treatment of skin cancer by exposure to X-rays was used as early as 1899; the discovery of the induction of skin cancer by X-rays was not made until 1902. Radium was first used therapeutically for the treatment of solid tumors in 1903; however both Marie Curie and Becquerel, who first described radioactivity (in uranium salts) in 1898, had reason to regret the casual handling of radium. She developed bone cancer and he developed an inflammation of the skin from carrying a piece of radium around in his waistcoat pocket.

Radioactivity was a sensation in the first decade of this century. The very concept of rays that could go right through you *without you feeling a thing* caused shivers of schadenfreud and for some the exposure of ones skeleton to outside view was uneasily akin to the exposure of ones soul. Conversely, the hair-loss associated with exposure to X-rays or radioactivity led to at least one clinic being set up, to which women flocked, using uranium salts as depilatory treatment.

In 1927, Müller had discovered that irradiation with X-rays caused mutations to appear in *Drosophila* at a higher rate than they did spontaneously and so X-rays became a tool for advancing the genetic studies of this organism (Müller, 1927). L.J. Stadler independently discovered the same phenomenon in barley, but published a year later (Stadler, 1928). These observations also provided the link between the inter-twined histories of research into DNA repair and cancer. To quote Müller:

> . . . moreover the converse effect of X-rays in occasionally producing cancer may also be associated with their action in producing mutation.

Müller also warned that in *Drosophila*, as in mammals, treatment is followed by a period of infertility, but mutations continue to appear in eggs produced after fertility is regained, contrary to the then current belief in radiotherapy that these eggs are "uninjured material."

Quantification of these observations of killing and mutagenesis began in the 1920s with the objective of pinning down the meaning of and gaining control over this terrible double-edged sword. The basis of the analysis was the "dose effect" or "dose response" curve in which some convenient biological end point, commonly the ability of a suspension of bacterial cells to form colonies, or the V_{max} of an enzyme activity, was measured as a function of an applied dose of radiation.

A theory was applied to these curves known as "hit theory." This purported to describe the physical action of the radiation, whatever its type—γ, α-particles, X-rays—in terms of quantum "hits" on a "target" in the cell such that after *n* or more hits the cell would die, whereas it would survive $n - 1$ hits. The distribution of hits in targets would be Poisson. The Poisson expression for the probability of receiving exactly n hits in a target of volume v is:

$$P(n) = \frac{(vD)^n \cdot e^{-vD}}{n!}$$

where the term vD represents the mean number of hits after dose D. The number of survivors, then, of that dose of irradiation is the sum of all cells in which the target has received $k = 0, 1, 2, .3, \ldots n - 1$ hits and is

$$N/N_0 = e^{-vD} \sum_{k=0}^{n-1} \frac{(vD)^k}{k!}$$

When $n = 1$, $k = 0$ and this expression reduces to

$$N/N_0 = e^{-vD} \qquad (\text{or } \ln N/N_0 = vD)$$

and the curve becomes a simple exponential. When the log of the fraction of survivors ($\log N/N_0$) is plotted against dose, therefore, it yields a nice straight line and this is the way dose effect curves have been plotted ever since, for purposes of evaluation and comparison.

Relatively simple modifications of these expressions deal with the imaginary situations where one needs to inactivate more than one target in a cell to kill it. The attraction of these modifications was that they promised a method of understanding the sensitive targets of ionizing radiations in cells by the effects of physically definable units of energy. For example, the term vD in the expressions which is the mean number of hits in a target, is meant to relate this to dose, D and the volume of the target, v. If you know the ionizing potential of, for example, an alpha-particle passing through water, then the volume of a target is revealed. So, at the dose of radiation which gives 37% survival, $N/N_0 = e^{-1}$, that is $vD = 1$ and $v = 1/D$. Likewise, dose effect curves can be used to predict the number of hits required to inactivate a cell or the number of targets that must be hit. In particular, a simple derivation of the multi-hit equation for the special case of single hits in m targets is

$$N/N_0 = 1 - (1 - e^{-vD})^m$$

This expression makes it possible to use dose effect curves to give both target number (m), related to the shoulder on sigmoid curves, and target "size" (v), related to the final exponential slope.

This mathematics became particularly beguiling since, almost without exception, dose-effect curves although various, often sigmoid and even complex, had exponential segments, thus conforming to theory. Furthermore, for simple biological systems, target theory gave very good measurements of the molecular weight of molecules—both enzymes and DNA. For example, the bacteriophage ϕX174 has a single-stranded DNA genome of 1.7×10^6 daltons; the D_{37} of its inactivation by ^{60}C γ-irradiation leads to a value of 1.8×10^6 daltons (Jung, 1968).

The search for and discovery of other agents, such as ultraviolet light and mustard gas, having very similar effects on living cells, together with the growth of nuclear physics, lent impetus to the development of target theory as a means of exploring the nature and behavior of genetic material. Very sophisticated mathematical treatments of the effects of radiations on a variety of biological systems—bacteria, viruses, or fruit flies and their survival or mutagenesis; or on sub-cellular components such as, *in vivo*, chromosome abnormalities, or, *in vitro*, enzyme activity and DNA breakage were developed, particularly by Lea (1946), by Zimmer (1961), Timoféeff-Ressovsky (1947) and others.

Unfortunately, the sophistication of the mathematics could not be matched by the accuracy of the data. For any given dose effect curve, a large number of parameters of hit number, target number, or target size could be chosen to give adequate fits to the data. Nor, by the early fifties, could talk of numbers and sizes of targets be reconciled with biological realities. For example, bacteria with very uniform sizes of cell clearly presented a limited range of potential target numbers and sizes between one species and another; yet survival curves could differ by several orders of magnitude in both respects. (Dean et al., 1966. Figure 1). Similar discrepancies in sensitivity were commonplace between organisms with apparently comparable genomes, or between different tissues of the same organism or when using different biological end-points in the same organism. Target theory provided good descriptions of the effects, but was a poor indicator of biological meaning.

Ironically, when in the thirties evidence of the nature of the genetic material was obtained by painstaking measurements of the mutagenic effects of UV on living tissue, no-one, least of all the investigators themselves believed a word of it (see Stadler, 1997).

Target theory was a long time dying, but the death knell was sounded well before it finally ceased to play a significant role in biological research. The course of research was completely altered by two sets of observations. The first was the discovery of mutants of bacteria and bacteriophages that had significantly different sensitivities to irradiation with either UV or X-rays. In 1958, Ruth Hill isolated a mutant of *Escherichia coli* B, B/r which was more resistant to UV than the original, and subsequently from B/r, another mutant, B_{s-1}, which was very much more sensitive to either UV or X-irradiation than the wild type (Hill, 1958; Figure 2). This observation in itself defied analysis in target theory terms. However, not only was the bacterium itself more sensitive to UV than *E. coli* B/r, but bacteriophage which had been irradiated in suspension separately from the host cell culture were much more sensitive when plaqued on B_{s-1} than on either B or B/r (Ellison et al., 1960; Harm, 1963). In the same paper, Harm showed that pre-irradiation of the bacterial host with low UV doses increased the plaquing efficiency of the UV-irradiated phage T3 (cf. Weigle reactivation, q.v.), and this property of the

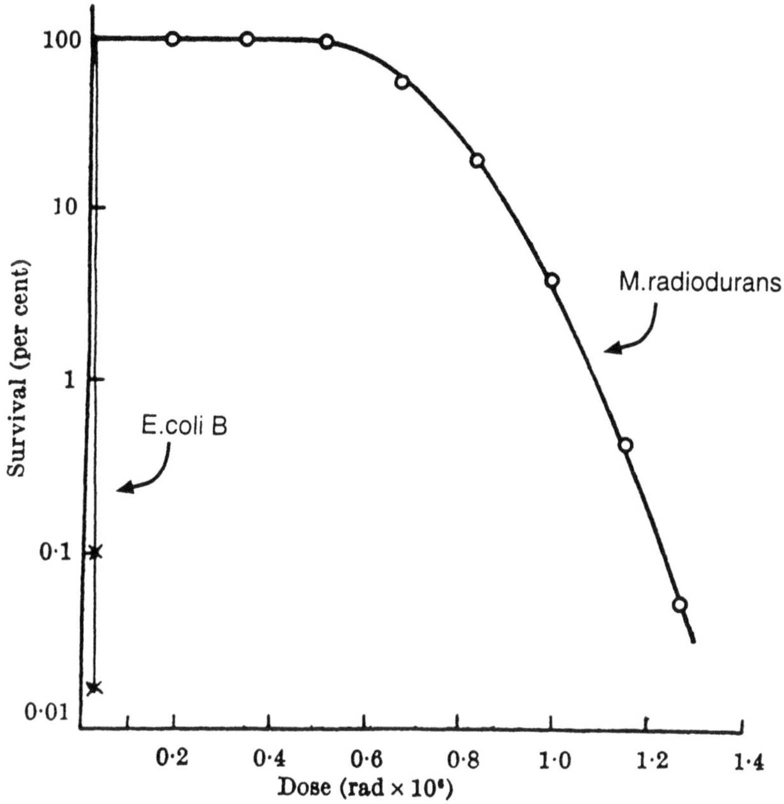

Figure 1. Survival curves of *M. radiodurans* and *E. coli B* irradiated under anoxia. The data for *E. coli B* have been replotted from Hill, 1958. (After Dean et al., 1966).

host cell was absent in the B_{s-1} mutant. No clearer indication that the resistance or sensitivity of cells to irradiation was a function of enzymatic activity in the damaged cell could be asked for.

The second kind of observation which rendered mathematics just about irrelevant to understanding what goes on in irradiated cells is typified by the detailed studies of mutagenesis carried out by Evelyn Witkin (1956). She discovered a phenomenon she called "mutation frequency decline" (mfd). Mutagenesis is the link between X-ray or UV damage and the genetic material and Witkin's work established that it, like survival, was the subject of metabolically controlled processes. Briefly, the observation is that mutation frequency at a given dose depends on the time delay before plating the treated cells and on the subsequent growth conditions.

Later, Patrick and Haynes (1964) made similar observations on UV-irradiated yeast cells, showing that survival was enhanced if, after irradiation, cells were

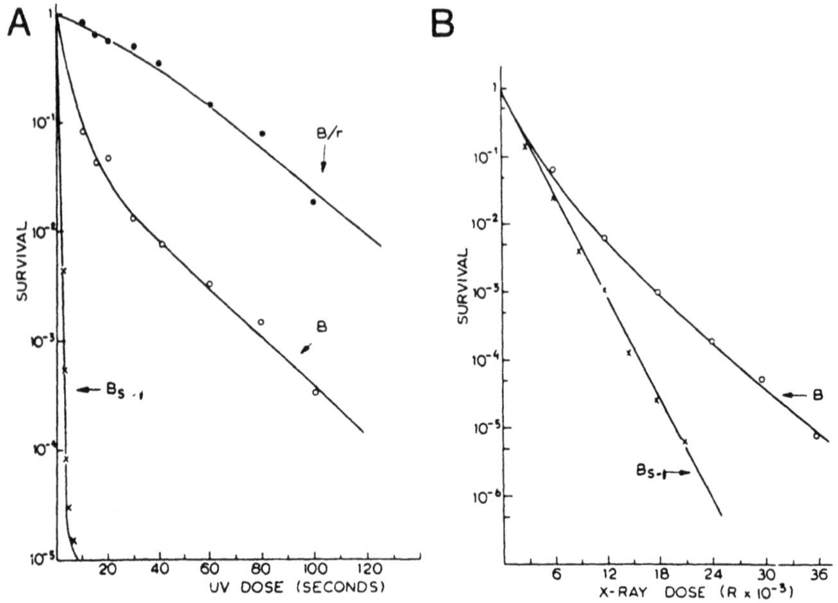

Figure 2. Survival curves for *E. coli B, B/r* and B_{S-1}. (**A**) UV. (**B**) X-rays. (Hill, 1958. Reprinted with kind permission of Elsevier Science—NL, Sara Burgerhartstraat 25, 1055 KV Amsterdam, The Netherlands.

allowed to incubate in buffer in the dark for several hours before plating (Patrick and Haynes, it seems, were enhancing their own survival by other means at the time). This was in fact a re-discovery of the effect which had been previously observed and extensively documented by Korogodin, but effectively was hidden from Western eyes by being published, in Russian journals, in Russian (Korogodin and Malumina, 1959).

The papers published concerning the photobiology and radiation biology of bacteria, bacteriophage and yeast in these few years constitute a genuine revolution in ideas about the effects of ionizing radiation, UV and what were then called "radiomimetic chemicals" in living systems. This was when DNA was recognized as the only significant target for the damaging actions of these treatments in cells and when the determinant of the contingent effects was seen to be the operation of repair processes in DNA. Nevertheless, it took 10 years from the publication of the Watson and Crick paper for the revolution to be completed. The final act of valediction or, perhaps, benediction, as people turned their minds from mathematics to biochemistry was given by Haynes in 1966, who showed that the concept of DNA repair processes could be included in the Poisson statistical formulations of target theory and provide satisfying dose-effect curves for both the wild-type and mutant forms of *E. coli* (Haynes, 1966). Having said that, it would be wrong to underestimate the role that target theory played and still plays in research. It ensured discipline in obtaining and presenting data and still provides a methodology by which comparisons between treatments can be meaningfully made. It was also used to test to self-destruction the relevance of target theory to explanations of cell survival and mutagenesis (for example, Augenstein et al., (1966).

Substrates

Of course, the chromosomes—and more specifically the DNA they contain—had been seen to be the significant target of radiation damage for decades before the idea of DNA repair burst upon the scene. From the first experiments of Müller, it had been clear that X-ray-induced mutations in *Drosophila* were often physically manifest as deletions, large and small, in appropriate regions of the salivary-gland chromosomes. Indeed, this correlation was one of the foundations of the chromosome theory of heredity. From there, it was a short step to the use of chromosomes, especially of plants, where the increasing skills of cytologists meant that both meiotic and mitotic material in all stages could be treated and readily examined. Thus the effects of both ionizing radiation and radiomimetic chemicals could be assayed (Wolff, 1961). The advantage of these biological systems was that the immediate consequences of the insults could be studied. There was no need to infer the occurrence of cellular events in dead cells from the behavior of a surviving fraction of the population. It did not matter whether the cells with damaged chromosomes lived or died, the damage could be seen anyway.

The studies showed that radiation and chemicals broke chromosomes and that broken ends rejoined, not necessarily in the same configuration as they previously held. These may be the earliest intimations of DNA repair, although no such interpretation was or could have been put upon them at the time. Chromosomes are composed of both protein and DNA and nothing in their microscopic appearance indicates what it is that holds them together, or is broken by X-rays. Indeed in the early years of radiobiology probably more effort was put into studies of the inactivation of other cell components, particularly enzymes, than of DNA. Having said that, Müller himself suggested that chromosome rejoining constituted a process of "repair."

That DNA was likely to be the specific target of genetic damage was first indicated by the action spectrum of UV-induced mutagenesis of various organisms. Three different groups of workers reported their results with three different species of plant at the 7th International Congress of Genetics at Edinburgh in 1939. Larry Stadler reported the results of mutagenetic studies with irradiated maize pollen, Alexander Hollaender worked with spores of the fungus *Trichophyton* and Knapp and Schreiber did their experiments with spermatozoa of the liverwort *Sphaerocarpus*. In each case, the peak mutagenic activity of ultraviolet light coincides with the peak absorption wavelength of nucleic acid—260 nm. (Figure 3) (Hollaender,

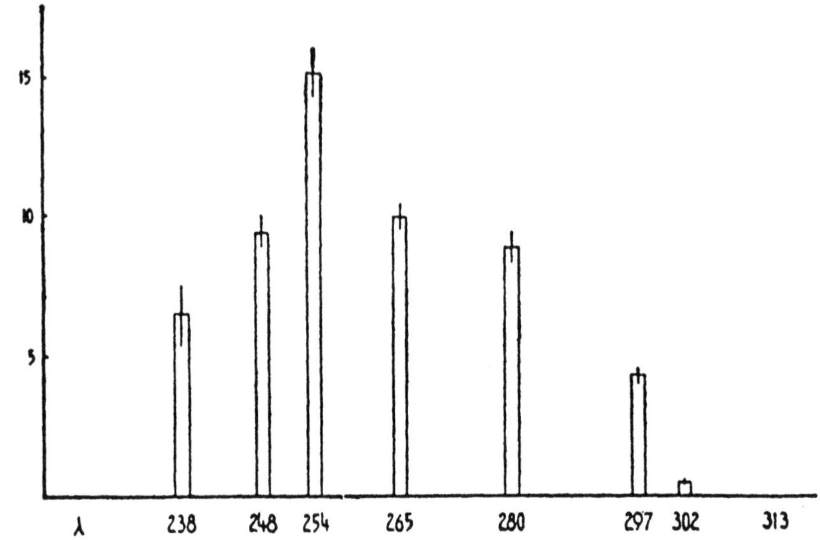

Figure 3. Wavelength dependence of maize pollen endosperm deficiency to monochromatic UV irradiation. (Stadler, 1939).

1939; Knapp and Schreiber, 1939; Stadler, 1939). However, serious study of the effects of radiation on DNA did not really begin until its role in heredity was established in the early 1950s. It pivoted crucially around the genetics and physiology involved in infection of bacteria by bacteriophages examined by Luria, Hershey and Delbruck (see Cairns et al., 1966). The key is that treatment and physical measurements of the easily distinguished components of phage could be separated from the treatment and measurements of the host bacterium, information could be separated from its consequences. The weight of evidence that DNA was the target of UV or X-ray inactivation of bacteriophages would seem to be as compelling by 1960 as the evidence from host-cell reactivation that the damage was repaired. However there was by no means a consensus at that time that either was the case and the discussion sections of most papers hedged their bets. Perhaps phage were too simple to be true.

IONIZING RADIATION

The primary action of ionizing radiation is, of course, ionization. This may affect the structure of DNA directly or indirectly through the formation of free radicals from, for example, water. Much attention has been given to what are the important changes in DNA. There is an extensive literature dealing with the many possible experimental variables involving the irradiation of bases, nucleosides, nucleotides or DNA: dry or in aqueous solution.

One of the most interesting and consistently applicable principles to emerge from these studies was the "oxygen effect." This is that, dry or wet, molecular or biological, systems show greater sensitivity when irradiated in the presence of oxygen than not. The effect is related to the concentration of oxygen, and it saturates. No universally-applicable explanation has been offered to account for it, but Howard-Flanders and Alper (Alper, 1956; Howard-Flanders and Alper, 1957) offered a model which can be applied to both living and molecular effects. It is as if there are two categories of lesion induced by ionizing radiation: irreversible and reversible, with the probability of the latter class becoming fixed or irreversible enhanced in the presence of oxygen. This may be because of the scavenging of electrons by oxygen which would block both charge neutralization of an ionized molecule and restitution of a radical. It is significant that many of the so-called "radioprotective agents," such as cysteine and glutathione are reducing agents.

Chemical changes in the base moiety with subsequent destruction or modification of the base may be an important means by which mutations are induced in DNA by ionizing radiation. However, this aspect of the effect of ionizing radiation tended in the past to receive less attention than the consequences of changes in the sugar-phosphate backbone of DNA which lead to strand breaks. Perhaps this is partly because assays for the complex variety of base changes were more troublesome than the several simple assays of the molecular weight of DNA treated either in solution or *in vivo* which were available; and partly because later it was found that DNA repair fell naturally into two broad categories—the repair of base damage and the repair of strand breaks. It seems to have been forgotten, as attention has shifted from the study of substrates to that of repair processes, that the consensus of the enormous variety of studies on damage induction is that close to 60%–80% of ionizing radiation damage is to bases and only 20%–40% is to the sugar-phosphate backbone, leading to strand breaks (see reviews by Ward, 1988; 1990). This ratio was confirmed by measurement of the decline in MW of naked ϕX174 DNA with irradiation, compared with its inactivation curve (Lytle & Ginoza, 1969). It is, however, also the case that the relative biological importance of these two areas of chemical change suggested by irradiating a single-stranded bacteriophage such as ϕX174 is not repeated when it comes to irradiating organisms containing double-stranded DNA.

Here we have to take note of two areas of research. One concerns the effects of irradiating bacteriophage and the other, yeast. Whereas inactivating single-stranded viruses by ionizing radiation leads to a close correlation between the estimated molecular weight of its DNA (or RNA) and its molecular weight determined by other means—which implies a one-to-one correspondence between lesions and inactivation—only about 5%–10% of lesions induced in double-stranded viruses, or in transforming DNA cause inactivation (Ginoza, 1967). Freifelder, working with bacteriophage T_7, showed that there was a 1:1 correlation between the number of double-strand breaks, measured by sedimentation in sucrose gradients, in T_7 DNA and its inactivation with X-rays (Freifelder, 1965). This relationship held both for irradiation in aerobic conditions in phosphate buffer, when the phage were extremely sensitive to irradiation, and anaerobic irradiation in the presence of cysteine, when they were very resistant. However, irradiation in other solutions produced an intermediate level of sensitivity and the sedimentation analysis indicated that only about 40% of the inactivation was attributable to double-strand breaks.

The historic experiments with yeast were those of Beam et al. (1954) and of Mortimer (1955), working on yeast cells of different ploidies. Haploid yeast cells have a simple exponential inactivation curve, but there is always a resistant "tail." Beam et al. showed that this resistant tail was the survival curve of those cells in the population which were budded. Target theory was invoked to show that the resistance was not a trivial effect of the presence of two nuclei or two sets of chromosomes—indeed budded haploid cells are more resistant than diploid cells. The conclusion was that ionizing radiation yielded two kinds of lethal damage—recessive and dominant. Mortimer went on to show, by mating irradiated haploid cells with unirradiated ones, that this was indeed the case. Dominant lethal damage accounted for the shape of survival curves of polyploid series of yeast strains. This was, ironically, deduced from target theory, for it was found that target size was exactly proportional to ploidy for genomes of $2n$ and above (Latarjet and Ephrussi, 1954; Lucke and Sarachek, 1953; Mortimer, 1958).

It was surmised that X-ray-induced base damage would lead mostly to recessive lesions and that the significant damage in cells other than haploids was, therefore, of a different nature. In due course, it was shown that the principal agent of lethality was the formation of double-strand breaks, since in strains defective in double-strand break repair, inactivation and the induction of double-strand breaks were in a 1:1 ratio (Ho, 1975).

The dominance of the double-strand breaks is due to another very important phenomenon discovered by Weinert and Hartwell (1988): The presence of any unrepaired double-strand breaks causes an arrest of cell-division at the G2 stage of the cell cycle. This was the first evidence that DNA repair and cell cycle controls were coordinated to maximize the viability of the damaged cell and was the harbinger of modern research leading to a holistic account of cell life. It is worth remarking that it is the stage of the cell cycle, rather than mere ploidy, that is important for survival. This is implicit in Beam et al.'s observation (1954) that budded haploid cells are more resistant than a mixed population of diploid cells, and was presaged by the observation in 1930 by Holweck and Lacassagne that irradiated yeast cells died at the G2 stage of the cell cycle, after DNA replication but before cell division.

Whatever happens to the lesions induced by ionizing radiation other than double-strand breaks, from the late 1960s onward, almost all research on repair of ionizing radiation damage has focussed on the repair of double-strand breaks; or perhaps one should state rather that ionizing radiation has been the principal means of inducing double-strand breaks with a view to understanding the repair of this substrate in DNA.

ULTRAVIOLET LIGHT

The first systematic investigation of the mutagenic effects of ultraviolet light was instigated by Stadler (1939) who, from the earliest stages argued that, since the absorption spectra of many biological compounds was well known, then systematic investigation of the action spectrum of ultraviolet light in mutagenesis should reveal the identity of the genetic material. He chose to work with corn pollen and was somewhat disappointed to find that the significant material was such a dull molecule as DNA. He and the others drew the conclusion that DNA was merely acting as a kind of chromophore for absorbing and passing on the energy. Larry Stadler became one of the most respected geneticists of his time for his analysis of the nature of the gene, but an article entitled *The Gene* published posthumously in 1954 (Stadler, 1954), makes it poignantly clear that he had missed the implications of the work with phage and of the paper on the Double Helix (Watson and Crick,1953) (which he may not have even seen). Revolutions are never so blindingly obvious as they appear 40 years later.

Following the work of Beukers and Berends (1960) on the products of UV-irradiation of purines and pyrimidines in frozen solution, it became accepted that the only photoproduct with biological significance in DNA was the 5,6 cyclobutane ring formed between adjacent pyrimidines, more specifically the thymine dimer which is twice as readily formed as either of the other two kinds (see Rupert, 1975). There is a certain amount of wishful thinking in this assumption, since thymine dimers are very stable chemically and survive violent chemical extraction procedures which makes them very easy to assay quantitatively. The ability to quantify thymine dimers by chromatography allowed the very first direct demonstration of the action of DNA repair in any living system (Boyce and Howard-Flanders, 1964; Setlow and Carrier, 1964). Other important photoproducts such as 6–4 dimers have been neglected in repair assays until quite recently, when it became apparent that they might be both metabolically and environmentally relevant.

CHEMICAL DAMAGE

The mutagenic effects of nitrogen mustard were discovered by Auerbach and Robson in 1940, but their results were not published until 1947 because of the war (Auerbach and Robson, 1947). Lotte Auerbach had been encouraged in the search for mutagenic chemicals by Müller, who felt that this was the road to directed mutation, as opposed to the essentially random nature of mutations induced by X-rays. Robson's role was to help in the handling of the mustard gas, but it seems they were rather too sanguine over this, since it was generated by heating the appropriate mixture over a flame in an open vessel and exposing the flies (*Drosophila*) to it. Both got severe skin burns during this operation and resorted to wearing gloves.

Mutagenesis by chemicals soon became a popular instrument for examining the biological effects of a wide (and controllable) range of chemical changes in DNA (Brookes and Lawley, 1961). The principal change brought about by nitrogen mustard is the alkylation of purines. Because mustard gas is so difficult to handle, in due course a number of substitute alkylating agents were found, including nitrosoguanidine and ethyl methanesulphonate. Chemicals with other effects also came into use: base analogues such as 6-aminopurine (adenine is 2-aminopurine) and bromo-uracil which substitute for their analogues in DNA and cause higher rates of mispairing during DNA synthesis; chemicals such as acridines causing frame-shifts to occur during DNA synthesis; and chemicals like methyl methanesulphonate which cause strand breaks and are in that sense more genuinely radiomimetic than other alkylating agents.

The link between mutagenesis and carcinogenesis led to a rapid rise in interest in the occurrence of mutagens in the environment through the use of chemicals in industry, pharmaceuticals, cosmetics, and food and to the development of testing procedures for their occurrence. One of the foundations of these procedures was the ingenious "Ames Test," devised by Bruce Ames (see Ames et al., 1973) using the knowledge gained of repair in *Escherichia coli* to develop strains of the closely related *Salmonella* which would simultaneously detect a variety of types of chemical mutagen. Thousands of chemicals have been screened by this and other derivative tests for their potential carcinogenicity. Many of these chemicals have provided insights into the nature and development of cancer (Ames, 1979).

FROM SUBSTRATES TO ENZYMES

Photoreactivation

The simplest example of an enzymic repair process is that of photoreactivation. This phenomenon was rediscovered in 1949 by Kelner, who found that UV-irradiated spores of *Streptomyces* showed higher survival if exposed to white light after the UV treatment (Kelner, 1949). At the time, this was not interpreted as an enzymic repair process but it was so. Enzymic action became demonstrable when it was shown that it could occur *in vitro* if, for example, transforming DNA of *Haemophilus influenzae* was incubated with cell-free extracts of yeast or *E. coli* after UV-treatment and then exposed to 400 nm light. (Rupert, 1958; 1960). This assay made it possible to purify the enzyme from a variety of sources (e.g., Sutherland, 1974; Rupert, 1975). Once an assay for thymine dimers became available, it was shown that the action of the enzyme was to restore the dimers to monomeric thymidine *in situ*.

Excision Repair[1]

Although the example of photoreactivation is the simplest, the earliest demonstration of the repair of DNA damage *in vivo* was by Boyce and Howard-Flanders and by Setlow and Carrier, working with *E. coli*. Independently, they showed that UV-induced thymine dimers were removed from DNA during incubation over about 20 min in the dark after induction. This process did not occur in *hcr*- mutants (*uvrA*-, *uvrB*- or *uvrC*-). At the same time it was shown that for every dimer removed, about 40 bases were lost from the DNA and later, Pettijohn and Hanawalt (1964) showed that the process was accompanied by a comparable amount of conservative DNA synthesis (i.e. incorporation of radioactive nucleotides into both parental strands of DNA).

These macromolecular events suggested a process which came to be known as excision-repair, the model being described in glorious technicolor by Hanawalt and Haynes in *Scientific American* (1967). This is notable for the prescience of the publisher's artist who depicted the damage-detecting and incision complex as a close-fitting sleeve around the DNA molecule. More importantly, the model suggested the existence of at least four kinds of enzymic activity: an endonuclease, an exonuclease a polymerase, and a ligase. Efforts to identify the relevant enzymes, in spite of the existence of mutant genes were, however, frustrated for many years. It is probably unfair to suggest that this was because almost everything about the model was wrong (except the artist's sleeve), but it is certainly true that research continued to yield surprises. It turned out that incision was a double event, nicks being made in sequence 3′, by UVRB and 5′ by UVRC in a kind of rolling complex:

uvrA::DNA → uvrA::DNA::uvrB → DNA::uvrB →
uvrC::DNA::uvrB → 3′nick → 5′nick → release by uvrD

UvrD codes not for an exonuclease but for DNA helicase II. The repair synthesis is carried out by PolI but, in spite of the extreme UV-sensitivity of mutants of PolA defective in its 5′ → 3′ exonuclease activity, the whole process is completed effectively without benefit of this activity. To add to the difficulties in this research, it turns out that PolI is not even essential to it, and either PolII or PolIII will do. It is worth remarking that with many of the proteins playing in these dramas there is a discrepancy between the activities that can be demonstrated *in vitro* and what they do in the organized cell. It is interesting that the beautiful and fascinating properties *in vitro* of PolI beguiled biochemists into overestimating first its role in DNA replication and, later, in excision repair.

The key papers in the dramatic remodelling of the Hanawalt and Haynes model and the elucidation of enzymatic activities were probably those by Seeberg and his colleagues in 1976, describing the UV-repair activity of a cell-free extract from *E. coli*. This led to the subsequent analysis of its components by Seeberg and the discovery of bimodal excision by Sancar and Rupp in 1983. In due course, recombinant DNA technology made purifying the protein products of the intransigent *uvr* genes possible, so that their activity could be properly followed *in vitro*.

It is interesting that the original excision repair model is showing a Lazarus effect in explaining mismatch repair and a second excision pathway in the fission yeast, *Schizosaccharomyces pombe*.

Excision Repair in Eukaryotes

Eukaryotic studies were converted from radiobiology to repair shortly after the biochemical demonstration of the reality of excision repair in *E. coli.* The occasion was the search for mutants of simple eukaryotes that were sensitive to radiation damage and the stimulus was the isolation by Robin Holliday of three such mutants in the smut fungus *Ustilago maydis*. Following the publication of his model for recombination in fungi, he argued that some or many of the enzymes used would also be required for DNA repair processes, especially if, in *Ustilago* as in bacteriophage and *E. coli*, this should involve recombination. Indeed, the three mutants he chose to work with—from the many he isolated—all affected recombination in one way or another (Holliday, 1965a, 1965b).

This example inspired searches for radiation sensitive mutants in yeast. The way was led by Nakai and Matsumoto (1967) who isolated one mutant, UV^s_1 which was very sensitive to ultraviolet and a second, X^s_1 sensitive to X-rays. They went one step further than isolation and survival curves by making the double-mutant and showing that like double mutants of *recA* and *uvrA* in *E. coli* it was much more sensitive to UV than either single mutant alone. This was the first demonstration of the existence of more than one type or pathway of DNA repair of UV damage in yeast, and inspired the later work of Game and Cox (1972; 1973; 1974), Brendel and Haynes (1973) and Louise Prakash (1993) in the genetic analysis of pathways of repair in yeast. This led to the classification of the many mutant loci into "epistasis groups," which are defined as those mutants which, when combined in the same strain, are no more UV-sensitive than the most sensitive of the two when alone.

The list of genes affecting repair in yeast was expanded by Snow (1967) and Cox and Parry (1968) who focussed on the phenotype of UV-sensitivity and by Resnick (1969), who isolated X-ray-sensitive mutants. By the time the roll-call of mutants had reached 29, and a common nomenclature had been agreed upon, it was clear that the three "epistasis groups" defined by Game and Cox at least had the merit of internal consistency (all mutants tested within any one group were epistatic with one another and additive or synergistic with those in either of the other two groups) and corresponded with their physiological effects. For example, mutants in one group (the "rad3 group") conferred only UV-sensitivity, in the second ("rad6 group"), mutants were sensitive to both UV and X-rays and the third group of seven genes, most of them isolated by Resnick, seemed only to be involved in conferring resistance to X-rays (the "rad52 group"). The epistasis model of repair pathways was also successful in predicting that a strain triply mutant with one mutation from each epistasis group would be completely deficient for UV repair (Game and Cox, 1974).

Although workers went on adding genes to this triple pathway model for some time, it is doubtful if this was more than a displacement activity for want of anything better to do until good biochemical assays for repair could be devised. It may be that the model served a useful purpose as a kind of filing system, keeping yeast repair research neatly compartmentalized. Biochemical assays in these years proved to be a relatively intransigent problem. It could be shown, once the technical difficulties of labelling thymine residues in yeast had been overcome, that at least five of the genes in the "rad3" group (only UV-sensitive) were all necessary for dimer excision (Unrau et al., 1971; Wheatcroft, 1972) and that at least two of the genes in the X-ray-sensitivity ("rad52") group were required for repairing double-strand breaks (Contopoulou et al., 1987). Demonstration of the enzyme activities involved, however, was long delayed in spite of the plethora of repair genes which had been identified. Research in yeast DNA repair was, for a decade or more, largely phenomenological. The first repair enzyme to be identified from yeast was the product of the cdc9 gene, DNA ligase (Johnston and Nasmyth, 1978), This is also the ligase involved in DNA replication and appears to be the only DNA ligase in yeast.

The real release from this impasse arrived, as in *E. coli*, with cloning technology. The key to a remarkable passage of research with a wholly unpredictable and exciting outcome was the finding the the product of the yeast RAD3 gene had a sequence suggesting it might be a helicase (Sung et al., 1987) and also that the gene was essential; disruption mutants were lethal. This led to the understanding that the essential process interupted in rad3 mutants was transcription (Feaver et al., 1993; Guzder et al., 1994). In due course, its role as a component of the transcription factor, TFIIH, was established (Buratowski, 1994; Conaway and Conaway, 1993). When combined with the products of other genes from this epistasis group, the complex probably forms a special version of the transcription factor responsible for locating DNA damage and directs the first step in repair in the 90% of yeast DNA that is not being transcribed at any one time.

Excision Repair in Mammalian Cells

The understanding of excision repair in mammalian cells has, in the last two or three years, come to a remarkable convergence with the work done in yeast. Of the 13 yeast genes presently known to be or suspected of being involved in excision repair, 11 have their counterparts in mammals. Furthermore, the association of yeast genes with transcription factor TFIIH is echoed by their mammalian homologues. This convergence is quite gratifying for those who embarked on the isolation of repair mutants in yeast 30 years ago and would alone be sufficient justification for the use of yeast as a model organism for the study of cell biochemistry of relevance to man.

Nevertheless, research with mammalian systems made the switch from radiobiology to DNA repair from a different starting point altogether—with an assay. This must have been directly inspired by the demonstration of excision in *E. coli*, and particularly of repair synthesis, although the paper by Rasmussen and Painter (1964) came out at almost the same time as that by Pettijohn and Hanawalt (1964). This assay took advantage of the fact that DNA replication in mammalian cells takes place in a narrowly-defined period in the cell cycle, occupying only 5% or less of the total time. Thus a mammalian cell culture offered a pulse label of tritiated thymidine shows only about 5% of its cell nuclei labelled—as visualized in an autoradiogram. These nuclei are very densely and evenly labelled. However, if the culture has been UV-irradiated before the pulse of tritium, not only are these densely-labelled nuclei seen, but scattered silver-grains appear in all the others, the intensity depending on, among other things the dose of UV. This was named "unscheduled DNA synthesis" or UDS for short.

Many other assays for the presence of dimers, or of strand-breaks incurred during repair or dimer loss have involved procedures *in vitro* and address specific details of the repair processes. However, UDS has been the assay of choice for general inferences about DNA repair. The assay became a powerful tool for identifying genes involved in repair. These were deliberately sought in tissue cultures from small mammals such as hamsters or mice, screened for sensitivity to UV. The UDS assay was also applied to tissues of human beings suffering from an assortment of syndromes which included some sort of adverse response to sunlight—a sign expected if a person is defective in some aspect of the repair of UV-induced DNA damage.

Probably nearly 200 rodent mutant lines have now been isolated and tested by genetic complementation for their genetic identity. Complementation at first involved cell-fusion techniques and this finally established the existence of 11 complementation groups, each representing a single gene.

In humans, it was found that tissue from sufferers of the rare hereditary disease xeroderma pigmentosum—where sunlight causes multiple brown pigmented patches on exposed skin and causes a range of other symptoms—was unusually sensitive to UV and failed to carry out UDS. Complementation studies, developed by Dirk Bootsma and his colleagues (DeWeerd-Kastelein et al., 1972), showed that xeroderma was not monotypic in origin; at first four and finally seven complementation groups were identified, XPA through to XPG. With the advent of cloning technology this list of genes was added to (and some re-discovered) by transfection of mutant rodent cell lines.

The homology of excision repair in yeast and mammalian cells has been established by the discovery that transcription factor TFIIH in mammals contains the XP homologues of the RAD gene proteins in TFIIH of yeast (Buratowski, 1994). The core complex contains XPD (RAD3) and XPB (RAD25/SSL2) as well as homologues of Tfb1 and Ssl1, and associates with XPG (RAD2), which confers the 5′ endonuclease function. ERCC4/XPF associates with ERCC1 to form the 3′ endonuclease as do their yeast homologues RAD1 and RAD10. Finally, it has been found that XPA protein, like yeast RAD14, binds to UV-irradiated DNA, but is specific to 6–4 adducts. The final twist to this story of convergence of research in different fields will not have escaped the attentive reader. It is that, although the homology does not extend to protein sequences, nor to the transcription connection, excision-repair processes in these eukaryotes and in *E. coli* are homologous since the molecular events appear to involve a binary incision 3′ and 5′ of the damage, triggered by the recognition of the damage by a complex of proteins traversing DNA.

The discovery of the involvement of a transcription factor in excision repair was prompted by the observation by Bohr et al. (1985) that excision-repair was differentiated between fast repair coupled with transcription and a slower mode observed in non-transcribing regions of DNA. One of the human genes, XPC, is required only for the latter mode of repair. Its homologue, RAD4 in yeast, where similar differentiation has been observed, is found to associate with the TFIIH complex, suggesting that these proteins may be involved in associating the complex specifically with inactive chromatin.

Mismatch Repair

Although the Hanawalt-Haynes model may not be exactly appropriate to excision-repair, it does seem to provide a good description of another repair system found in both bacteria and eukaryotes—namely mismatch repair. It is a probably a universal faculty for identifying base-pairs that are mismatched, excising one of the strands and re-synthesizing it, using the opposite strand as a template. It is an aspect of DNA metabolism which is independent of the need to repair damage induced by outside agencies, being more concerned with fidelity in replication and recombination. It very probably plays an important role in the reproductive isolation of species (Chambers et al., 1996; Hunter et al. 1996; Radman et al., 1980; Resnick et al., 1989).

Mismatch repair was an essential component of the Holliday model for meiotic recombination in fungi (Holliday, 1965a). This states that recombination depends on the formation of relatively short stretches of hybrid DNA between homologous chromatids within which, if they spanned a region of heterozygosity, a mispaired base-pair would appear. Mismatch repair would determine whether or not aberrant segregations would be found in the tetrads formed on completion of meiosis.

However, the first mutants which were recognized as affecting mismatch repair were found in bacteria. The *hex* mutant of *Streptococcus (Diplococcus) pneumoniae*, which increases the transformation rate of certain markers a hundred-fold, was found to be a mutator. Transformation in *S. pneumoniae* involves the uptake of a single strand of donor DNA and efficiency is limited by the correction of the mutational difference between donor and recipient. In *hex*-mutants, the directionality of this correction is abolished (Lacks, 1970). Subsequently a number of mutator genes were identified in *E.coli*—mutH, mutS and mutL, as well as uvrD (Cox, 1976). These were shown to be defective in mismatch repair by an assay in which *E.coli* cells were transfected with phage λ hybrid DNA molecules. This assay was also used to show that the mismatch correction was strand specific and depended on the absence of methylation at the 6-position of adenine residues in GATC sequences. It was proposed that the function of methylation was for the purpose of strand recognition, allowing discrimination of newly-synthesized DNA from parental strands (Radman et al. 1980). This would allow mismatches to be corrected by excision of newly-synthesized DNA, thus providing a proof-reading function additional to that built into DNA polymerases. The same system acts to reduce recombination between DNA molecules having 10% or more of heterologous sites (Shen and Huang, 1993).

That the identical editing system exists in eukaryotes was shown by identifying and cloning genes of yeast or man that were homologous to the *E.coli* genes (MLH, MSH, for MutL homologue etc.). In yeast, these have been found to be necessary for keeping a low frequency of aberrant segregations of markers in meiosis and for promoting infertility of inter-specific hybrids because, as in bacteria, extensive mis-matches between pairing partners in meiosis causes, under the influence of these gene products, failure of the reciprocal crossing-over which in yeast is essential for meiosis to be successful (Chambers et al., 1996; Hunter et al, 1996; Resnick et al., 1989). Furthermore, msh2 of yeast is a strong mutator, implicating it in promoting replication fidelity. Defects of these genes in man lead to a failure to correct instability of tracts of short repeated sequences and this may lead to a hereditary cancer called nonpolyposis colorectal cancer (HNPCC).

Recombination Repair

As Ruth Hill observed, the survival of irradiated bacteriophage from UV depends on whether the host bacterium is proficient in excision repair or not, plaque-forming ability being much higher in the former kind of host. This was called host cell reactivation and the bacteria designated *hcr+* or *hcr–*.

A second form of host cell-mediated reactivation depended on the multiplicity of infection. X-ray or UV-irradiated phage were more resistant when co-infecting than when singly infecting the host, or when irradiated after the latent period, when DNA synthesis had begun. This came to be known as multiplicity reactivation and was independent of the excision-repair status of the host cell (Symonds et al., 1962). The interpretation was that there is present in the host cells a second system of DNA repair, requiring homologous molecules and probably involving recombination between them. In 1965, the first mutants which controlled this process were isolated in *E. coli* by Clarke and Margulies (1965). They were in fact isolated as mutants of F– cells deficient in Hfr-mediated genetic exchange and three loci were described: Rec A, RecB and RecC. All three were deficient in multiplicity reactivation, and were extremely sensitive to X-rays. In addition, RecA turned out to be extremely sensitive to UV as well; the other two, while sensitive, were less so. Investigation of the fate of DNA in these strains after UV-irradiation showed that in the RecA mutant DNA was rapidly and extensively degraded, while that in the other two was stable. The mutants were nicknamed "rec-less" and "cautious" respectively.

Later, Rupp et al. (1971) were able to demonstrate a physical exchange of material between molecules of DNA in irradiated cells defective in excision-repair. In RecB or Rec C mutants, this did not occur. The phenomenon could not be assayed in a rec-less mutant.

The ensuing history of the understanding of DNA repair in *E.coli* and a number of other apparently unrelated phenomena is closely involved with the array of puzzles that the RecA gene and its mutant alleles presented over the next ten or fifteen years. It was 12 years before the protein coded by this gene could be isolated in sufficient quantities to allow studies *in vitro* of its enzymic properties. These include a single-strand binding activity together with an ATP-dependent helicase which is capable of incorporating single-stranded DNA into homologous ds DNA molecules (Radding, 1988); even then it continued to yield surprises. The unravelling of this complexity revealed yet another form of DNA damage repair operating in *E. coli*.

SOS Repair

As with photoreactivation, excision repair and recombination repair, the story starts with yet another form of reactivation of UV-irradiated phage in host cells. This phenomenon was first observed by Luria(1947), but came to be known as Weigle reactivation. He observed that UV-irradiated bacteriophage λ survives better when plaqued on *E. coli* which has been previously irradiated with a small dose of UV (Weigle, 1953). This paper also showed that whatever repair process is induced in the host cells by these small doses of UV was mutagenic. For a time it came to be known as "error-prone repair," distinguishing it from excision repair which was said to be error-free. The distinction was dramatically demonstrated by Howard-Flanders, who found that the rate of induction of mutation by UV was 100–1000-fold higher in strains of *E. coli* defective in excision repair.

Other apparently unrelated UV-induced phenomena came to be associated with RecA and with another gene, LexA. LexA mutants constitutively express phage λ, that is loss of LexA function causes prophage λ to be induced. Of course, prophage λ is also induced by low levels of UV-irradiation. Like RecA, LexA mutants are sensitive to both UV and ionizing radiation. LexA differs in that mutants appear to have a high rate of spontaneous mutation. Another UV-inducible phenomenon in *E.coli* is filament formation; after UV, growth continues, but septation becomes impaired, resulting in the formation of long filamentous cells in the culture. In 1967, Witkin proposed that both these UV-induced phenomena (and Weigle reactivation) were the result of the induction of inducers of gene activity which were sensitive to the presence of damage in the DNA and which included an error-prone repair pathway (Witkin, 1967). She implicated RecA and LexA in the sensing and induction pathway. In due course another UV-induced phenomenon was observed, namely the appearance in the cell of an unknown protein of 40kDa molecular weight. Like λ prophage, this protein, protein X, was constitutively expressed in LexA mutants. Also present in LexA mutants was a proteolytic activity which cleaved λ cI repressor, reinforcing the conclusion that LexA coded for a repressor, affecting among others the gene for this protease and for protein X, which is destroyed in cells after UV.

In 1975, it was possible for Miroslav Radman to propose a unifying hypothesis: the SOS Repair Hypothesis (Radman, 1975). The essentials of this were (it has been much refined since, or rather made much fuzzier with the addition of detail after detail) that when the replication fork of DNA synthesis arrives at a damaged site such as a pyrimidine dimer, it stalls. The consequence is the induction of an activity which over-rides the editing function of PolIII (PolC) and permits the replication complex to continue, inserting random bases opposite the dimer. The whole system is dependent for its control on LexA and RecA. In LexA mutants it is constitutive and in RecA mutants it is absent altogether. LexA is the key repressor as it is of a host of other so-called "SOS functions."

The hypothesis acquired a neat closure with the discovery that "protein X" was itself RecA protein (Emmerson and West, 1977; Gudas and Mount, 1977; Little and Kleid, 1977; McEntee, 1977) and that this was the protease that cleaves λ cI protein and allows the phage to enter the lytic phase (Craig and Roberts, 1980). RecA, although present in low levels constitutively, is itself heavily repressed by LexA protein. It acquires its proteolytic activity as a result of interaction with unreplicated DNA at stalled replication forks and among the proteins it cleaves is the LexA repressor. This versatile protein thus embodies a damage-sensing activity, a protease activity triggered by the presence of DNA damage which releases a number of genes from repression by the LexA repressor (not to mention λ repressor) and, given the appropriate substrate consequent on the original damage, manipulates DNA to perform steps in recombination repair. The RecA-mediated inducible repair and SOS functions are conserved in a large number of bacteria, gram-negative and gram-positive, with local variations, and RecA itself is well-enough conserved for the *E. coli* gene to function appropriately when introduced into other species. Its long history has obviously allowed parasites like λ to evolve to tap into the system, giving them an escape route when there are signs of trouble with the host. It has also become clear recently that many eukaryotes have genes involved in repair or recombination or both that have homology with RecA.

Recombination Repair in Eukaryotes

In yeast and in *Ustilago*, ionizing radiation, UV and radiomimetic chemicals all induce recombination in mitotic cells (Holliday, 1961; Roman and Jacob, 1957). It is likely that the same is true of mammalian cells, since Wolff et al. (1975)

demonstrated multiple sister-strand exchange in cultured mammalian cells after UV.

Since the increase in recombination is photoreactivable when induced by UV, the cause is the production of pyrimidine dimers, as for killing and mutagenesis. Other processes are involved besides excision repair, since the rate of induction as a function of UV dose is enhanced 50–100 fold in excision-defective mutants (Hunnable and Cox, 1971). Recombinants accumulate for several hours when UV-treated cells are held in non-growth conditions. No mutation has been shown to abolish UV-induced recombination in yeast, although some in the RAD52 epistasis group reduce it. It is presumed that the substrate for the process is the presence, perhaps transiently, of single-strand nicks or gaps, but an attempt to demonstrate the physical exchange of DNA between homologous DNA strands after UV gave negative results (Resnick et al.,1981). The phenomenon remains virtually unexplored, but it is clear that it contributes to the repair of pyrimidine dimers and other base damage because of the extreme synergism of the RAD52 epistasis group mutants when combined with mutants of either of the other two groups.

RAD52 genes are involved in recombination. After exposure to X-rays, it is clear that recombination, in the sense of exchange of genetic material between homologous DNA molecules, is the principal if not the only means of repairing double-strand breaks. The first intimation of this came from the work of Carl Beam, who showed that haploid cells of yeast had two distinct populations, one of cells very sensitive to X-rays and one of very resistant cells. (Beam et al., 1954). The resistant subset were the budded cells and the survival curve of this subset is sigmoid like that of diploid cells, with a shoulder and an exponential portion. Budded yeast cells are those which have reached the stage in the cell cycle where DNA synthesis has begun, and later Brunborg and Williamson using synchronous cultures, were able to show that resistance of the cells to X-rays was directly proportional to the amount of DNA synthesis completed.

In due course, Ho (1975)—using neutral sucrose gradients to measure the molecular weight of yeast DNA molecules—showed that double-strand breaks induced by X-rays were repaired within three hours and this repair did not take place in a diploid rad52/rad52 mutant. Twelve years later, the new technique of orthogonal field gel electrophoresis (OFAGE) devised by Olson (which separates yeast chromosomes by size) was used to show the same phenomenon. Repair required diploidy and the activities of the RAD51, RAD52 and RAD54 genes (Contopolou et al., 1987).

What is called recombination in yeast is really a complex of systems, including meiosis, mitotic recombination, plasmid integration, unequal crossing-over, ectopic recombination between homologous repeats, developmentally-induced mating-type switching by recombination, and very rare non-homologous recombinations. Genes identified by their repair defects play overlapping roles in these various events, some being common to several, some absolutely required for one but not another and so on.

A similar range of activities is found in plant and mammalian cells. Mutants that are sensitive to ionizing radiation have also been isolated in mammalian cells. About nine complementation groups have been identified in rodent cell lines and many are defective in double-strand or single-strand break repair or both. The activities together or severally affect all kinds of molecular interactions in DNA, including non-homologous "end joining" which may be involved in the formation of the assortment of chromosome abberations (inversions, translocations, and deletions) which have been observed from the earliest days of treatment of cells with X-rays, as well as in the commonly observed non-homologous integration of transfecting DNA. The impression one gets is that it is so important not to lose fragments of DNA that attaching loose ends to any chromosome, regardless of aptness, is preferable to losing them at the next cell division.

NOTE

[1]See Appendix for gene nomenclature.

APPENDIX

Nomenclature

It has been the practice among geneticists to name any gene they identify by some abbreviation which usually hints at the phenotype of the mutant which defines it. Thus Ruth Hill's mutant of *Escherichia coli* B_{s-1} she called "**hcr**-" because it could not accomplish *h*ost *c*ell *r*eactivation. When Boyce and Howard-Flanders isolated their UV-sensitive *E. coli* mutants they called them "**uvr**" (for UV resistance). Because they found there was more than one gene which could mutate to give this phenotype, they appended capital letters to distinguish them, according to the current convention among *E. coli* geneticists: **uvrA**, **uvrB**, **uvrC** and **uvrD**. For a time, these were also known as "**hcr**," since they too were host cell reactivation negative; however, in due course, the "**hcr**" designation was dropped and **uvr** became the conventional name. In like manner, Clarke and Margulies and Howard-Flanders and his colleagues, when they isolated mutants unable to carry out genetic recombination, called them **recA**, **recB**, **recC** and so on.

A similar convention exists among yeast geneticists, but there was not at first any consensus about names; **uvs**, **xs**, **uxs** or X^s were all used. In due course, however, radiation-sensitive mutants as a class were designated "**rad**" and numbers added to distinguish genes and alleles. In yeast, there is a further convention which is that the dominant allele of a gene (usually the wild-type) is written in capitals. Thus **rad3.1** and **rad3.2** are mutant alleles of the **RAD3** gene.

In mammalian systems, different conventions again apply, which have to do with how the genes were found. Human genes are often named after the syndrome which results from mutation. Hence **XPA**, **XPB**, **XPC** (etc.) are genes affecting xeroderma pigmentosum, and **CS**, Cockaigne's Syndrome. Rodent mutants, once assigned to complementation groups (originally by cell-fusion techniques) are designated **RCG1–11**. However, in recent years, genes have been isolated by cloning, again usually by complementation of mutant rodent cell lines. When this happens, the clone or gene is designated **ERCC** (for excision repair cross-complementing). **ERCC1** through **6** are synonymous with **RCG1** through **6**. Since there is such good homology both at the functional and protein sequence level between yeast and mammalian excision repair genes, it is worth tabulating them:

YEAST	*MAMMALS*
RAD1	XPF; RCG4
RAD2	XPG; RCG5
RAD3	XPD; RCG2
RAD4	XPC;
RAD10	RCG1
RAD14	XPA
RAD23	HHR23A; HHR23B
RAD25 (SSL2)	XPB; RCG3
RAD26	CSB; RCG6

REFERENCES

Alper, T. (1956). The modification of damage caused by primary ionization of biological targets. Radiation Res. 5, 573–586.

Ames, B.N., Durston, W.E., Yamasaki, E., & Lee, F.D. (1973). Carcinogens are mutagens: a simple test system combining liver homogenates for activation and bacteria for detection. Proc. Natl. Acad. Sci. USA 70, 2281–2285.

Auerbach, C. & Robson, J.M. (1947). The production of mutations by chemical substances. Proc. Roy. Soc. Edinburgh, B 62, 27–32.

Augenstein, L.G., Masom, J., & Zelle, M.R. (1966). Advances in Radiation Biology. Academic Press, New York.

Beam, C.A., Mortimer, R.K., & Tobias, C.A. (1954). The relation of radioresistance to budding in *Saccharomyces cerevisiae*. Arch. Biochem. Biophys. 49, 110–122.

Beukers, R. & Berends, W. (1960). Isolation and identification of the irradiation product of thymine. Biochem. Biophys. Acta 41, 550–551.

Bohr, V.A., Smith, C.A., Okumoto, D.S., & Hanawalt, P.C. (1985). DNA repair in an active gene: removal of pyrimidine dimers from the DHFR gene of CHO cells is much more efficient than in the genome overall. Cell 40, 359–369.

Boyce, R.P. & Howard-Flanders, P. (1964). Release of ultraviolet light-induced thymine dimers from DNA in *E. coli*. Proc. Natl. Acad. Sci. USA 51, 293–300.

Brendel, M. & Haynes, R.H. (1973). Interactions among genes controlling sensitivity to radiation and alkylation in yeast. Molec. Gen. Genetics 125, 197–216.

Brookes, P. & Lawley, P.D. (1961). The reactions of mono- and di-functional alkylating agents with nucleic acids. Biochem. J. 80, 496–503.

Brunborg, G. & Williamson, D.H. (1978). The relevance of the nuclear division cycle to radiosensitivity in yeast. Molec. Gen. Genet. 162, 277–286.

Buratowski, S. (1994). The basics of basal transcription by RNA polymerase II. Cell 77, 1–3.

Cairns, J., Delbruck, M., Stent, G., & Watson, J.D. (1966). Phage and the origins of molecular biology. Cold Spring Harbor Laboratory, Press, New York.

Chambers, S.R., Hunter, N., Louis, E.J., & Borts, R.H. (1996). The mismatch repair system reduces meiotic homologous recombination and stimulates recombination-dependent chromosome loss. Molec. Cell. Biol. 16, 6110–6120.

Clarke, A.J. & Margulies, A.D. (1965). Isolation and characterisation of recombination-deficient mutants of *Escherichia coli* K12. Proc. Natl. Acad. Sci. USA 53, 451–459.

Conaway, R.C. & Conaway, J.W. (1993). General initiation factors for RNA polymerase II. Annu. Rev. Biochem. 62, 161–190.

Contopolou, C.R., Cook, V.E., & Mortimer, R.K. (1987). Analysis of DNA double-strand breakage and repair using orthogonal field alternation gel electrophoresis. Yeast 3, 71–76.

Cox, B.S. & Parry, J.M. (1968). The isolation, genetics and survival characteristics of ultraviolet light-sensitive mutants in yeast. Mutation Res. 6, 37–55.

Cox, E.C. (1976). Bacterial mutator genes and the control of spontaneous mutation. Annu. Rev. Genetics 10, 135–156.

Craig, N.L. & Roberts, J.W. (1980). *E. coli* recA protein-directed cleavage of phage lambda requires polynucleotide. Nature (London) 283, 26–30.

De Weerd-Kastelein, E.A., Keijzer, W., & Bootsma, D. (1972). Genetic heterogeneity of xeroderma pigmentosum demonstrated by somatic cell hybridization. Nature New Biol. 238, 80–83.

Dean, C.J., Feldschreiber, P., & Lett, J.T. (1966). Repair of X-ray damage to the deoxyribonucleic acid in *Micrococcus radiodurans*. Nature (London) 209, 49–52.

Ellison, S.A., Feiner, R.R., & Hill, R.F. (1960). A host effect on bacteriophage survival after ultraviolet irradiation. Virology 11, 294–296.

Emmerson, P.T. & West, S.C. (1977). Identification of protein X of *Escherichia coli* as the recA+/tif+ gene product. Molec. Gen. Genetics 155, 77–85.

Feaver, W.J., Svejstrup, J.Q., Bardwell, L., Bardwell, A.J., Buratowski, S., Gulyas, K.D., Donahue, T.F., Friedberg, E.C., & Kornberg, R.D. (1993). Dual roles of a multiprotein complex from *S. cerevisiae* in transcription and DNA repair. Cell 75, 1379–1387.

Freifelder, D. (1965). Mechanism of inactivation of coliphage T7 by X-rays. Proc. Natl. Acad. Sci. USA 54, 128–134.

Game, J.C. & Cox, B.S. (1972). Epistatic interactions between four rad loci in yeast. Mutation Res. 16, 353–362.

Game, J.C. & Cox, B.S. (1973). Synergistic interactions between *rad* mutants in yeast. Mutation Res. 20, 35–44.

Game, J.C. & Cox, B.S. (1974). Repair systems in *Saccharomyces*. Mutation Res. 26, 35–44.

Ginoza, W. (1967). The effects of ionizing radiation on nucleic acids of bacteriophages and bacterial cells. Annu. Rev. Microbiol. 21, 325–368.

Gudas, L.J. & Mount, D.W. (1977). Identification of the recA(tif) gene product of *Escherichia coli*. Proc. Natl. Acad. Sci. USA 74, 5280–5284.

Guzder, S.N., Qiu, H., Sommers, C.H., Sung, P., Prakash, L., & Prakash, S. (1994). DNA repair gene RAD3 of *S. cerevisiae* is essential for transcription by RNA polymerase II. Nature (London) 367, 91–94.

Hanawalt, P.C. & Haynes, R.H. (1967). The repair of DNA. Scientific American 216(2), 36–43.

Harm, W. (1963). In: Repair from Radiation Damage (Sobels, F.H., Ed.), p. 107. Pergamon Press, New York.

Haynes, R.H. (1966). The interpretation of microbial inactivation and recovery phenomena. Radiation Res. Suppl. 6, 1–29.

Hill, R.F. (1958). A radiation-sensitive mutant of *Escherichia coli*. Biochem. Biophys. Acta 30, 636–637.

Ho, K. (1975). Induction of double-strand breaks by X-rays in a radiosensitive strain of yeast. Mutation Res. 30, 327–334.

Hollaender, A. (1939). Wave-length dependence of the production of mutations in fungus spores by monochromatic ultra-violet radiation (2180-3650A). Proc. 7th. Int. Cong. Genetics 153–154.

Holliday, R. (1961). Induced mitotic crossing-over in *Ustilago maydis*. Genetical Res. 2, 231–248.

Holliday, R. (1965a). A mechanism of gene conversion in fungi. Genetical Res. 5, 282–304.

Holliday, R. (1965b). Radiation sensitive mutants of *Ustilago maydis*. Mutation Res. 2, 557–559.

Holweck, F. & Lacassagne, A. (1930). Action sur les levures des rayons Xmous (K der Fer). Comptes Rendus Soc. Biol. 103, 60–62.

Howard-Flanders, P. & Alper, T. (1957). The sensitivity of microorganisms to irradiation under controlled gas conditions. Radiation Res. 7, 518–540.

Hunnable, E.G. & Cox, B.S. (1971). The genetic control of dark recombination in yeast. Mutation Res. 13, 297–309.

Hunter, N., Chambers, S.R., Louis, E.J., & Borts, R.H. (1996). The mismatch repair system contributes to meiotic sterility in an interspecific yeast hybrid. EMBO J. 15, 1726–1733.

Johnston, L.H. & Nasmyth, K.A. (1978). *Saccharomyces cerevisiae* cell cycle mutant cdc9 is defective in DNA ligase. Nature (London) 274, 891–893.

Jung, H. (1968). Habilitationsschrift. Heidelberg: University of Heidelberg.

Kelner, A. (1949). Effect of visible light on the recovery of *Streptomyces griseus* conidia from ultraviolet irradiation injury. Proc. Natl. Acad. Sci. USA 35, 73–79.

Knapp, E. & Schreiber, H. (1939). Quantitative Analyse der mutationsauslosenden Wirkung monochromatischen UV Lichtes in Spermatzoiden von *Sphaerocarpus*. Proc. 7th. Int. Cong. Genetics, 175–176.

Korogodin, V.I. & Malumina, T.S. (1959). Recovery of viability of irradiated yeast cells. Priroda 48, 82–85 (in Russian).

Lacks, S.A. (1970). Mutants of *Diplococcus pneumoniae* that lack deoxyribonucleases and other activities possibly pertinent to genetic transformation. J. Bacteriol. 101, 119–131.

Latarjet, J. & Ephrussi, B. (1954). Courbes de survie de levures haploides et diploides soumises aux rayons. Comptes Rend. Acad. Sci. Paris 229, 306.

Lea, D.E. (1946). Actions of Radiations on Living Cells. Cambridge University Press, Cambridge.

Little, J.W., & Kleid, D.G. (1977). *Escherichia coli* protein X is the recA gene product. J. Biol. Chem. 252, 6251–6252.

Lucke, W.H. & Sarachek, A. (1953). X-ray inactivation of polyploid *Saccharomyces*. Nature (London) 171, 1014–1015.

Luria, S.E. (1947). Reactivation of irradiated bacteriophage by transfer of self-reproducing units. Proc. Natl. Acad. Sci. USA 33, 253–264.

Lytle, C.D. & Ginoza, W. (1969). Frequency of single-strand breaks per lethal γ-ray hit in ϕX174 DNA. Int. J. Radiation. Biol. 14, 553–560.

McEntee, K. (1977). Protein X is the product of the recA gene of *Escherichia coli*. Proc. Natl. Acad. Sci. USA 74, 5275–5279.

Mortimer, R.K. (1955). Evidence for two types of X-ray-induced lethal damage in *Saccharomyces cerevisiae*. Radiation Res. 2, 361–368.

Mortimer, R.K. (1958). Radiobiology and genetic studies on a polyploid series (haploid to hexaploid) of *Saccharomyces cerevisiae*. Radiation Res. 9, 312–332.

Müller, H.J. (1927). Artificial transmutation of the gene. Science 66, 84–87.

Nakai, S. & Matsumoto, K. (1967). Two types of radiation-sensitive mutants in yeast. Mutation Res. 4, 129–136.

Patrick, M.H. & Haynes, R.H. (1964). Dark recovery phenomena in yeast; 1. Comparative effects with various inactivating agents. Radiation Res. 21, 144–163.

Pettijohn, D.E. & Hanawalt, P.C. (1964). Evidence for repair-replication of ultra-violet damaged DNA in bacteria. J. molec. Biol. 9, 395–410.

Prakash, L. & Prakash, S. (1979). Three additional genes involved in pyrimidine dimer removal in *Saccharomyces cerevisiae*: RAD7, RAD14 and MMS19. Molec. Gen. Genet. 176, 351–359.

Radding, C.M. (1988). Homologous pairing and strand exchange promoted by *Escherichia coli* RecA protein. In: Genetic recombination, G.R. Smith (Ed.) pp. 193–229. American Society for Microbiology, Washington, D.C.

Radman, M. (1975). SOS repair hypothesis: phenomenology of an inducible DNA repair which is accompanied by mutagenesis. In: Molecular Mechanisms of Repair of DNA (Hanawalt, P. & Setlow, R.B., Eds.), pp. 335–367. Plenum Pub., New York.

Radman, M., Wagner, B.W., Glickman, B.W., & Meselson, M. (1980). DNA methylation, mismatch correction and genetic stability. In: Progress in Environmental Mutagenesis (Alacevic, M., Ed.), pp. 121–130. Elsevier/Nth. Holland Biomedical, Amsterdam.

Rasmussen, R.E. & Painter, R.E. (1964). Evidence for repair of ultra-violet damaged deoxyribonucleic acid in cultured mammalian cells. Nature (London) 203, 1360–1362.

Resnick, M.A. (1969). Genetic control of radiation sensitivity in *Saccharomyces cerevisiae*. Genetics 62, 519–531.

Resnick, M.A., Boyce, J., & Cox, B.S. (1981). Post-replication repair in *Saccharomyces cerevisiae*. J. Bacteriol. 146, 285–290.

Resnick, M.A., Skaanild, M., & Nilsson-Tillgren, T. (1989). Lack of DNA homology in a pair of divergent chromosomes greatly sensitizes them to loss by DNA damage. Proc. Natl. Acad. Sci. USA 86, 2276–2280.

Roman, H. & Jacob, F. (1957). Effet de la lumiere ultraviolette sur la recombination genetique entre alleles chez la levure. C. R. Acad. Sci. (Paris) 245, 1032–1034.

Rupert, C.S. (1958). Photoreactivation *in vitro* of ultraviolet irradiated *Haemophilus influenzae* transforming factor. J. Gen. Physiol. 41, 451–471.

Rupert, C.S. (1960). Photoreactivation of transforming DNA by an enzyme from bakers' yeast. J. Gen. Physiol. 43, 573–595.

Rupert, C.S. (1975). Enzymatic photoreactivation: overview. In: Molecular Mechanisms for Repair of DNA. (Hanawalt, P. & Setlow, R.B., Eds.), pp. 73–87. Plenum Publishing, New York.

Rupp, W.D., Wilde, C.E.I., Reno, D.L., & Howard-Flanders, P. (1971). Exchanges between DNA strands in ultraviolet-irradiated *Escherichia coli*. J. Molec. Biol. 61, 25–44.

Sancar, A. & Rupp, W.D. (1983). A novel repair enzyme: UVRABC excision nuclease of *Escherichia coli* cuts a DNA strand on both sides of the damaged region. Cell 33, 249–260.

Seeberg, E., Nissen-Mayer, J., & Strike, P. (1976). Incision of ultraviolet-irradiated DNA by extracts of E. coli requires three different gene products. Nature (London) 263, 524–526.

Setlow, R.B. & Carrier, W.L. (1964). The disappearance of thymine dimers from DNA: an error-correcting mechanism. Proc. Natl. Acad. Sci. USA 51, 226–231.

Shen, P. & Huang, H.V. (1993). Effect of base-pair mismatches on recombination via the RecBCD pathway. Molec. Gen. Genetics 218, 358–360.

Snow, R. (1967). Mutants of yeast sensitive to ultraviolet light. J. Bacteriol. 94, 571–575.

Stadler, D.J. (1997). Ultraviolet-induced mutation and the chemical nature of the gene. Genetics 145, 863–865.

Stadler, L.J. (1928). Mutations in barley induced by X-rays and radium. Science 68, 186–187.

Stadler, L.J. (1939). Genetic studies with ultraviolet radiation. Proc. 7th. Int. Cong. Genetics 269–276.

Stadler, L.J. (1954). The gene. Science 120, 811–819.

Sung, P., Prakash, L., Matson, S.W., & Prakash, S. (1987). RAD3 protein of *Saccharomyces* is a DNA helicase. Proc. Natl. Acad. Sci. USA 84, 8951–8955.

Sutherland, B.M. (1974). Photoreactivating enzyme from human leukocytes. Nature 248, 109–112.

Symonds, N. (1962). Multiplicity reactivation and radiation stability. J. Molec. Biol. 4, 319–321.

Timofeeff-Ressovsky, N.W. & Zimmer, K.G. (1947). Biophysik. I. Das Trefferprinzip in der Biologie. Leipzig: Hirzel.

Unrau, P., Wheatcroft, R., & Cox, B.S. (1971). The excision of pyrimidine dimers from DNA of ultraviolet-irradiated yeast. Molec. Gen. Genetics 113, 359–362.

Ward, J.F. (1988). DNA damage produced by ionizing radiation in mammalian cells: identities, mechanisms of formation and repairability. Prog. Nucl. Acid Res. and Molec. Biol. 35, 95–125.

Ward, J.F. (1990). The yield of double-strand breaks produced intracellularly by ionising radiation: a review. Int. J. Radiation Biol. 57, 1141–1150.

Watson, J.D. & Crick, F.C. (1953). Molecular structure of nucleic acids: a structure for deoxyribose nucleic acid. Nature (London) 171, 740–741.

Weigle, J.J. (1953). Induction of mutation in a bacterial virus. Proc. Natl. Acad. Sci. USA 39, 628–636.

Weinert, T.A. & Hartwell, L.H. (1988). The RAD9 gene controls the cell cycle response to DNA damage in Saccharomyces cerevisiae. Science 241, 317–322.

Wheatcroft, R. (1972). Genetic control of repair in *Saccharomyces*. D. Phil. Thesis. University of Oxford, Oxford.

Witkin, E. (1956). Time, temperature and protein synthesis: a study of ultraviolet-induced mutation in bacteria. Cold Spring Harbor Symp. Quant. Biol. 21, 123–140.

Witkin, E.M. (1967). The radiation sensitivity of Escherichia coli B: a hypothesis relating filament formation and prophage induction. Proc. Natl. Acad. Sci. USA 57 1275–1279.

Wolff, S. (1961). In: Mechanisms in Radiobiology (Errera, F. & Forssburg, W., Eds.), p. 419. Academic Press, New York.

Wolff, S., Bodycote, J., Thomas, G.H., & Cleaver, J.E. (1975). Sister chromatid exchanges in xeroderma pigmentosum cells that are defective in DNA excision repair or post-replication repair. Genetics 81, 349–355.

Zimmer, K.G. (1961). Studies on Quantitative Radiation Biology. Oliver & Boyd, Edinburgh & London.

HANS SPEMANN[1]

ERIC WEISCHAUS[2]

CHRISTIANE NUSSLEIN-VOLHARD[3]

EDWARD LEWIS[4]

LEWIS WOLPERT[5]

CHARLES DARWIN[6]

TOM CECH[7]

MANFRED EIGEN[8]

SOL SPIEGELMAN[9]

JOHN BENET LAWES[10]

JOSEPH HENRY GILBERT[11]

JEAN BAPTISTE BOUSSINGAULT[12]

JUSTUS VON LIEBIG[13]

HERMAN WILFARTH[14]

HERMANN HELLRIEGEL[15]

DIMITRI NELJUBOV[16]

ANTON LANG[17]

JAMES BONNER[18]

FRITZ WENT[19]

EISHII KUROSAWA[20]

PHILIP WAREING[21]

FREDERICK STEWARD[22]

JENS SKOU[23]

KENNETH THIMANN[24]

ALAN HODGKIN[25]

RICHARD KEYNES[26]

BERT SAKMANN[27]

ERWIN NEHER[28]

THOMAS HUNT MORGAN[29]

RUDOLPH PETERS[30]

DAN KOSHLAND[31]

MELVIN CALVIN[32]

Chapter 4

EVOLUTION IN AN RNA WORLD

Peter Schuster

INTRODUCTION

Until the eighties biochemists had an empirically well established but nevertheless prejudiced view on natural and artificial functions of proteins and nucleic acids. Proteins were thought to be nature's unbeatable universal catalysts, highly efficient as well as ultimately specific, and as in the case immunoglobulins even tunable to recognize previously unseen molecules. After Watson[2] and Crick's[2] famous discovery of the double helix, DNA was considered as the molecule of inheritance capable of encoding genetic information and sufficiently stable to allow for essential conservation of nucleotide sequences over many replication rounds. Mutations are rare cases of replication errors and provide the basis for innovation in genetics. RNA's role in the molecular concert of nature was reduced to the transfer of

Foundations of Modern Biochemistry,
Volume 4, pages 159–198.

ISBN: 0-7623-0351-4

sequence information from DNA to protein, be it as mRNA or as tRNA. Ribosomal RNA and some rare RNA molecules did not fit well into this picture; To them a kind of scaffolding function was attributed like holding supramolecular complexes together or bringing protein molecules into the correct spatial positions required for activity. This conventional picture was based on the idea of complete division of labor. Nucleic acids, DNA as well as RNA, were the templates, ready for replication and read-out of genetic information and accordingly unable to do catalysis. Proteins were the catalysts and thus not capable of template function. In both cases these rather dogmatic views turned out to be wrong.

Tom Cech[7] and Sidney Altman discovered RNA molecules with catalytic functions (Cech, 1983, 1986, 1990; Gurrier-Takada et al., 1983). Both examples were dealing with RNA cleavage reactions catalyzed by RNA. The name "ribozyme" was created for this new class of biocatalysts because they combine properties of **ribo**nucleotides and en**zymes**. The first example is concerned with the self-splicing reaction of group I introns: Without the help of a protein catalyst a non-coding region of an RNA transcript cuts itself out during mRNA maturation. The second example concerns the enzymatic reaction of RNase P which catalyzes tRNA formation from the precursor poly-tRNA. For a long time, biochemists knew that this enzyme consists of a protein and an RNA moiety. It was tacitly assumed that the protein is the catalyst and the RNA has some backbone function. As it turned out in the eighties the converse was true: The RNA acts as the catalyst and the protein carries the scaffold function.

Specificity of conventional protein enzymes is provided by precise molecular fit. The mutual recognition of an enzyme and is substrate is the result of various intermolecular forces which are almost always strongly dominated by hydrophobic interaction. In contrast, specificity of catalytic RNAs is provided by base pairing (see for example the hammerhead ribozyme in Figure 1) and to a lesser extent by tertiary interactions. Both are the results of hydrogen bond specificity. Metal ions too, in particular Mg^{2+}, are often involved in RNA structure formation and catalysis. Catalytic action of RNA on RNA is exercised in the cofolded complexes of ribozyme and substrate. Since the formation of a ribozyme's catalytic center which operates on another RNA molecule requires sequence complementarity in parts of the substrate, ribozyme specificity is thus predominantly reflected by the sequence and not by the three-dimensional structure of the isolated substrate.

RNA catalysis is not only concerned with RNA cleavage; non-natural ribozymes that show ligase activity (Bartel and Szostak, 1993) were obtained and many (so far not yet successful) efforts have been undertaken to prepare a ribozyme with RNA replicase activity. RNA catalysis does not only operate on RNA, nor do nucleic acid catalysts require the ribose backbone. Ribozymes were trained by evolutionary techniques to process DNA rather than their natural RNA substrate (Beaudry and Joyce, 1992), and catalytically active DNA molecules were evolved as well (Breaker and Joyce, 1994; Cuenoud and Szostak, 1995). Systematic studies revealed many other examples of RNA catalysis on non-nucleic acid substrates (see also the section on optimization of RNA properties). A spectacular finding in this respect was that oligopeptide bond cleavage and formation is catalyzed by ribosomal RNA and more than 90% of the protein fraction can be removed from ribosomes without loosing the catalytic effect on peptide bond formation (Green and Noller, 1997).

The second prejudice was disproved only last year by the demonstration that oligopeptides can act as templates for their own synthesis and thus show autocatalysis (Lee et al., 1996; Severin et al., 1997). In this very elegant work, Reza Ghadiri and his coworkers at the Scripps Institute in La Jolla showed that template action does not necessarily require hydrogen bond formation. Two smaller oligopeptides of chain lengths 17 (E) and 15 (N) are aligned on the template (T) by means of the hydrophobic interaction in a coiled coil of the leucine zipper type and the 32-mer is produced by spontaneous peptide bond formation between the activated carboxy-group and the free amino residue (see Figure 1):

$$E + T + N \leftrightarrow E{\cdot}T{\cdot}N \rightarrow T{\cdot}T \leftrightarrow 2T$$

The hydrophobic cores of template and ligands consist of alternating valine and leucine residues and show a kind of knobs-into-holes type packing in the complex. The capability for template action of proteins is a consequence of the three-dimensional structure of the protein α-helix which allows the formation of coiled coils. It requires that the residues making the contacts between the helices fulfill the condition of space filling and thus stable packing. Modification of the oligopeptide sequences alters the interaction in the complex and thereby modifies the specificity and efficiency of catalysis. Coiled coil interaction is not restricted to two helices. Triple helices and even higher helical aggregates are readily formed by polypeptide-α-helices under suitable conditions. The triple helix was shown to be active as template for ligation:

$$E + T{\cdot}T + N \leftrightarrow E{\cdot}T{\cdot}T{\cdot}N \rightarrow T{\cdot}T{\cdot}T \leftrightarrow T + T{\cdot}T$$

Triple helix formation expands the catalytic repertoire of template induced oligopeptide formation beyond simple autocatalysis since reaction networks of the hypercycle type (Eigen[8] and Schuster, 1979) can be designed with this reaction too.

Older results and recent findings have provided a revised and comprehensive picture of those properties of RNA and protein molecules which are relevant for evolution. They will be discussed further in the forthcoming sections. Here we recall several apparent differences. Although very stable special RNAs are known, for example the RNAs of viroids (Riesner and Gross, 1985), the vast majority of RNA molecules is much less stable against hydrolytic degradation than proteins. Both classes of molecules comprise representatives showing catalytic functions as well as those having template properties. There are, nevertheless, fundamental differences. Apart from special cases, RNA molecules are more flexible than proteins and this is a disadvantage for forming active sites that are spatially defined with

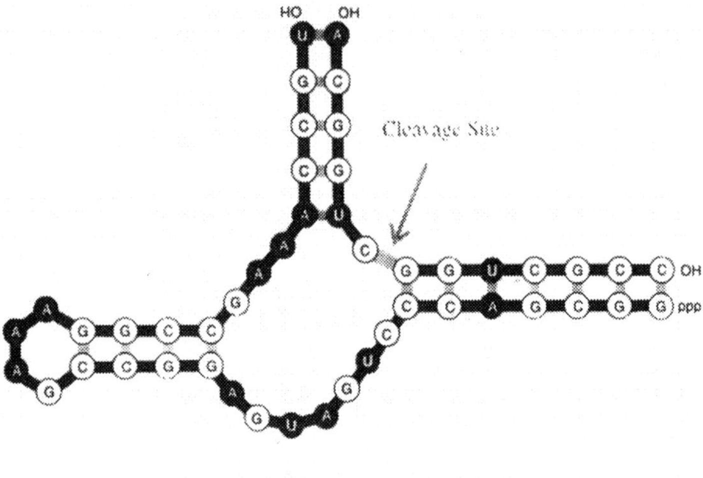

The "Hammerhead" Ribozyme

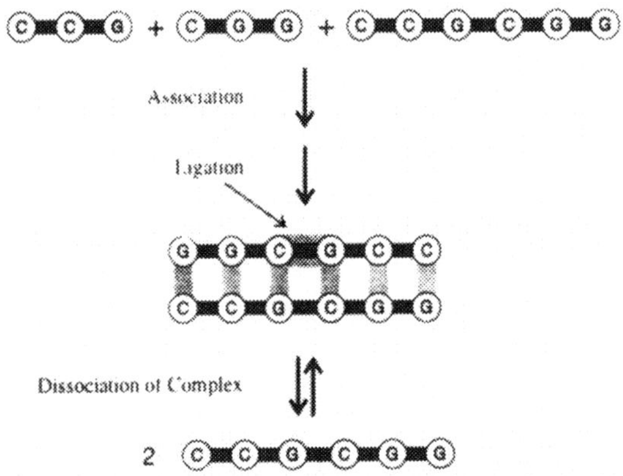

Autocatalytic Replication of a Hexanucleotide

Figure 1. Catalysis and template action of RNA and proteins. Catalytic action of one RNA molecule on another one is shown in the simplest case, the "hammerhead ribozyme." The substrate is a tridecanucleotide forming two double-helical stacks together with the ribozyme ($n = 34$) in the confolded complex. Tertiary interactions determine the detailed structure of the hammerhead ribozyme complex and are important for the enzymatic reaction cleaving one of the two linkages between the two stacks. Substrate specificity of ribozyme catalysis is caused by secondary structure in the cofolded complex between substrate and catalyst. Autocatalytic replication of oligonucleotide and nucleic acid is based on $G \equiv C$ and $A = U$ complementarity in the hydrogen bonded complexes of nucleotides forming a Watson-Crick type double helix. Günter von Kiedrowski's experi-

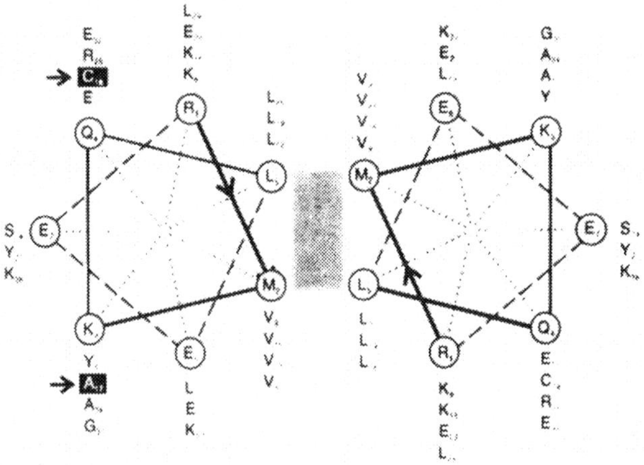

Helical-Wheel Diagram of Template-Ligand Interactions

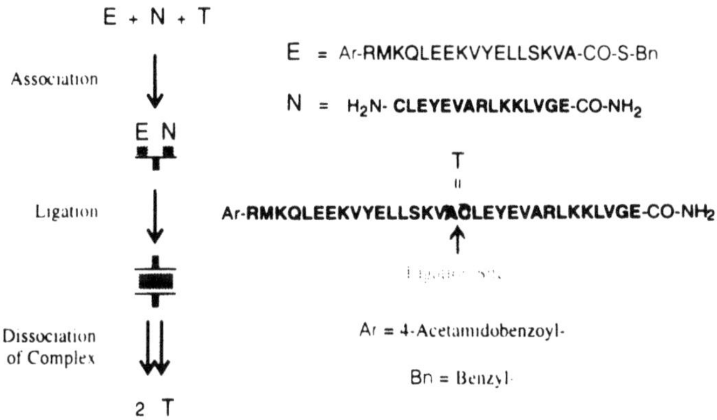

Autocatalytic Replication of an Oligopeptide

Figure 1. (continued) ment on enzyme-free, template induced, synthesis of a hexanucleotide (von Kiedrowski, 1986) starts out from two trinucleotides that associate specifically with a complementary hexanucleotide to form a complex which facilitates ligation. Dissociation of the double helical complex yields two identical hexanucleotides. In the case of larger oligomers the autocatalytic reaction is limited by a kind of product inhibition: the ligated product does not dissociate sufficiently fast from the duplex. Reza Ghadiri succeeded in creating oligopeptide molecules that are capable of template-induced autocatalytic synthesis. Two suitably modified and activated molecules forming α-helices associate with an α-helical template through forming a coiled coil structure of the leucine-zipper type.

high precision. In addition, the catalytic repertoire of RNA molecules is rather poor compared to that of proteins (For a recent and very detailed comparison of RNA and protein enzymes see Narlikar and Herschlag, 1997). The shortcomings of RNA catalysis can be overcome at least in part when ribozymes operate on RNA molecules. Then, base pairing in the cofolded complexes between catalysts and substrates provide the basis for the high specificity of ribozyme catalysis (Figure 1).

The capability to act as template is a rare property within the class of proteins since it requires special tertiary structures. In contrast, all RNA molecules share this property because the potential for template action is a consequence of their backbone structure and thus does not dependent on the sequence. Therefore, we may consider nucleic acid molecules as "obligatory" templates. The molecular basis for template action of RNA and DNA is the complementarity of bases in the Watson-Crick double helix of nucleic acids (**G** ≡ **C** and **A** = **U** or **A** = **T** in RNA or DNA, respectively). Given this property one can easily conceive a mechanism for replication: "It has not escaped our notice that the specific pairing we have postulated immediately suggests a copying mechanism for the genetic material" as Watson and Crick say in their famous closing phrase of their milestone paper in *Nature* (Judson, 1979). The logic of RNA replication is particularly simple and can be visualized as an analogue of the conventional photographic process where the positive is obtained by copying a negative and vice versa. Template-induced RNA synthesis starts at the 5′-end and proceeds in stepwise manner by addition of complementary nucleotides one at a time to the growing chain. One strand called the plus strand ("plus" in Figure 2) acts as the template on which the complementary minus strand ("minus" in Figure 2) is synthesized. The minus strand in turn is the template for plus strand synthesis. **Complementary replication**, as this two step autocatalytic process is commonly characterized, is catalyzed *in vivo* by specific enzymes called RNA replicases and occurs in nature when viral RNA is replicated the host. High template specificity is required because the replicase has to recognize viral RNA among the enormous variety of other RNA molecules in the host cell. Complementary replication can be easily studied *in vitro* (see the section on molecular evolution experiments) and represents the basis for early evolution experiments in the test-tube.

The fundamental problem of complementary replication with or without enzyme catalysis is the separation of plus and minus strands during the course of replication. Long plus-minus duplexes, once formed, are too stable to dissociate under the conditions of replication. The enzyme prevents double strand formation by separating the two strands during the replication processes. Then, the two strands can form their individual secondary structures which, in general, are sufficiently stable to prevent the strands from recombination to the duplex. An evolutionarily important and universal feature of nucleic acid replication is a direct consequence of being "obligatory" templates: A replication error or mutation will be transferred automatically to future generations because of base complementarity in the copying process. Autocatalytic proteins, being occasional templates, have no intrinsic and hence do not share this universal property with the RNA molecules. Consequently, they cannot mutate simply by single events of error copying.

The statement in the last paragraph saying that all RNA molecules have the intrinsic capacity for template action, however, does not imply that enzyme-free RNA replication is a straightforward process and thus easy to achieve. In contrary, the impressive series of elegant experiments carried out by Leslie Orgel and coworkers (Orgel, 1987; 1992) revealed the difficulties of this reaction. So far no solution for RNA replication using activated mononucleotides as building blocks but without enzymes, has been found. The two presumably most prohibitive problems are: (i) single strands rich in purine bases (**G** and **A**) form complexes through self-interaction and are thus not available as templates and (ii) duplexes of already very moderate chain lengths do not dissociate readily so that the first products of template induced synthesis cannot undergo further replication cycles (Figure 2). Autocatalytic template-induced synthesis of oligonucleotides from smaller oligonucleotide precursors, however, was successful: a hexanucleotide was made by ligation of two trideoxynucleotides (von Kiedrowski, 1986). His system (Figure 1) is the oligonucleotide analogue of the autocatalytic, template-induced, ligation of oligopeptides discussed above. In contrast to the latter system, the oligonucleotides do not form triple-helical complexes. Isothermal, autocatalytic, template-induced synthesis, however, cannot be used to prepare longer oligonucleotides because of the duplex dissociation problem. Julius Rebek and co-workers (Tjivikua et al., 1990; Nowick et al., 1991) use other classes of rather complex templates carrying the complementary units, and obtain replication under suitable conditions. Their molecules, like nucleic acids, contain a backbone whose role is to bring "molecular digits" into sterically appropriate positions, so that they can be read by their complements. Complementarity is based on essentially the same principle as in nucleic acids: specific hydrogen bonding patterns allow the recognition of the complementary digit and discriminate between all other letters of the alphabet. As with Ghadiri's coiled coils of oligopeptide α-helices, the hydrogen bonding pattern in these model replicators may be assisted by opposing electric charges carried by the complements.

THE CONCEPT OF AN RNA WORLD

A large number of successful experimental studies which tried to work out plausible chemical scenarios for the origin of life have been conducted in the past (Mason, 1991). A sketch of a possible sequence of events in prebiotic evolution is shown in Figure 3. Most of the building blocks of present day biomolecules are available from different prebiotic sources, from extraterrestrial origins as well as from processes taking place in the primordial atmosphere or near hot vents in deep oceans. Condensation reactions and polymerization reactions formed non-instructed polymers, for example random oligopeptides of the protenoid type (Fox

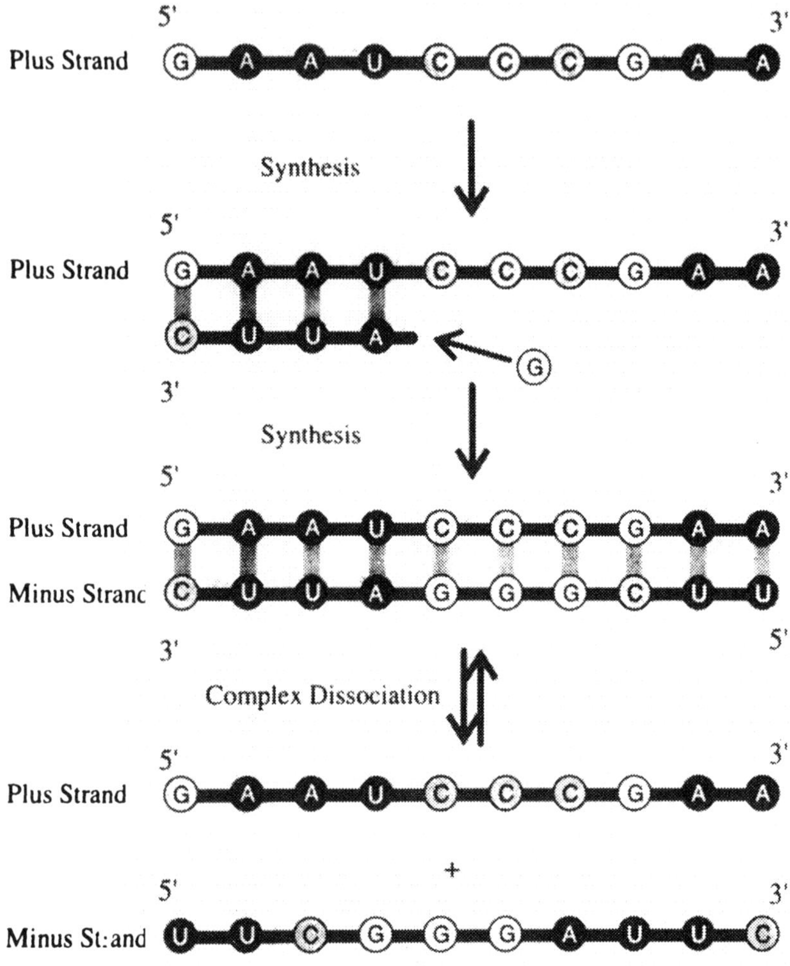

Complementary Replication

Figure 2. The logic of complementary replication and mutation. Template-induced synthesis of RNA is based on nucleotide complementarity (G ≡ C and A = U) in the double helix. The synthesis starts at the 5′-end of the template and adds nucleotide after nucleotide to the growing chain. In this way a negative copy is obtained, being the minus- or plus-strand when a plus- or a minus-strand was the template, respectively. Dissociation of the double-helical plus-minus-duplex completes complementary replication. Three classes of

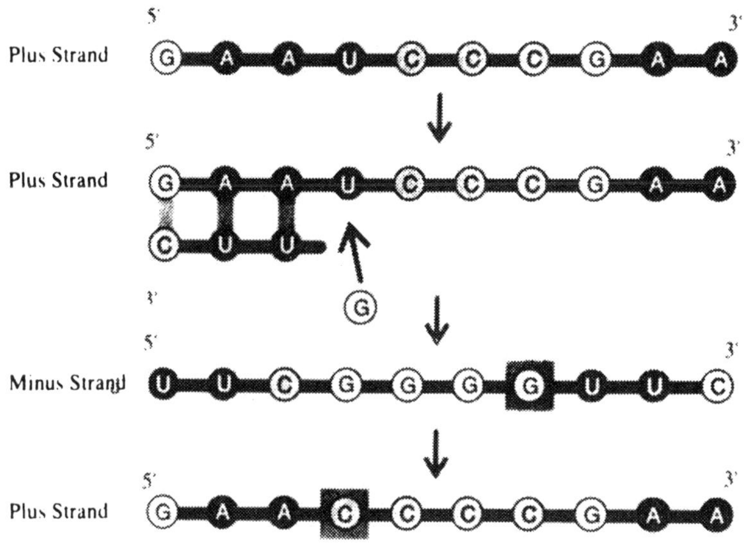

Point Mutation

GAAUCCCGAA → GAAUCCCGUCCCGAA

Insertion

GAAUCCCGAA → GAAUCCA

Deletion

Common Mutations

Figure 2. (continued) mutants are common: (i) point mutations, (ii) insertions and (iii) deletions. Point mutations are tantamount to incorporation of an incorrect nucleotide into the growing chain and thus leave the chain length constant. Insertions consist of double or multiple copies of part of the sequence. In the case of deletions part of the sequence is omitted during replication. The chain length increases in the case of insertions and becomes smaller with deletions.

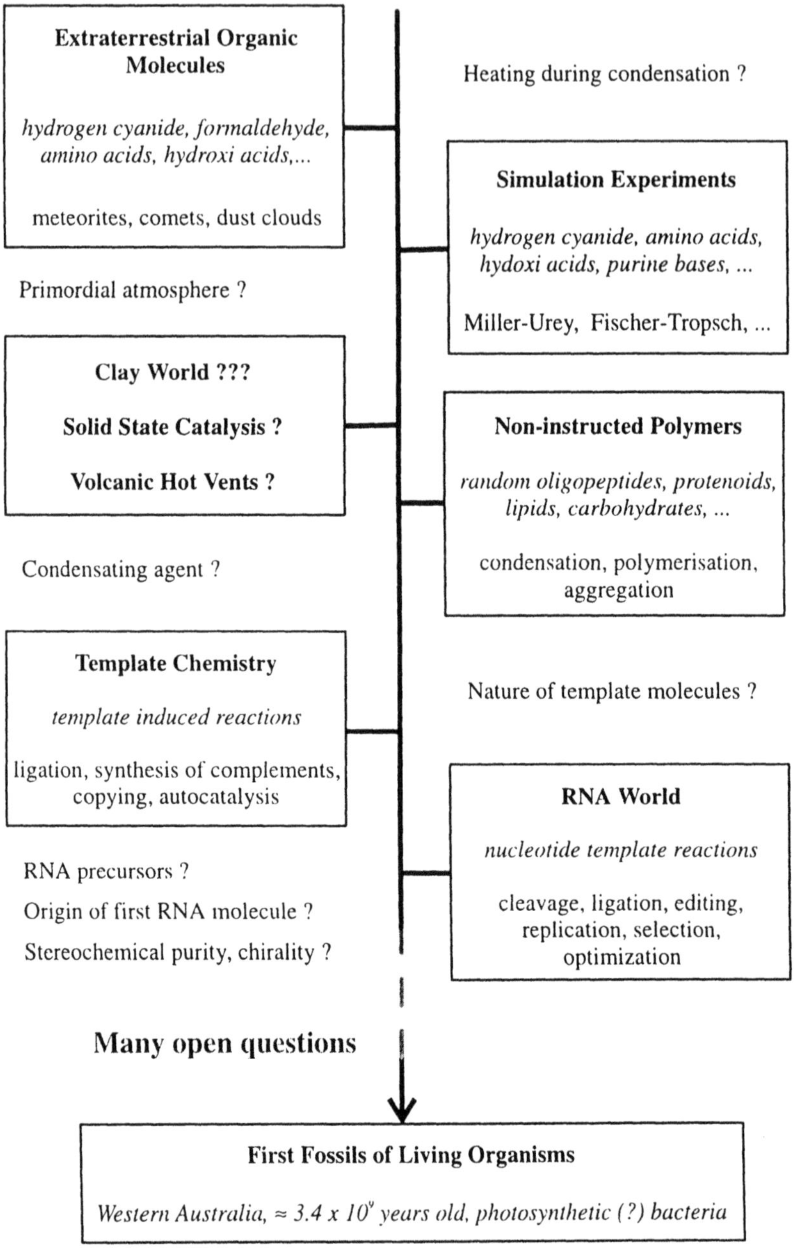

Figure 3. Some facts and open questions about the origin of life. The concept of an RNA world preceding present-day genetics based on DNA, RNA and protein, has been conceived by chemists and molecular biologists interested in problems of prebiotic chemistry and early biological evolution. Several pathways which are plausible under

and Dose, 1977). A first major transition leads from a world of simple chemical reaction networks to autocatalytic processes that are able to form self-organized systems which are capable of replication and mutation as required for Darwinian evolution. This transition can be seen as the interface between chemistry and biology since an early Darwinian scenario is tantamount to the onset of biological evolution. Two suggestions were made in this context: (i) autocatalysis arose in a network of reactions catalyzed by oligopeptides (Kauffman, 1993) and (ii) the first autocatalyst was a representative of a class of molecules with "obligatory" template function (Eigen, 1971; Orgel, 1987). The first suggestion works with molecules that are easily available under prebiotic conditions but lacks plausibility because the desired properties, conservation and propagation of mutants, are unlikely to occur with oligopeptides. The second concept suffers from opposite reasons: It is very hard to obtain the first nucleic acid like molecules but they would fulfill all functional requirements. Many further and not yet understood steps (Eigen and Schuster, 1982) are required to arrive at the first organisms that formed the earliest identified fossils which were found in Western Australia and are about 3.4×10^9 years old.

For two obvious reasons, RNA was chosen as candidate for the leading molecule in a simple scenario at the interface between chemistry and biology: (i) RNA is thought to be capable of storing retrievable information because it is an obligatory template and (ii) it has catalytic properties. Although the catalytic properties of RNA are less universal than those of proteins, they are apparently sufficient for processing RNA. RNA molecules operating on RNA molecules form a kind of self-organizing system that can develop a form of molecular organization with emerging properties and functions. This scenario has been called the RNA world (see e.g. the collective volume by Gesteland and Atkins, 1993 as well as Joyce, 1991). It has been speculated that functionally correlated RNA molecules have developed a primitive translation machinery based on an early genetic code. After such a relationship between RNA and proteins had been established, the stage was

---

Figure 3. (continued) prebiotic conditions lead to the synthesis of the building blocks of present day biopolymers. A commonly made assumption suggests that these building blocks are turned into (non-instructed) biolpolymers by means of condensation reactions. Order in the great diversity of random polymers is brought about by means of template-induced autocatalytic processes. Darwinian selection and evolution is occurring first in the RNA world. Catalytic properties in addition to the template action of RNA molecules make this idea particularly attractive. Problems and open questions on the way to an RNA world are indicated by question marks. At the current stage of our knowledge the further development of primordial replicators to the first cells (not shown in detail) leaves many questions open. The first fossils of living organisms, presumably precursors of present-day cyanobacteria, were dated to 3.4×10^9 years ago.

set for concerted evolution of proteins and RNA. Proteins may induce vesicle formation in lipid-like materials and eventually lead to the formation of compartments. After a number of steps such an ensemble might have developed a primitive metabolism and thus led to the first protocells (Eigen and Schuster, 1982). DNA now being the backup copy of genetic information is seen as a late comer in prebiotic evolution. A successful experimental approach to self-reproduction of micelles and vesicles has been discussed with respect to its relevance for prebiotic structure formation (Bachmann et al., 1992). The basic reaction leading to autocatalytic production of amphiphilic materials is the hydrolysis of ethyl caprylate. The combination of vesicle formation with RNA replication represents an particularly important step towards the construction of a kind of minimal synthetic cell (Luisi et al., 1994).

Despite the experimental studies and the attempts to build comprehensive models, a satisfactory, detailed, and complete answer to the origin of replication problem is still not at hand. Chemists working in prebiotic chemistry see a number of problems for an RNA world as a plausible direct successor of the functionally unorganized prebiotic chemistry (see e.g. the reviews Orgel, 1987, 1992; Joyce and Orgel, 1993; Schwartz, 1997): (i) no convincing prebiotic synthesis of RNA building blocks has been demonstrated, (ii) materials for successful RNA synthesis require high degree of purity that can be hardly achieved under prebiotic conditions, (iii) RNA is a highly complex molecule whose stereochemically correct synthesis (3′-5′ linkage) requires an elaborate chemical machinery, and (iv) enzyme-free template-induced synthesis of RNA molecules from monomers has not been achieved so far. In particular, the dissociation of duplexes into single strands (see next section) and the optical asymmetry problem are of major concern. Template induced synthesis of RNA molecules requires pure optical antipodes. Enantiomeric monomers (containing L-ribose instead of the natural D-ribose) are "poisons" for the polycondensation reaction on the template since their incorporation causes a truncation of the growing polynucleotide chain. Several suggestions postulating more intermediates between chemistry and biology were made. Most of these intermediates were thought to be more primitive and easier to synthesize than RNA but, nevertheless, still having the capability of template action (Schwartz, 1997). Glycerol, for example, was suggested as a substitute for ribose because it is structurally simpler and it lacks chirality. However, no successful attempts to use such less sophisticated molecules together with the natural purine and pyrimidine bases attached to the backbone for template reactions have been reported yet.

Despite the current problems concerning RNA at the origin of life—in particular, the origin of the first RNA molecules and enzyme-free replication—the idea of an RNA world turned out to be a very useful concept. It initiated the search for molecular templates and created an entirely new field which may be characterized as **template chemistry** (Orgel, 1992). A series of systematic studies were performed, for example, on the properties of nucleic acids with modified sugar moieties (Eschenmoser, 1993). This studies revealed the special role of ribose and provided explanations why this molecule is basic to all life processes. The difficulties with RNA replication can be overcome with the help of natural enzymes like virus specific replicases and then the door is open for studies on evolution *in vitro* which will be considered in the remaining part of this chapter.

MOLECULAR EVOLUTION EXPERIMENTS

Evolution has been in the core of biological thought ever since the first publication of Charles Darwin's[6] "Origin of Species" in 1859. Theodosius Dobzhansky cast this fact into the famous phrase "Nothing in biology makes sense except in the light of evolution" (Dobzhansky, 1973). Despite its central role in biology, evolution (not in the historical sense but as a scientific discipline) suffers from a serious drawback. Apart from very few examples, evolutionary change cannot be observed directly and thus has to be inferred from the inevitably incomplete fossil record of extinct species. In the first half of this century it was apparently out of question to do conclusive and interpretable experiments on evolving populations because of two severe problems: (i) Time scales of evolutionary processes are usually prohibitive for laboratory investigations and (ii) the numbers of possible genotypes are outrageously large and thus only a negligibly small fraction of all possible sequences can be realized and evaluated by selection. This is true for both evolution *in vitro* and *in vivo*. If however generation times could be reduced to a minute or less, thousands of generations, numbers sufficient for the observation of evolutionary phenomena, could be recorded in the laboratory. Experiments with RNA molecules fulfill the time scale criterion for evolution in the test-tube since the time required for the replication of RNA molecules can be reduced to less than one minute. With respect to the "combinatorial explosion" of the numbers of possible genotypes the situation is less clear. Population sizes of nucleic acid molecules of 10^{15} to 10^{16} individuals can be produced by random synthesis in conventional automata. These numbers cover roughly all sequences up to chain lengths of $n = 27$ nucleotides. These are only short RNA molecules but as we shall see in the penultimate section, it is unnecessary to cover all sequences in order to find solutions to given evolutionary optimization problems.

Studies on the Qβ Replication Assay

The first successful attempts to study RNA evolution *in vitro* were carried out in the late sixties by Sol Spiegelman[9] and his group at Columbia University (Spiegelman, 1971). They made use of an RNA replicase isolated from *Escherichia coli* cells infected by the RNA bacteriophage Qβ and prepared a medium for replication by adding the four ribonucleoside triphosphates (**GTP**, **ATP**, **CTP**, and **UTP**) in a suitable buffer solution. Qβ RNA, when transferred into this medium, instantaneously started to replicate. Evolutionary experiments were carried out by means of the serial transfer technique (Figure 4). Materials consumed in RNA replication

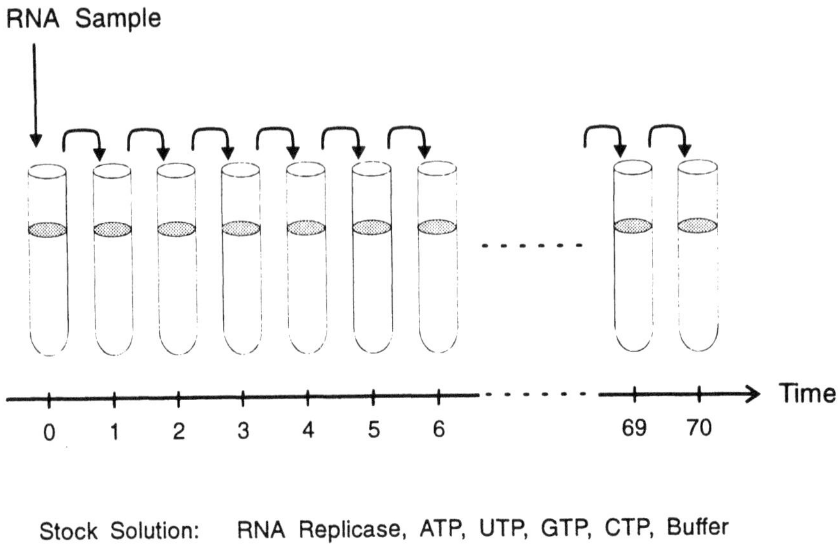

Figure 4. The technique of serial transfer. An RNA sample which is capable of replication in the assay is transferred into a test-tube containing stock solution. This medium contains the four nucleoside triphosphates (ATP, UTP, GTP and CTP)and a virus specific RNA polymerase, commonly Qβ-replicase because of the stability of this protein, in a suitable buffer solution. RNA replication starts instantaneously. After a given period of time a small sample is transferred to the next test-tube and this procedure is repeated about one hundred times. The transfer has two consequences: (i) the material consumed in the replication is replaced, and (ii) the distribution of RNA variants is subjected to a constraint selecting for the fastest replicating species. Indeed, the rate of replication is increased by several orders of magnitude in serial transfer experiments starting out from natural Qß RNA and leading to variants that are exclusively suited for fast replication and hence are unable to infect their natural hosts, *Escherichia coli.*

were replenished by transfer of small samples of the current solution into fresh stock medium. The transfers were made after equal time intervals. In series of up to one hundred transfers the rate of RNA synthesis increased by orders of magnitude. The increase in the replication rate occur in steps and not continuously as one might have expected. Analysis of the molecular weights of the replicating species showed a drastic reduction of the RNA chain lengths during the series of transfers. Initially the Qβ RNA was 4220 nucleotides long but the species finally isolated contained little more than 200 bases. What happened during the serial transfer experiments was a kind of degradation due to suspended constraints on the RNA molecule. In addition to being suitable for replication the RNA has to code for four different proteins in the host cell and presumably also needs a suitable structure in order to facilitate packing into the virion. In test-tube evolution these constraints are released and the only remaining requirement is recognition of the RNA by Qβ replicase. The rate of replication is increased by shortening the viral RNA through deletions during replication. Evidence for a non-trivial evolutionary process came a few years later when the Spiegelman group published the results of another serial transfer experiment that gave evidence for adaptation of an RNA population to environmental change. The replication of an optimized RNA population was challenged by the addition of ethidium bromide to the replication medium (Kramer et al., 1974). This dye intercalates into DNA and RNA double helices and thus reduces replication rates. Further serial transfers in the presence of the intercalating substance led to an increase in the replication rate until an optimum was reached. A mutant was isolated from the optimized population which differed from the original variant by three point mutations.

Extensive studies on the reaction kinetics of RNA replication in the Qβ replication assay were performed by Christof Biebircher in the laboratory of Manfred Eigen in Göttingen (Biebricher and Eigen, 1988). These studies revealed consistency of the kinetic data with the two-cycle many-step mechanism shown in Figure 5. Depending on the concentration, the growth of template molecules allows one to distinguish three phases of the replication process: (i) At low concentration all free template molecules are instantaneously bound by the replicase which is present in excess and therefore the template concentration grows exponentially, (ii) excess of template molecules leads to saturation of enzyme molecules, when the rate of RNA synthesis becomes constant and the concentration of the template grows linearly, and (iii) very high template concentrations impede dissociation of the complexes between template and replicase, and the template concentration approaches a constant in the sense of product inhibition. Despite the apparent complexity of RNA replication kinetics the mechanism fulfills a simple over-all rate law provided the activated monomers, ATP, UTP, GTP, and CTP, as well as Qβ replicase, are present in access. Then, the rate of increase for the concentration c_i of RNA species $\mathbf{I}_i$ follows the simple relation

$$dc_i/dt_i = k_i \cdot c_i - c_i \cdot \Phi,$$

which, in absence of constraints ($\Phi = 0$), leads to exponential growth. This growth law is identical to that found for asexually reproducing organisms and hence replication of molecules in the test-tube leads to the same principal phenomena that are found with evolution proper.

RNA Amplification Assays

RNA replication in the Qβ replication systems requires specific recognition by the enzyme which implies sequence and structure restrictions. Accordingly, only RNA sequences that fulfill these criteria can be replicated. In order to be able to amplify RNA free of such constraints many-step replication assays have been developed. Two of them are sketched in Figure 6. The discovery of the DNA

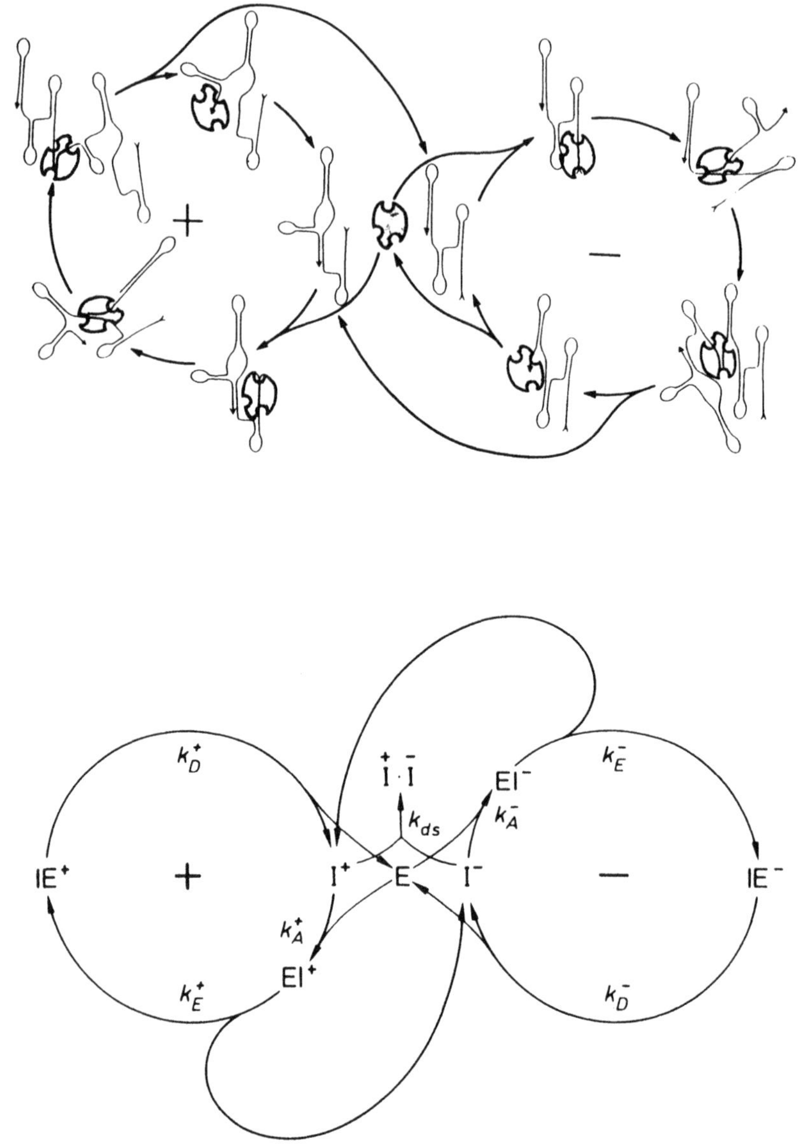

Figure 5. The mechanism of RNA replication by means of viral-specific RNA replicases. The sketch shows a simplified version of the mechanism of RNA replication by Qβ replicase. The mechanism within each of the two cycles consists of binding of the RNA (I^+ and I^- standing for plus- or minus-strand respectively) to the enzyme (E), elongation of the growing chain (see Figure 2) and, eventually, dissociation of the enzyme-RNA complexes. Rate constants for association, elongation and dissociation are indicated by k_A^+, k_E^+, k_D^+, k_A^-, k_E^-' and k_D^-, respectively. According to complementar-

Figure 5. (continued) ity in the replication process, the plus-strand (I^+) is synthesized in the right-hand cycle starting with complex formation between the replicase and minus-strand, whereas newly synthesized minus-strand (I^-) results from the left-hand cycle. The two cycles are complemented by double-strand formation representing a dead end of RNA replication. The upper part of the Figure indicates how the enzyme prohibits double strand formation by allowing template and product to form their specific secondary structures during synthesis. Reproduction of the Figure is by courtesy of the authors (Biebricher and Eigen, 1988).

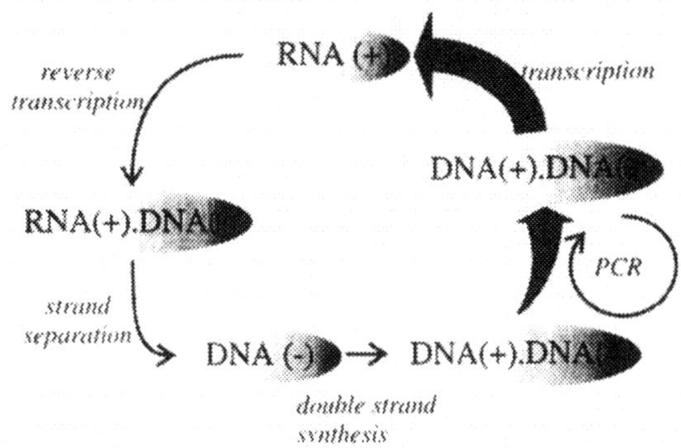

Polymerase Chain Reaction in RNA Amplification

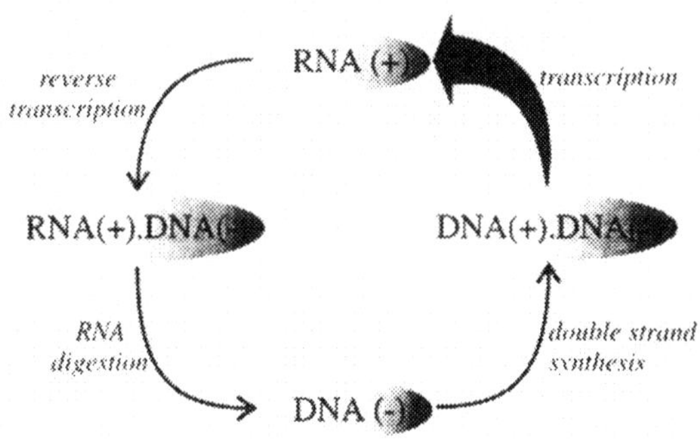

The Self-Sustained Sequence Replication Reaction

Figure 6. Amplification of RNA molecules by assays that are sequence- insensitive. The first assay (upper part) combines the polymerase chain reaction (PCR) of DNA templates with reverse transcription and transcription. Commonly used enzymes are TAQ-polymerase, HIV reverse transcriptase and bacteriophage T7 RNA polymerase. The assay requires a temperature program applying higher temperatures for double strand dissociation. The second assay (lower part) shows the self-sustained sequence replication reaction (3SR) which can be carried out isothermally because double strand dissociation is replaced by enzymatic digestion of the RNA strand in the RNA-DNA duplex. The enzymes used are HIV reverse transcriptase, RNase H and T7 RNA polymerase.

polymerase chain reaction (PCR; Mullis,[2] 1990) was a milestone towards sequence independent amplification of DNA sequences. However, it has one limitation: double helix separation requires higher temperatures and conventional PCR works with a temperature program. In the first assay shown in Figure 6, PCR is combined with reverse transcription and transcription by means of bacteriophage T7 RNA polymerase in order to yield a sequence independent amplification procedure for RNA. This assay contains two possible amplification steps: PCR and transcription.

The second assay makes use of the isothermal self-sustained sequence replication reaction of RNA (3SR; Fahy et al., 1991). Instead of double strand melting to yield single strands the RNA•DNA hybrid obtained through reverse transcription is converted into single stranded DNA by RNA digestion making use of RNase H. DNA double strand synthesis and transcription complete the cycle. Here, transcription by T7 polymerase represents the amplification step.

Artificially enhanced error rates needed for the creation of sequence diversity in a population can be achieved readily with PCR. Reverse transcription and transcription are also susceptible to increase of mutation rates. These two and other new techniques for RNA amplification provide universal and efficient tools for the study of molecular evolution under laboratory conditions and make the usage of viral replicases with their undesirable sequence specificities obsolete.

Evolution of Ribozymes

Evolutionary methods were also used to modify catalytic RNA molecules (Beaudry and Joyce, 1992; Lehman and Joyce, 1993): Group I introns were trained to cleave DNA instead of their genuine RNA substrates or they were reprogrammed to work in presence of Ca^{2+} rather than Mg^{2+} cations. The RNA amplification system used in these experiments is closely related to the PCR based assay shown in Figure 6. Successful variations in the ribozyme are monitored by chemical tagging. A chemical tag consisting of a stretch of DNA is attached to the group I ribozyme or its variant, it replaces the part of the exon that is cleaved off in the natural reaction. Only those variants that are free of the DNA end and hence were able to cleave DNA rather than natural RNA are retained in the evolutionary assay. Gerry Joyce and his group created several group I ribozymes with different predefined properties in this way and demonstrated that molecular evolution can be used to derive novel molecules with properties of technological interest.

OPTIMIZATION OF RNA PROPERTIES

The desire to create RNA molecules with predefined properties and to optimize their efficiencies and specificities has led to a new technique called "evolutionary biotechnology" or "applied molecular evolution." Natural selection or its analogue in test-tube evolution optimizes fitness or replication rate constants, respectively. High replication rates, however, are neither required nor wanted in the search for molecules with optimal properties other than reproduction efficiency. Natural selection has to be replaced by another selection procedure that allows optimization of arbitrary properties. The problem is completely analogous to the task of applied plant geneticists or animal breeders. They are searching for variants with desired properties irrespective of their fitness and achieve success through consecutive intervention with reproduction in the sense that they eliminate individuals with undesired properties. "Molecule breeders," as one could characterize the experimentalists in applied molecular evolution, do precisely the same thing except that different molecules cannot be separated by hand and more sophisticated techniques are required. The commonly applied concept of selection cycles in evolutionary biotechnology is sketched in Figure 7. Three steps are carried out one after another: (i) amplification of genotypes, (ii) diversification through mutation or random synthesis, and (iii) selection of desired variants. Optimization is achieved in successive selection cycles. Amplification and diversification are routinely achieved in replication assays. If very large diversity is desirable, as is the case at the beginning of selection experiments, random oligonucleotide synthesis rather than mutation is applied. The essential step in the cycles is successful selection which requires chemical intuition as well as experimental skill.

Molecules with optimal binding properties to given target molecules can be selected by the so-called SELEX techniques that were invented by Larry Gold and Jack Szostak and their groups at Boulder and Boston, respectively (Gesteland and Atkins, 1993). Target molecules are attached to a chromatographic column and the molecules with desired binding properties are selected through retention from a solution containing all created variants (Figure 8). Application of a different solvent allows the bound molecules to be released and used in subsequent selection cycles. Changes in the solvents applied in the course of optimization experiments improve the binding constants until eventually the optimal binders are obtained. The evolutionary design of optimally binding RNA molecules, so-called **aptamers**, has already found wide-spread application (Ellington, 1994 a,b). An illustrative example is the selection of an aptamer that allows discrimination between the two ligands theophylline and caffeine through a ratio of 10^5 in the binding constants although the two molecules differ only by a single methyl group.

An elegant series of experiments aiming at evolutionary design of ribozymes with RNA ligase function has been carried out by Jack Szostak and his group at Boston (Ekland et al., 1995). A random stretch of 220 nucleotides was attached to the 3′-end of leader sequence. The search for molecules with ligase activity yielded two classes of catalytic RNA: (i) molecules which are capable of self-ligation and (ii) truly catalytic molecules which ligate a 5′-primer to a hairpin loop with a dangling 3′-end that is partially complementary to the primer. Suitable molecules were identified by means of chemical tags. Many different molecules with catalytic activities but little sequence homology were obtained and optimized by selection cycles (Figure 6). These results suggest that particular catalytic activities of RNA molecules are not restricted to certain sequences. In the contrary many and very different se-

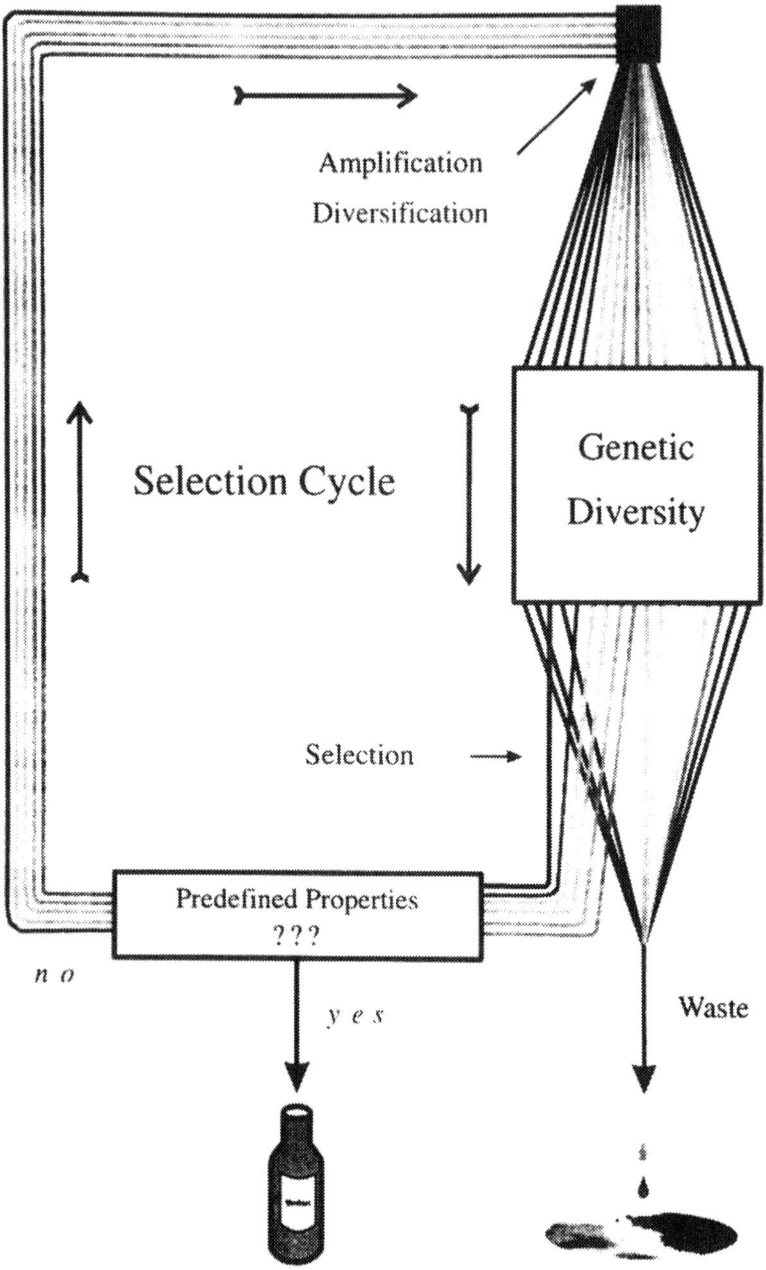

Figure 7. An optimization technique for properties and functions of biopolymers based on molecular evolution with intervention. The goal is achieved by means of selection cycles. Each cycle consists of three phases: (i) amplification of initial or

Figure 7. *(continued)* pre-selected molecules, (ii) creation of genetic diversity and (iii) selection of suitable candidates through intervention, for example by the SELEX technique (Figure 8). Molecular properties are tested after each selection phase and the cycles are terminated when either the desired result has been achieved or no further improvement of molecular properties has been observed.

quences can lead to the same function (see also the subsection on selective neutrality). Ribozymes were also designed for other reactions. Reactions such as alkylation, acylation and aminoacylation of nucleotides and oligonucleotides are efficiently catalyzed by optimized RNA molecules.

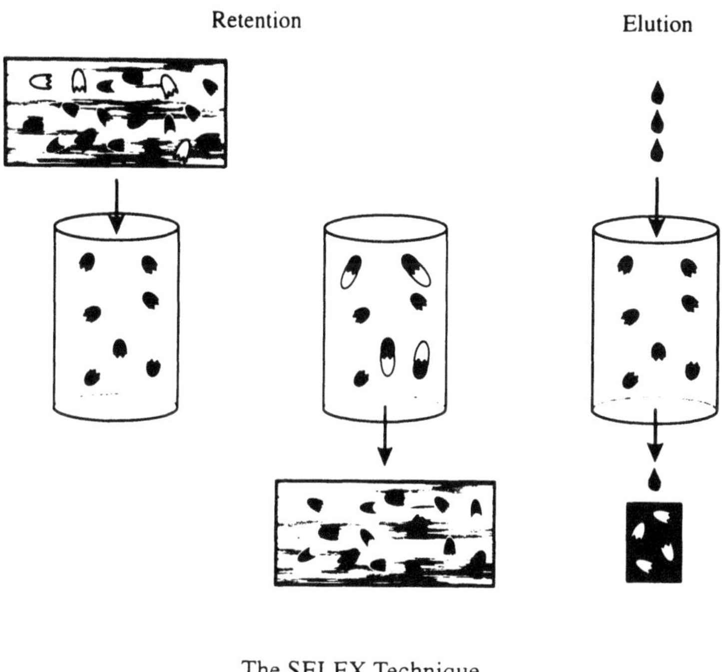

The SELEX Technique

Figure 8. A sketch of the SELEX technique used to select for molecules with optimal binding constants to predefined target molecules. The SELEX procedure selects for molecules with sufficiently high binding constants, so called aptamers, in two steps. Target molecules are attached to a chromatographic column which allows for selective retention of sufficiently strong binders. A different solvent is applied to release the binders and canalize them to the next selection round. Commonly some tens of selection cycles (Figure 7) are sufficient to isolate optimal binding RNA molecules.

MODELING EVOLUTION

At almost at the same time Sol Spiegelman presented the results of his first experiments on RNA evolution in the test-tube, Manfred Eigen published a seminal study on self-organization and evolution of biological macromolecules (Eigen, 1971). His investigations aimed at a mathematical analysis of the processes going on in ensembles of replicating molecules. The major concern of Eigen was to show how biological information originates from self-organization in populations of molecules. Indeed, populations of RNA molecules can "learn" about their environments by means of trial-and-error strategies just as populations of organisms do. The mechanism by which populations "learn" is Darwin's principle of evolution, creation of variants followed by selection of the fittest. Genetic diversity in populations is indeed regulated by mutation and selection, two counteracting processes. Mutation introduces new molecular variants into populations by changing the polynucleotide sequences of genotypes whereas selection operates in opposite direction through evaluation of variants according to replication efficiencies and elimination of those that fall below the population average. Fitness, being Darwin's measure of the number of viable offspring in future generations, can be expressed in terms of replication and degradation rate constants as well as mutation frequencies (see next subsection). The changes selection introduces into evolving populations result in a time series of non-decreasing mean values of fitness. Mean fitness, thus, is the quantity that is maximized through variation and selection. It can be and is commonly used to monitor evolutionary processes. In fact, Spiegelman's experiments and Eigen's theoretical analyses have shown that Darwinian evolution and adaptation to changes in the environment are no privileges of cellular life. Molecules which like polynucleotides are obligatory templates, form populations and undergo evolution when brought into environments that are suitable for replication. These populations usually consist of the fittest genotype and a cloud of mutants surrounding it. They evolve and adapt according to Darwin's principle just as populations of cellular organisms do.

Molecular Quasispecies

The model that was used in Eigen's early paper and in the following studies (Eigen and Schuster, 1979; Eigen et al., 1989) is based on chemical reaction kinetics. Error-free replication and mutation are treated as parallel chemical reactions. Materials consumed by RNA synthesis are replenished by continuous flow or discontinuously by serial transfer (Figure 4). Considering all possible mutation pathways as independent reaction channels provides a great challenge since by single-point mutations every RNA sequence of chain length **n** can give rise to 3**n** different mutants. The entire diversity of sequences can be illustrated by means of an abstract space denoted as **sequence space**. For nucleic acids it can be viewed as a generalized hypercube (Figure 9). Because of combinatorics the genetic diversity of DNA, RNA or protein sequences is hyperastronomically large. In total there are $k^{\mathbf{n}}$ possible different genotypes with k being the number of monomer classes, i.e. $k = 4$ for nucleic acids and $k = 20$ for proteins, respectively. These numbers reach 10^{100} for polynucleotides of chain length $\mathbf{n} = 167$ and polypeptides of chain length $\mathbf{n} = 77$. Sequence space is so large that populations can cover only negligibly small fractions of it. Nevertheless, the concept of sequence space turned out to be very useful for understanding evolution. It is an arrangement of sequences that reflects the mutational distance between genotypes which is the (minimal) number of mutations that are required to interconvert the two genotypes. The common assumption is that single point mutations dominate. Then the Hamming distance $d(j,k)$, which is tantamount to the number of point mutations between two sequences represents the natural measure of distance in sequence space.

The usage of hypercubes for the representation of all possible genotypes goes back to the population geneticist Sewal Wright (Wright, 1932). The distance between two sequences, $\mathbf{I}_j$ and $\mathbf{I}_k$, in the hypercube (generalized to the natural four letter alphabet or not) is the Hamming distance $d(j,k)$. Wright introduced the notion of **fitness landscape** in a metaphor illustrating evolutionary optimization as a hill-climbing process in sequence space. The fitness landscape is obtained by assigning fitness values to individual sequences in sequence space. The evolutionary process is then modeled as hill-climbing of populations on the landscape (see Figure 14, pp. 194). For a long time, Wright's metaphor was nothing but a heuristic because in reality it is very hard if not impossible to determine the fitness values. In recent years, the concept of fitness landscapes saw a revival in the sense that model landscapes were conceived that allowed evolutionarily relevant quantities to be derived from few parameters. The "single-peak" landscape discussed below and the Nk-model introduced by Stuart Kauffman may serve as examples (Kauffman, 1993). Realistic fitness landscapes were derived from the properties of biopolymers, in particular RNA (Schuster, 1997a), and they were applied in modeling of test-tube evolution (see forthcoming sections).

In order to show how a model can be extended to full sequence space we derive all mutation rates from a few parameters by means of the uniform error rate approximation. This approach assumes that mutation rates are independent of the nature of the nucleotide exchange (for example, **G→A**, **G→U**, **G→C**, etc.) and the position in the sequence. The uniform error rate model thus does not consider the existence of "hot spots" at the sequence where mutations occur with higher frequency than in other position. It neglects also deletions and insertions. What we gain is a simple expression which allows analysis of error propagation through replication and mutation by means of standard mathematical techniques. Within this approximation the frequency at which the variant $\mathbf{I}_k$ is obtained as an error copy of $\mathbf{I}_j$, denoted by Q_{kj}, becomes a simple function of the mutation rate per replication round and nucleotide site, p, and the Hamming distance between the two sequences, $d(k,j)$. In order to produce $\mathbf{I}_k$ from $\mathbf{I}_j$ the incorporation of $n-d(k,j)$ correct digits and $d(k,j)$ erroneous digits is required:

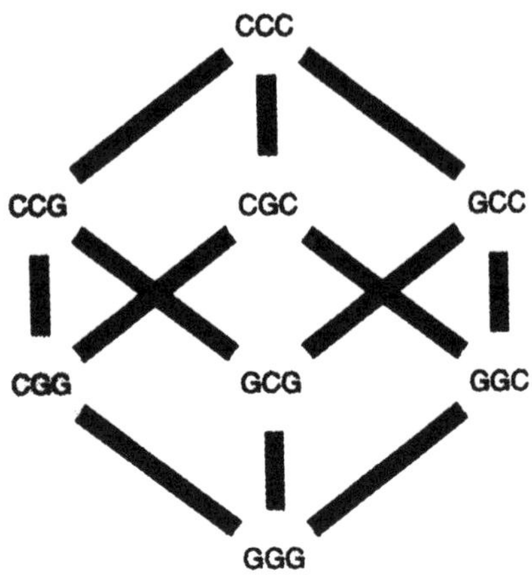

Error Class	Number of Sequences
0	1
1	3
2	3
3	1

Hypercube in Dimension n=3

Error Class	Number of Sequences
0	1
1	9
2	27
3	27

AUGC Sequence Space in Dimension n=3

Figure 9. Sequence spaces of biopolymers and hypercubes. Genotypes may be ordered according to their Hamming distances. The sequence spaces of binary

$$Q_{kj} = (1 - p)^{n-d(k,j)}\, p^{d(k,j)}.$$

Accordingly, the frequency of correct copies of genotype $\mathbf{I}_k$ is given by $Q_{kk} = (1 - p)^n$ and the fitness or the selective value of the phenotype associated with genotype $\mathbf{I}_k$, w_k, can now be expressed in terms of the replication rate constants, f_k, the replication accuracy, Q_{kk}, and the degradation rate constant d_k:

$$w_k = f_k\, Q_{kk} - d_k.$$

The conditions under which a population approaches a stationary, i.e. time independent, mutant distribution were derived from the kinetic differential equations. In this stationary distribution called **quasispecies**, the most frequent genotype of highest fitness, the master sequence, is surrounded by closely related mutants[1] (Figure 10).

Increasing the mutation rate implies creation of more diversity in populations and accordingly, the relative weights between master sequence and mutants decrease in the quasispecies (Figure 11). The mutation rate, however, cannot be enlarged

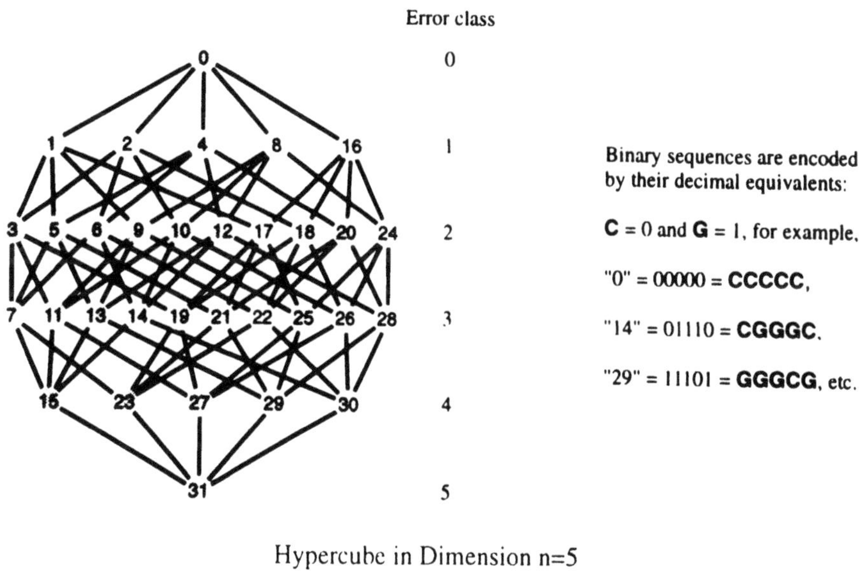

Figure 9. (continued) sequences are particularly simple objects. When nearest neighbors, represented by pairs of sequences of Hamming distance 1, are connected by straight lines, hypercubes of dimension *n* are obtained (see upper and lower part of the illustration which are dealing with (hyper)cubes of dimensions $n = 3$ and $n = 5$, respectively). In four letter alphabets the corresponding sequence spaces are generalized hypercubes (the example with $n = 3$ is shown in the middle of the figure).

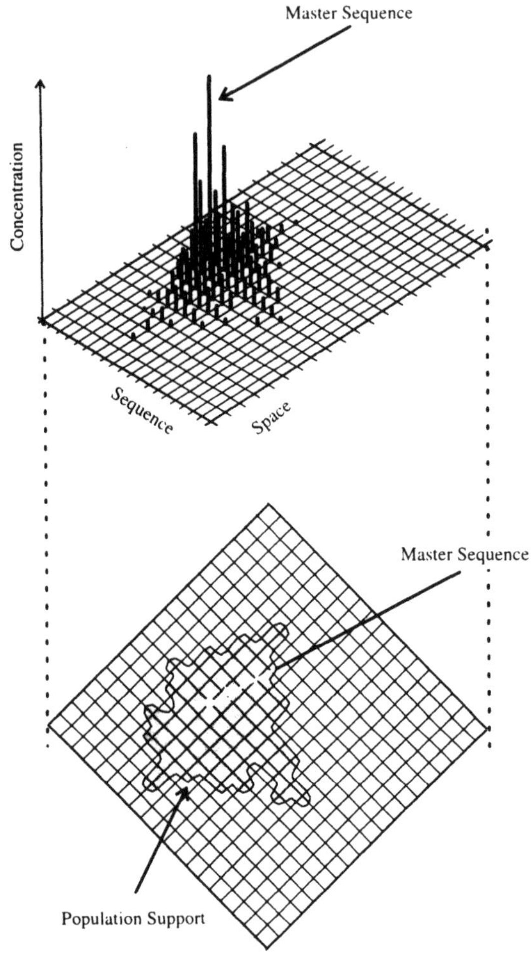

Figure 10. The molecular quasispecies and its support in sequence space. Due to unavoidable non-zero mutation rates, replicating populations form distributions of genotypes or polynucleotide sequences. As shown in the sketch these distributions are centered around a most frequent genotype called the master sequence. A population thus occupies a connected region in sequence space which, according to usual mathematical terminology, is called the support of the population.

indefinitely without losing the capacity for inheritance. Making too many errors per replication causes finite lifetimes for the master sequence and the mutants. Then, no stationary distribution can be sustained and the population migrates through sequence space. The mathematical analysis of the kinetic differential equations suggests the existence of an error threshold which separates the regime of stationary mutant distributions from non-stationarity (Figure 11). The error threshold has been studied first by means of an approximation based on the neglect of reverse mutations and was then verified for the "single-peak" and other model fitness landscapes.[2] It permits the derivation of two simple expressions for the threshold value. A stationary mutant distribution requires

$$Q_{mm} = (1-p)^n > Q_{min} = \sigma_m^{-1} \quad \text{or} \quad n < n_{max} = \ln \sigma_m / p.$$

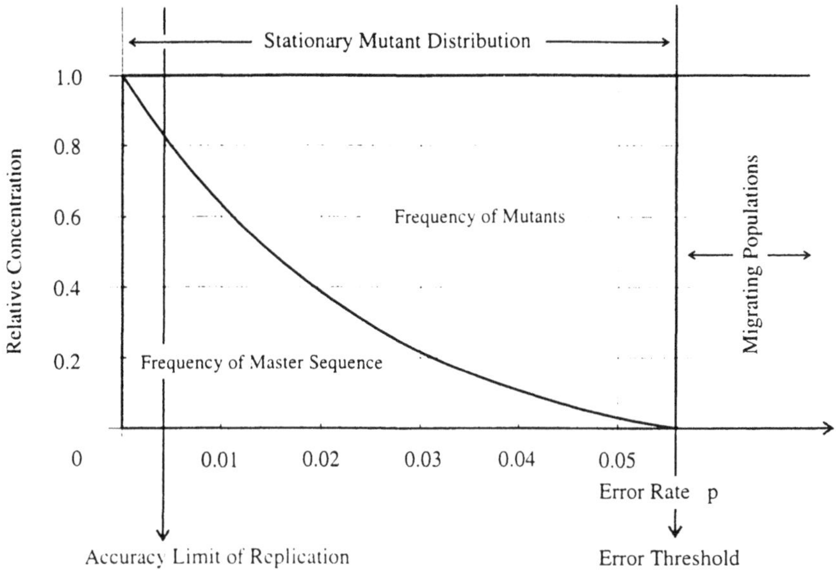

Figure 11. The error threshold of replication and mutation in genotype space. Asexually reproducing populations with sufficiently accurate replication and mutation, approach stationary mutant distributions which cover some region in sequence space. The condition of stationarity leads to a (genotypic) error threshold. In order to sustain a stable population the error rate has to be below an upper limit above which the population starts to drift randomly through sequence space. In case of selective neutrality, i.e. the case of equal replication rate constants, the superiority becomes unity, $\sigma_m = 1$, and then stationarity is bound to zero error rate, $p_{max} = 0$. Polynucleotide replication in nature is confined also by a lower physical limit which is the maximum accuracy which can be achieved with the given molecular machinery. As shown in the illustration, the fraction of mutants increases with increasing error rate. More mutants and hence more diversity in the population imply more variability in optimization. The choice of an optimal mutation rate depends on the environment. In constant environments populations with lower mutation rates do better, and hence they will approach the lower limit. In highly variable environments those populations which approach the error threshold as closely as possible have an advantage. This is observed for example with viruses, which have to cope with an immune system or other defence mechanisms of the host.

The superiority of the master sequence is denoted by σ_m and represents a measure for its advantage in selection. It can be expressed (in the case of equal degradation rate constants) as the ratio of the replication rate constant of the master genotype, f_m, and the mean replication rate constant of the rest of the population, $\bar{f}$:

$$\sigma_m = f_m / \bar{f}; \qquad \bar{f} = \sum_{i, i \neq m} x_i f_i / (1 - x_m)$$

The variables x_i denote the frequencies of the genotypes $\mathbf{I}_i$ (i = 1, . . . , N and $\Sigma_{i=1}^{N} x_i = 1$) in the population. The superiority of the master sequence thus is always larger than one ($\sigma_m > 1$) except in the case of selective neutrality, $f_1 = f_2 = \ldots = f_N = f$, where we have $\sigma_m = 1$ (see forthcoming sections). A larger value of the superiority implies that lower accuracy of replication can be tolerated. Alternatively, longer sequences can be replicated at constant replication accuracy without losing stationarity of the quasispecies. Although the model that has been used in the derivation of the molecular quasispecies is rather simple, the results are also representative for replication and mutation in real populations.

The error threshold corresponds to a kind of phase transition in an abstract space of replication and mutation parameters. It becomes sharper with increasing chain length n and reminds one of cooperative conformational changes in biopolymers.[3] The error threshold also limits the genetic information that can be stored in genotypes. For a given accuracy of the replication process the chain length of the polynucleotide must not exceed a maximal value n_{max} which to a good approximation is about the reciprocal error rate (p^{-1}). This relation is fulfilled since $\ln\sigma_m$ is commonly not very different from one.

Virus populations, in particular those of RNA viruses, were found to replicate close to the error threshold. RNA replication in viruses occurs at much lower fidelity than DNA replication in pro- and eucaryotes: the error rates are commonly between 10^{-3} and 10^{-4} compared to values between 10^{-8} to 10^{-10} observed in the DNA cases. The main reason for lower fidelity seems to be caused by the lack of an error correction mechanism in the replication of RNA viruses. Indeed all organisms except viruses make use of proof-reading in order to improve the accuracy of replication. In some cases, for example with foot-and-mouth disease, influenza virus or HIV, the rate of production of identifiable mutants is so high that RNA genotypes change even within one infected individual. It should be mentioned that it is generally difficult to determine mutation rates *in vivo*. The number of isolated mutants enables one to compute only a lower limit to the mutation rate since non-viable variants cannot be found in this way. Often, the frequency of mutations in virus particles is very high: in influenza A, for example, only one out of approximately 30 virus particles is infectious. Experiments with virus-specific replicases *in vitro* yielded more reliable information on replication error rates. These data give clear evidence that mutation rates for most viruses are close to the reciprocal genome length (n^{-1}) as required for an operation at the error threshold. Analysis of mutant distributions in populations provide additional information. As predicted by the quasispecies concept, virus populations replicating near the error threshold show very high sequence heterogeneity. This has been verified by cloning individual genotypes and comparing their sequences with the consensus sequence of the population. Almost no identified sequence was identical with the consensus.

Genotypes and Phenotypes

Phenotypes are created by unfolding of genotypes in a suitable environment. In general, phenotypes and their formation are so complex that there is currently no chance to interpret phenotype evolution at the molecular level. Development, cell differentiation, cellular metabolism, and biopolymer folding have to be understood in sufficient detail before even decent guesses can be made on the relations between changes in the genomes and their consequences for the properties of phenotypes. Even in the very simple case of RNA viruses, virion self-assembly and the encoding of regulatory events for viral life cycles into RNA structures are yet not fully understood. The rapidly increasing number of complete genomic sequences of viruses and bacteria will provide a wealth of information on life cycles and metabolism. Successful analysis of these data requires efficient models for genotype-phenotype relations and to develop these represents a great challenge for theorists in biology.

Sol Spiegelman (Spiegelman, 1971) was the first to suggest that three-dimensional RNA structures are the phenotypes in the molecular evolution of RNA. Indeed, all properties that are relevant for RNA replication *in vitro*, and thus determine fitness in test-tube experiments, result from the interaction of the RNA with the replicase. In particular, the RNA has to carry structural features that allow specific recognition by the enzyme. For this purpose a short stretch of RNA is apparently sufficient. Christof Biebricher (Biebricher, 1987) has shown that RNA molecules with chain length of only 25 nucleotides are readily recognized and replicated by Qβ replicase. The conserved regions of the accepted RNA molecules contain a few clustered **C**-residues together with a hairpin loop structure at the 5′-ends. In the molecular evolution of RNA investigation of genotype–phenotype mapping boils down to the search for relationships between sequences and molecular structures. Although much simpler than in any other evolutionary system, predictions of three-dimensional structures of RNA molecules from known sequences represent, nevertheless, a difficult and still unsolved problem. What is possible at the present state of the art are investigations at the level of secondary structures. Such studies have revealed several regularities which apparently are of much more general validity than the secondary structures of RNA would suggest (Schuster, 1997a; 1997b). These results are summarized in four statements: (i) The numbers of possible sequences are larger by far than the numbers of stable structures[4] and this implies that there is redundancy in the mapping of sequences onto structures in the sense that many sequences form identical structures. (ii) A relatively small number[5] of common structures is in contrast with the great number of rare structures. A straightforward conjecture says that only the common structures play a role in evolution *in vivo* and *in vitro* because rare structures have too low a probability to be found by search techniques. (iii) In order to find a sequence that folds into a common structure one need not search the whole sequence space. It is sufficient to explore a spherical environment of an arbitrarily chosen reference sequence whose diameter is much smaller than the chain length. (iv) Sequences folding into common structures form extended **neutral networks** spanning almost the entire sequence space. These four regularities have direct consequences for evolutionary optimization. We shall discuss them in the forthcoming subsection on the dynamics of evolutionary optimization.

Selective Neutrality

Comparative analysis of sequences and molecular structures has revealed an astonishingly high degree of redundancy in the sequence-structure relations of biopolymers. Sequence homologies in proteins are often hardly detectable and, nevertheless, the molecules form the same "structure." At this point it is necessary to be precise about what is meant by a biologically relevant notion of structure. Structures at atomic resolution as obtained by X-ray crystallography of proteins or nucleic acids are, of course, never identical for different sequences. Structural information at this high level of resolution provides the details required for the architectures of active sites, for the mechanisms of catalytic reactions as well as for molecular recognition at regulatory DNA sequences. For some parts of biomolecules, much less detail is required to retain function and then very different sequences may serve the same purpose. Thus, biologically adequate notion of structure is based on function and other features relevant to fitness. It needs to be fussy wherever details are required but at the same time it has to be coarse-grained where more details would be physiologically irrelevant. When such a notion of structure is adopted the redundancy in sequence–structure relations becomes evident. Since structures may also be redundant with respect to function, an even higher degree of neutrality than is inferred from sequence-structure relationships is operative in evolution where fitness is the only relevant property of phenotypes.

Many genotypes give rise to the same phenotype. In addition, different phenotypes are often not distinguished by selection because they have approximately the same fitness. In cases of neutrality, populations may drift randomly through the set of neutral variants instead of improving fitness through adaptive selection. This fact has been pointed out with remarkable clarity by Charles Darwin in his "Origin of Species" (Darwin, 1859):

> "Variations neither useful nor injurious would not be affected by natural selection, and would be left either a fluctuating element, as perhaps we see in certain polymorphic species, or would ultimately become fixed, owing to the nature of the organism and the nature of the conditions."

When Motoo Kimura (Kimura, 1983) introduced the "neutral theory of evolution" in the late sixties which was based on his stochastic theory of population genetics and the molecular data of comparative sequence analysis, he and his colleagues from molecular biology (King and Jukes, 1969) were confronted with heavy criticism and negative comments by the community of conventional selectionists of the neo-Darwinian school. Despite the undeniable success of molecular evolution in the reconstruction of phylogenetic trees from molecular data and the overwhelming evidence for the existence of selectively neutral alleles reflected in vast sequence heterogeneity of populations, the neutralist–selectionist debate is unending (Ohta and Kreitman, 1996). It seems that macroscopic biology is unable to end this controversy because of the inherent difficulty in obtaining unambiguously interpretable data. (For a new interpretation of the role of random drift in evolutionary optimization based on data from test-tube evolution and computer simulation see next section).

Quasispecies theory in its original form is unable to deal with the existence of selectively neutral genotypes. The superiority of the master sequence σ_m becomes one and the error threshold converges to zero mutation rate ($p = 0$) in the case of neutrality. This result of the kinetic approach is easily interpreted by non-stationarity and migrating populations even for arbitrarily low error rates of replication in the sense of Kimura's theory. Conditions for the stationarity of phenotypes rather than genotypes, however, can be derived readily by a simple extension of the original quasispecies concept. All genotypes forming the same phenotype are lumped together.

The phenotype of highest fitness is called the master phenotype in complete analogy to the fittest genotype, the master sequence. A derivation that is closely related to those for the conventional (genotypic) error threshold yields the conditions for the existence of stationary phenotype distributions which is tantamount to a phenotypic error threshold occurring at a minimum accuracy Q_{min} (Figure 12; Schuster, 1997b):

$$Q_{mm} = (1 - p)^n > Q_{min} = \frac{1 - \bar{\lambda}_m \sigma_m}{(1 - \bar{\lambda}_m)\, \sigma_m}$$

The quantity $\bar{\lambda}$, called the degree of neutrality, represents the mean fraction of neutral neighbors of the genotypes forming the master phenotype. The two extreme situations are: $\bar{\lambda} = 0$ corresponding to no neutrality and $\bar{\lambda} = 1$, the completely neutral case in which all genotypes form the same phenotype. In the limit of vanishing neutrality, $\lim \bar{\lambda}_m \rightarrow 0$, the original genotypic error threshold is obtained. The existence of selectively neutral genotypes allows to tolerate more replication errors and accordingly the minimum accuracy required for a stable phenotype distribution decreases with increasing degree of neutrality. Interestingly, there is a critical degree of neutrality, $(\bar{\lambda}_m)_{cr} = \sigma_m^{-1}$, above which a population can tolerate unlimited mutation rates without loosing the master phenotype.

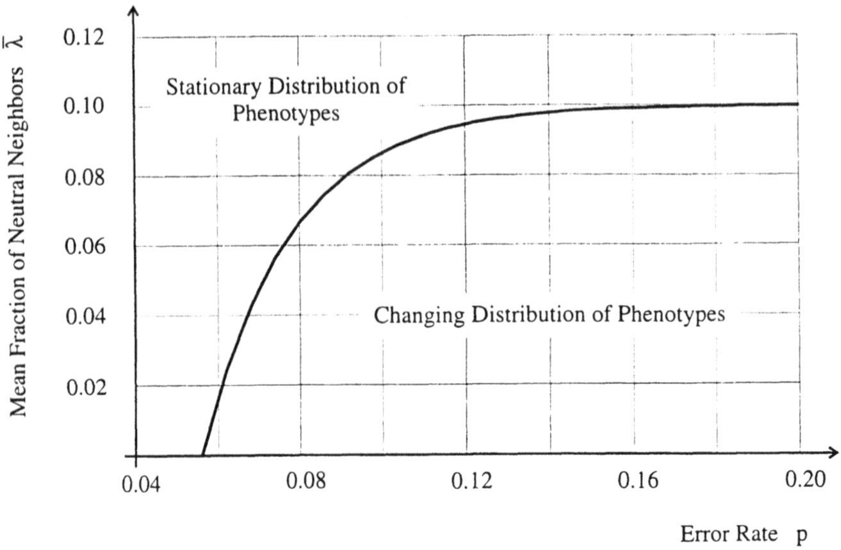

Figure 12. The error threshold of replication and mutation in phenotype space. The genotypic error threshold approaches zero in the case of selective neutrality. Despite changing genotypes a phenotype may be conserved in evolution whenever it has higher fitness than the other phenotypes in the population. The concept of error threshold can easily be extended to competition between phenotypes. The distribution of phenotypes is stationary provided the error rate does not exceed the maximum value p_{max} which is a function of the mean fraction of nearest neighbors, $\bar{\lambda}$, and the superiority of the master phenotype, σ. The illustration shows the position of the phenotypic error threshold in the $\bar{\lambda}$, p plane. Selective neutrality allows more errors to be tolerated and p_{max} increases accordingly with increasing $\bar{\lambda}$. If $\bar{\lambda}$ approaches the inverse superiority, $\bar{\lambda} \rightarrow \sigma^{-1}$, the tolerated error may grow to $p_{max} = 1$, and this means the phenotype will never be lost, no matter how many errors are made in replication.

The existence of neutral networks has important evolutionary consequences and strong implications for the development of a theory of sequence–structure mappings. In recent work attempts were made to derive mathematical models which are able to handle such redundant mappings. One model is based on sequences derived from some alphabet and the relations defined by their embedding in sequence space (Reidys et al., 1996). The mathematical concept chosen is random graph theory since it allows evolutionarily relevant problems to be addressed, for example the question whether or not the neutral network belonging to a given structure extends through sequence space. Depending only on the degree of neutrality, structures may form large networks connecting all neutral sequences or networks containing many disjointed small patches in sequence space. In analogy to percolation problems there is a critical degree of neutrality at which the appearance of the networks changes instantaneously,

$$\bar{\lambda}_{cr} = 1 - \sqrt[\kappa-1]{\frac{1}{\kappa}}$$

with κ being the size of the alphabet (κ = 4 for the natural **A**, **U**, **G**, **C** alphabet): Above the threshold value, $\bar{\lambda} > \bar{\lambda}_{cr}$, neutral neutral networks consist of a single component that spans whole sequence space and below threshold, $\bar{\lambda} < \bar{\lambda}_{cr}$, the network is partitioned into a great number of components. This difference in the shape of neutral networks is crucial for the efficiency of evolutionary optimization.

Comprehensive Dynamics of Molecular Evolution

The quasispecies concept is derived from the over-all kinetics of replication and mutation and corresponds to population genetics in asexually replicating populations. In essence, the ansatz of population genetics describes genotypic frequencies and their changes in time. This kinetic approach, however, is not sufficient for a comprehensive description of evolution since it does not provide a self-contained derivation of the input parameters for the rate equations. Fitness, in general, is a function of rate and equilibrium constants that are properties of the phenotypes which are the molecular structures in the test-tube evolution of RNA. In the simplest known case, the relations between genotypes and phenotypes are thus reduced to the mapping of RNA sequences onto secondary structures and hence can be included in models of evolutionary dynamics. An example of such a model is shown in Figure 13. The complex process of evolution is partitioned into three simpler phenomena: (i) population dynamics, (ii) population support dynamics, and (iii) genotype–phenotype mapping. Conventional population dynamics is extended by two more aspects. Population support dynamics[6] describes the migration of populations through sequence space and genotype-phenotype mapping provides the source of the parameters for populations dynamics. The model is self-contained in the sense that it is based on the rules of RNA structure formation, the kinetics of replication and mutation as well as the structure of sequence space, and thus needs no other source of inputs. The three processes shown in Figure 13 are connected by a cyclic relationship in which each process is driven by the previous one in the cycle and provides the input for the next one: (i) folding sequences into structures yields the input for population dynamics, (ii) population dynamics describes the arrival of new genotypes through mutation and the dying of old ones through selection, and determines thereby how and where the population migrates, and (iii) migration of the population in sequence space finally defines the new genotypes that are to be mapped into phenotypes and thus completes the cycle.

The model of evolutionary dynamics has been applied to interpret the experimental data on molecular evolution and it was implemented for computer simulations (Huynen et al., 1996; Fontana and Schuster, 1998). The computer simulations allows one to follow the optimization process in full detail at the molecular level.

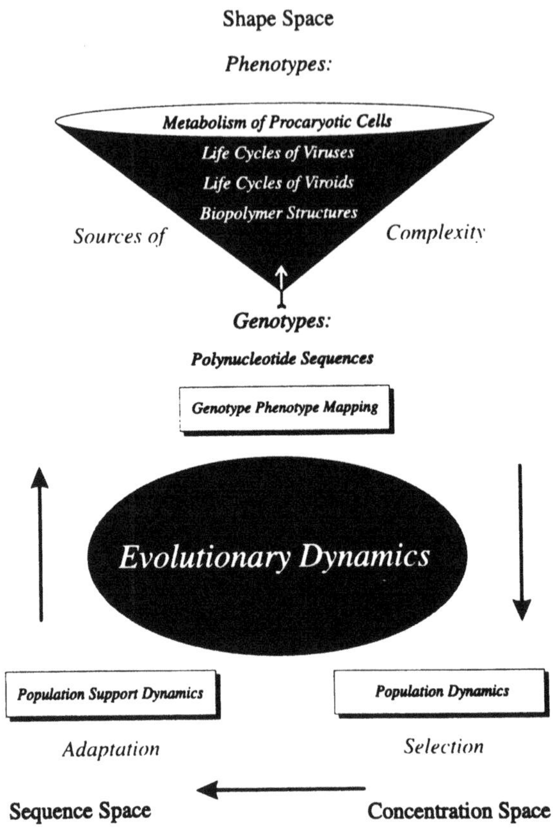

Figure 13. A comprehensive model of molecular evolution. The highly complex process of biological evolution is partitioned into three simpler dynamical phenomena, (i) population dynamics, (ii) support dynamics and (iii) genotype-phenotype mapping. Population dynamics describes how optimal genotypes with optimal genes are chosen by natural or artificial selection from a given reservoir. The basis of population dynamics is replication, mutation and recombination derived from chemical reaction kinetics. The variables are particle numbers or concentrations. In essence it is concerned with selection and other evolutionary phenomena occurring on short time-scales. Population support dynamics describes how the genetic reservoirs change when populations migrate in the huge space of all possible genotypes. Issues are the internal structure of the populations and the mechanisms by which the regions of high fitness are found in sequence or genotype space. Support dynamics deals with the long-term phenomena of evolution, for example with optimization and adaptation to changes in the environment. Genotype-phenotype mapping represents a core problem in evolutionary thinking since the dichotomy between the genotypes and phenotypes is the basis of Darwin's principle of variation and selection. All genetically

Individual runs are monitored as time series of structures which eventually lead to the optimized molecule. The simulations helped to clarify the role of neutral variants in evolution. Recording evolution experiments as well as computer simulation have shown that optimization does not occur continuously. Instead, stepwise increases of fitness are observed. The periods of increase are interrupted by long phases of almost constant fitness. Inspection of populations during the quasi-static phases revealed that constancy is restricted to the level of phenotypes or their properties. The genotypes are changing all the time and the apparent stasis is a result of selective neutrality or, in other words, populations drift randomly through sequence space but stay on neutral networks. As sketched in Figure 14 selective neutrality plays an active role in optimization. On a rugged landscape without neutrality, populations are regularly caught in evolutionary traps. Whenever a population reaches a local optimum in sequence space, i.e. a point that has no neighbors with higher fitness values, optimization comes to an end. If we are dealing with a sufficiently high degree of neutrality, however, the landscape consists of extended neutral networks for all common phenotypes. Almost all points having only non-higher neighbors belong to some neutral network. When a population reaches such a point at the end of an adaptive phase, it starts drifting randomly on the network until it comes to an area that also contains points of higher fitness. There the next adaptive period starts and the population continues the hill-climbing process. The role of neutral variants is to enable populations to leave local fitness optima and to proceed towards areas of higher fitness in sequence space. Optimization on realistic landscapes is a process on two time scales. Fast adaptive phases with substantial increase in fitness are interrupted by periods of random drift of neutral networks during which fitness is essentially constant. The combination of

Figure 13. (continued) relevant variation takes place on the genotypes, whereas the phenotypes are subjected to selection. Variations and their results are uncorrelated in the sense that a mutation yielding a fitter phenotype does not occur more frequently because of the increase in fitness. Any comprehensive theory of evolution thus has to include genotype-phenotype mapping. The problem however is the enormous complexity of the unfolding of genotypes that involves sophisticated processes from the formation of biopolymer structures to cellular metabolism, and the almost open-ended increase in complexity with the development of multicellular organisms. The three processes are related by a kind of cyclic causality in the sense that each process provides the input for the next one. Genotype-phenotype mapping produces the parameters for population dynamics. Population dynamics indicates which genotypes are dying out and what are the new variants. It determines thereby where populations are migrating in sequence space. Support dynamics finally defines the regions in sequence space from which the new phenotypes arise through unfolding of genotypes, and thus completes the cycle. Such cases of cyclic causality are characteristic for self-organizing systems like the ones with which evolutionary processes are concerned.

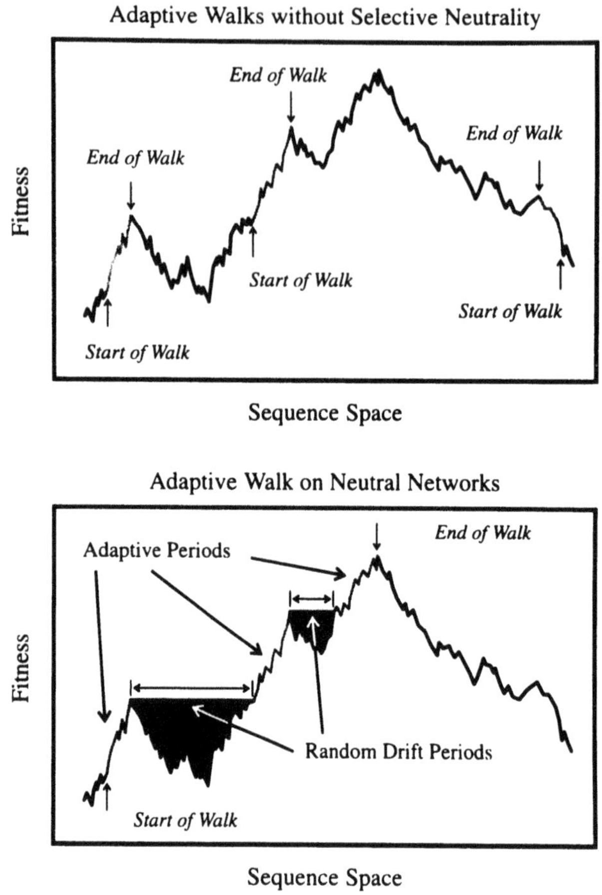

Figure 14. The role of neutral variants in evolutionary optimization. Evolution is often visualized as hill climbing by populations on an abstract fitness landscape (Wright, 1932). In the absence of selectively neutral genotypes the optimization corresponds to a single adaptive walk on the landscape. Adaptive walks allow the next step to be chosen arbitrarily from all directions where fitness is locally non-decreasing. Accordingly adaptive walks end whenever the populations reach local optima in fitness landscapes (upper illustration shows the non-neutral case). Populations containing mutant diversity in the sense of molecular quasispecies (Figure 11) can bridge over narrow valleys with widths of a few point mutations. In the absence of neutral neighbors they are however unable to span longer stretches in sequence space, and thus will approach only the next fitness peak. Since the walk ends there these local optima are considered to be evolutionary traps. Populations on rugged fitness landscapes with a sufficiently high degree of selective neutrality are not caught in local traps. They evolve by a combination of pure adaptive walks and random drift, along the network at roughly constant mean fitness (lower illustration showing the neutral case). Whenever a population reaches an area of higher fitness a new adaptive walk

Figure 14. (continued) starts. Eventually populations reach the global maximum of the landscape. Evolution thus occurs on two time scales: fast adaptive phases characterized by strongly increasing fitness are interrupted by long and seemingly stationary periods with almost constant mean fitness during which only the genotype changes randomly until an area of higher fitness in sequence space is found.

---

adaptation and drift allows escape from evolutionary traps and, depending on the degree of neutrality, eventually leads to the global optimum of the landscape.

RNA PERSPECTIVES

Experiments in molecular evolution with RNA molecules made two important contributions to evolutionary theory: (i) The Darwinian principle of (natural) selection is not restricted to cellular life since it is valid also for serial transfer experiments and other laboratory assays, and (ii) evolution in molecular systems is faster than evolution in organisms by many orders of magnitude and thus allows one to observe optimization and adaptation on easily accessible time-scales, i.e. within days or weeks. The second finding made selection and adaptation subjects of laboratory investigations and thus permitted the design and preparation of biopolymers tailored for predefined purposes. The enormous potential for the application of molecular evolution to the preparation of biopolymers has been recognized within the last decade when successful pioneering experiments raised hopes for a novel kind of biotechnology that copies the most successful principle from nature, namely evolution. In order to be able to create and optimize biomolecules with an efficiency that is suitable for technological applications, more experimental data and further development of the theory of evolution are required. Just as chemical engineering would be doomed to fail without solid backgrounds in chemical kinetics and material science, evolutionary biotechnology cannot be successful without a comprehensive knowledge of molecular evolution and structural biology.

The work with RNA has had a kind of pioneering character. Both the experimental approach to evolution in the laboratory and the development of a theory of evolution are much simpler for RNA than for proteins or viruses. RNA, however, does not seem to be the final choice in the design of molecules for real tasks in biotechnology. Proteins are much more versatile as catalysts and have a much wider stability spectrum. Thus a vision of the future will have to go beyond RNA-based systems and focus on the optimization of proteins without sacrificing the advantages of simple and intelligible RNA genotypes. Display of proteins on the surface of filamentous phages is already an available technique for the selection of polypeptide libraries. It is however still lacking a mechanism for mutation and hence does not provide the possibility of performing multiple selection cycles as required for a truly evolutionary approach. The next logical step in theory and application consists of the development of a coupled RNA and protein system that makes use of both replication and translation. Thus division of labor is introduced into laboratory evolution: RNA is the genotype, protein the phenotype and genotype and phenotype are no longer housed in the same molecules. The development of a theory for RNA and protein evolution will require an extension of the results with RNA to sequence–structure relations in proteins. There a huge body of theoretical and empirical knowledge is already available and the growing sequence and structure databanks provide a substantial amount of information which is not yet exploited. Molecular evolution seems to be approaching the end of an RNA-only scenario since the principal possibilities for RNA molecules in binding assays and as catalysts appear to be well understood. Extensions to much richer "worlds" including instructed protein molecules, are at hand.

ACKNOWLEDGMENTS

The work on the theoretical approach towards molecular evolution reported here is the result of a long lasting cooperation with Professor Manfred Eigen at the *Max-Planck-Institut für Biophysikalische Chemie* in Göttingen (B.R.D.). The recent investigations were performed at the *Institut für Theoretische Chemie der Universität Wien* (Austria) together with Drs. Walter Fontana, Peter Stadler and Ivo Hofacker, and at the *Institut für Molekulare Biotechnologie* in Jena (B.R.D. together with Drs. Christian Forst and Christian Reidys. Financial support of our studies by the Austrian *Fonds zur Förderung der Wissenschaftlichen Forschung* is gratefully acknowledged.

NOTES

1. "Relatedness" refers to the Hamming distance between master sequence and mutant and is expressed by the number of mutation events that are required to produce the mutant from the master. The frequency of individual mutants in the quasispecies is determined by their fitness and the Hamming distance from the master sequence.

2. The result is true for most fitness landscapes and seems to hold for all realistic landscapes in molecular evolution. There are, however, very smooth distributions of fitness values sometimes used in population genetics for which the transition between stationary quasispecies and drifting populations is smooth. A simple landscape showing a sharp transition is the "single-peak" fitness landscape that assigns a higher fitness value to the master sequence and the same lower fitness value to all mutants. It has some similarity to mean field approximations often applied in physics.

3. Conformational changes in biopolymers are commonly described by a model that has been derived by an application of the one-dimensional Ising model to the problem of cooperative transitions from random coil states into ordered mostly helical conformations of (homo)biopolymers (see e.g. Cantor and Schimmel, 1980). Although the threshold is mostly of the cooperative transition type, landscapes can be constructed for which the threshold corresponds to a first order phase transition.

4. A stable structure may but need not be a structure of minimum free energy. Results obtained for kinetically determined structures are essentially the same.

5. "Relatively small" refers to the number of common structures in relation to the total number of structures. Both, the numbers of common structure as well as the total numbers of structures increase exponentially with the chain length.

6. "Support" is a term used in mathematics. It distinguishes the actual from the possible. Every genotype which is present in the population belongs to the support, no matter whether it is represented by a single individual or by a large number of copies. All other potential genotypes do not belong to the support which thus represents the area in sequence space that is occupied by a population of genotypes.

If genotypes are ordered in sequence space, the support forms an area which consists of one, two or more connected compounds. Two genotypes are connected when they are separated by a single point mutation, i.e. when they have Hamming distance one.

REFERENCES

Bachmann, P.A., Luisi, P.L., & Lang, J. (1992). Autocatalytic self-replicating micelles as models for prebiotic structures. Nature (London) 357, 57–59.

Bartel, D.P. & Szostak, J.W. (1993). Isolation of a new ribozyme from a large pool of random sequences. Science 261, 1411–1418.

Beaudry, A.A. & Joyce, G.F. (1992). Directed evolution of an RNA enzyme. Science 257, 635–641.

Biebricher, C.K. (1987). Replication and evolution of short-chained RNA species by Qβ replicase. Cold Spring Harbor Symposia on Quantitative Biology, Vol. 52, pp. 299–306. Cold Spring Harbor Laboratory, Cold Spring Harbor, New York.

Biebricher, C.K. & Eigen, M. (1988). Kinetics of RNA replication by Qβ replicase. In: RNA genetics. Vol.1: RNA Directed Virus Replication (Domingo, E., Holland, J.J., Ahlquist, P., Eds.) pp. 1–21. CRC Press, Boca Raton, FL.

Breaker, R.R. & Joyce, G.F. (1994). A DNA enzyme that cleaves RNA. Chemistry & Biology 1, 223–229.

Cantor, C.R. & Schimmel, P.R. (1980). Biophysical chemistry. Part III: The Behavior of Biological Macromolecules, pp. 1041–1073. W.H. Freeman, San Francisco.

Cech, T.R. (1983). RNA splicing: Three themes with variations. Cell 34, 713–716.

Cech, T.R. (1986). RNA as an enzyme. Sci. Am. 255(5), 76–84.

Cech, T.R. (1990). Self-splicing of group I introns. Annu. Rev. Biochem. 59, 543–568.

Cuenoud, B. & Szostak, J.W. (1995). A DNA metalloenzyme with DNA ligase activity. Nature (London) 375, 611–614.

Darwin, C. (1859). The origin of species. Everyman's Library Vol. 811, p. 81 (1867) . J.M. Dent & Sons, London.

Dobzhansky, T. (1973). Nothing in biology makes sense except in the light of evolution. Amer. Biol. Teacher 35, 125–129.

Eigen, M. (1971). Self organization of matter and the evolution of biological macromolecules. Naturwissenschaften 58, 465–523.

Eigen, M. & Schuster, P. (1979). The Hypercycle. A Principle of Natural Self-Organization. Springer-Verlag, Berlin.

Eigen, M. & Schuster, P. (1982). Stages of emerging life. Five principles of early organization. J. Mol. Evol. 19, 47–61.

Eigen, M., McCaskill, J., & Schuster, P. (1989). The molecular quasi-species. Adv. Chem. Phys. 75, 149–163.

Ekland, E.H., Szostak, J.W., & Bartel, D.P. (1995). Structurally complex and highly active RNA ligases derived from random RNA sequences. Science 269, 364–370.

Ellington, A.D. (1994a). Empirical explorations of sequence space: Host-guest chemistry in the RNA world. Ber. Bunsenges. Phys. Chem. 98, 1115–1121.

Ellington, A.D. (1994b). Aptamers achieve the desired recognition. Current Biology 4, 427–429.

Eschenmoser, A. (1993). Hexose nucleic acids. Pure and Applied Chem. 65, 1179–1188.

Fahy, E., Kwoh, D.Y., & Gingergas, T.R. (1991). Self-sustained sequence replication (3SR): An isothermal transcription-based amplification system alternative to PCR. PCR Methods Appl. 1, 25–33.

Fontana, W. & Schuster, P. (1998). Continuity in evolution. On the nature of transitions. Science 280, 1451–1455.

Fox, S.W. & Dose, H. (1977). Molecular Evolution and the Origin of Life. Academic Press, New York.

Gesteland, R.F. & Atkins, J.F., Eds. (1993). The RNA World. Cold Spring Harbor Laboratory Press, Plainview, NY.

Green, R. & Noller, H.F. (1997). Ribosomes and translation. Annu. Rev. Biochem. 66, 679–716.

Gurrier-Takada, C., Gardiner, K., Marsh, T., Pace, N., & Altman, S. (1983). The RNA moiety of ribonuclease P is the catalytic subunit of the enzyme. Cell 35, 849–857.

Huynen, M.A., Stadler, P.F., & Fontana, W. (1996). Smoothness within ruggedness. The role of neutrality in adaptation. Proc. Natl. Acad. Sci. USA 93, 397–401.

Joyce, G.F. (1991). The rise and fall of the RNA world. The New Biologist 3, 399–407.

Judson, H.F. (1979). The eighth day of creation. The makers of the revolution in biology. Jonathan Cape Ltd., London.

Kauffman, S.A. (1993). The Origins of Order. Self-Organization and Selection in Evolution. Oxford University Press, New York.

Kimura, M. (1983). The Neutral Theory of Evolution. Cambridge University Press, Cambridge, U.K.

King, J.L. & Jukes, T.H. (1969). Non-Darwinian evolution: Random fixation of selectively neutral variants. Science 164, 788–798.

Kramer, F.R., Mills, D.R., Cole, P.E., Nishihara, T., & Spiegelman, S. (1974). Evolution *in vitro*: Sequence and phenotype of a mutant RNA resistant to ethidium bromide. J. Mol. Biol. 89, 719–736.

Lee, D.H., Granja, J.R., Martinez, J.A., Severin, K., & Ghadiri, M.R. (1996). A self-replicating peptide. Nature 382, 525–528.

Lehman, N. & Joyce, G.F. (1993). Evolution *in vitro*: Analysis of a lineage of ribozymes. Current Biology 3, 723–734.

Luisi, P.L., Walde, P., & Oberholzer, T. (1994). Enzymatic RNA synthesis in self-reproducing vesicles: An approach to the construction of a minimal synthetic cell. Ber. Bunsenges. Phys. Chem. 98, 1160–1165.

Mason, S.F. (1991). Chemical Evolution. Origin of the Elements, Molecules, and Living Systems. Clarendon Press, Oxford, U.K.

Mullis, K.B. (1990). The unusual origin of the polymerase chain reaction. Sci. Am. 262(4), 36–43.

Narlikar, G.J. & Herschlag, D. (1997). Mechanistic aspects of enzymatic catalysis: Lessons from comparisons of RNA and protein enzymes. Annu. Rev. Biochem. 66, 19–59.

Nowick, J.S., Feng, Q., Tjivikua, T., Ballester, P., & Rebek, J., Jr. (1991). Kinetic studies and modeling of a self-replicating system. J. Am. Chem. Soc. 113, 8831–8839.

Otha, T. & Kreitman, M. (1996). The neutralist-selectionist debate. BioEssays 18, 673–683.

Orgel, L.E. (1987). Evolution of the genetic apparatus. A review. Cold Spring Harbor Symposia on Quantitative Biology, Vol. 52, pp. 9–16. Cold Spring Harbor Laboratory, Cold Spring Harbor, NY.

Orgel, L.E. (1992). Molecular replication. Nature (London) 358, 203–209.

Reidys, C., Stadler, P.F., & Schuster, P. (1996). Generic properties of combinatory maps. Neutral networks of RNA secondary structures. Bull. Math. Biol. 59, 339–397.

Riesner, D. & Gross, H.J. (1985). Viroids. Annu. Rev. Biochem. 54, 531–564.

Schuster, P. (1997a). Landscapes and molecular evolution. Physica D 107, 351–365.

Schuster, P. (1997b). Genotypes with phenotypes: Adventures in an RNA toy world. Biophys. Chem. 66, 75–110.

Schwartz, A.W. (1997). Speculation on the RNA precursor problem. J. Theor. Biol. 187, 523–527.

Severin, K., Lee, D.H., Granja, J.R., Martinez, J.A., & Ghadiri, M.R. (1997). Peptide self-replication via template directed ligation. Chemistry 3, 1017–1024.

Spiegelman, S. (1971). An approach to the experimental analysis of precellular evolution. Quart. Rev. Biophys. 4, 213–253.

Tjivikua, T., Ballester, P., & Rebek, J., Jr. (1990). A self-replicating system. J. Am. Chem. Soc. 112, 1249–1250.

von Kiedrowski, G. (1986). A self-replicating hexadeoxynucleotide. Angew. Chem. Internat. Ed. Engl., 25, 932–935.

Wright, S. (1932). The roles of mutation, inbreeding, crossbreeding and selection in evolution. In: Proceedings of the Sixth International Congress on Genetics, (D.F. Jones, Ed.) Vol. 1, pp. 356–366. Ithaca, NY.

Chapter 5

BIOLOGICAL NITROGEN FIXATION

Phillip S. Nutman

THE NITROGEN FIXING MICROORGANISMS

It seems that nitrogen gas (dinitrogen) can be fixed (brought into combination with other elements at ordinary temperatures and pressures) only by a limited and rather odd assortment of procaryotes, either alone or in association with higher plants or animals. The free-living fixers are to be found amongst aerobic soil and water microorganisms such as *Azotobacter*, facultative and obligate anaerobes e.g. *Bacillus*, *Clostridium*, anaerobic sulfate-reducing bacteria e.g. *Desulfovibrio* and the photosynthetic bacteria—the blue-greens like *Anabaena* (Kennedy, 1997).

Of the symbiotic associations that fix nitrogen, that of the legumes is by far the most widespread and important. The microsymbiont *Rhizobium* that fixes the nitrogen, resides in specialized nodular outgrowths on the roots. Other root nodular symbioses that fix nitrogen at a high rate are induced by actinomycetes and occur on a wide assortment of trees and shrubs (*Frankia*-type symbioses). Structures inhabited by N-fixing blue-green algae are found in some cycads, ferns, etc.

Foundations of Modern Biochemistry,
Volume 4, pages 199–216.

ISBN: 0-7623-0351-4

The origin of the process of biological nitrogen fixation (BNF) is lost in the mists of geological time. The earliest forms of life on earth probably obtained their nitrogen from nitrogen oxides and ammoniacal compounds present in the primordial oceans. Volcanism and the weathering of plutonic rocks may have provided a small input of combined nitrogen, such as nitrides, and of gaseous nitrogen to the biosphere, and continue to do so.

The blue-green algae may have been among the earliest fixers, possibly associated with reef-forming organisms like the stromatolites still found locally today, as in Shark Bay, Western Australia. These stromatolites form the oldest known fossils from metamorphosed cherts of the Apex Basalt (Awramik et al., 1988; Schopf, 1993), Warrawoona group, Pilbara Craton, Western Australia, dated at 3450±16 Million years (Pidgeon, 1978). This unit contains actual carbonaceous microfossils whose morphotypes resemble some extant photo-autotrophic, chemo-autotrophic and chemo-organotrophic bacteria (see Schopf, 1993). Recently carbon isotope studies have documented traces of life in sediments at least 3850 million years old (Mojzsis et al., 1996; Nutman, A.P. et al., 1997). However, evidence of the nature and sophistication of these forms of life has been destroyed by high temperature metamorphism of the host rocks.

The origin of BNF in plant symbioses certainly came much later, probably coeval with the evolution of the macrosymbiont, and BNF eventually came to assume a major role in the global cycling of nitrogen. Of these, the largest contributors are the aforementioned leguminous symbioses; almost all legumes nodulate and fix nitrogen (Allen, O.N. & Allen, E.K., 1981). The *Leguminosae* is one of the largest families of flowering plants comprising more than 600 genera and nearly 20,000 species; only the *Orchidaceae* and *Compositae* are larger. Incidentally, the orchids also are symbiotic, but with fungi that do not fix nitrogen; the microsymbiont benefits its host in other ways. Endotropic mycorrhiza of the vesicular-arbuscular type, a symbiosis with fungi that does not fix nitrogen, are common on legume roots. This association enables the host plant to scavenge nitrogen compounds and other important nutrients such as phosphorus and minor elements from soil and plant litter. BNF has a high requirement for these elements and benefits especially from vesicular-arbuscular mycorrhiza which can occur on the same roots as nodules, and even within the nodules themselves.

Legumes are found in locations from the tropics to beyond the Arctic Circle and are most frequent and diverse in tropical rain forests and savannahs. They provide major sources of food, fibers, fodder, timber, drugs and many other products, and have done so since ancient times. Seeds of legumes have been found as tomb offerings in the earliest Egyptian and Tigris-Euphrates civilizations and from prehistoric and medieval lake dwelling sites in Europe.

Today the very large amounts of nitrogen fixed by crop and forage legumes probably exceeds the substantial input from artificial fertilizers worldwide. That fixed in natural ecosystems is probably of the same order of magnitude.

Quantities of nitrogen fixed on an area basis vary enormously. Free-living microorganisms fix up to a few tens of kilograms of nitrogen per hectare per annum whereas several hundreds kgN ha^{-1} may be fixed by a good leguminous forage or grain crop. The associated yield can even exceed the maximum response found in, for example, a modern cereal cultivar bred to respond to high rates of fertilizer nitrogen. BNF by nodulated plants is fuelled directly by the host's photosynthetic apparatus. This raises the question of how the high energy requirements for fixation are, in some cases, met without apparently drawing on those for growth, thus conserving energy? Legume-dependent production can be less polluting than that based on mineral fertilizers when rates of overall nitrification (conversion of ammoniacal nitrogen through nitrite to nitrate) do not exceed current uptake.

As a process, BNF is arguably on a par with carbon dioxide fixation as indispensable for the evolution of life. Both involve complex porphyrin chemistry and the participation of the trace metal ions Fe, Mg and Mo. Both also require special organelles, microorganisms for fixation and chloroplasts for photosynthesis, the latter possibly, like mitochondria, being originally of procaryotic origin (see Day and Poulton, 1996). The crucial importance of BNF and photosynthesis as life processes was recognized by the International Biological Programme (IBP) which, from 1960, has organized international symposia on these themes. Over the last half century much has been learned about the interactions between BNF and photosynthesis. It is not the intention of this chapter to deal with this in detail; texts in genetics, biochemistry and agriculture can be consulted. Our aim instead is to look at the early epoch-making discoveries in BNF and their background and to show how they depended upon a few individual scientists who, though constrained by the circumstances of the time, were able to ask the right questions and devise experiments to test their validity.

THE DISCOVERY OF BIOLOGICAL NITROGEN FIXATION IN 1886

During the half-century before 1886, agriculturists were deeply interested in the sources of nitrogen for plant growth. Chilean nitrate (from guano) was first imported into Europe in 1838 and, with other minerals, was being used in field trials, notably by Boussingault[12] at Bechelbronn in Alsace, and by Lawes[10] and Gilbert[11] at Rothamsted in England. Mineral fertilizers were not yet used in practical agriculture. Manuring was from the farmyard and stable and, near towns, from night soil. Farmyard manure was insufficient for all crops and fields not manured regained their fertility by fallowing and rotational cropping in which legumes played an important part. The restorative properties of legumes were recognized, though not understood, from ancient times. It was only with some knowledge of chemistry that agriculturists were able to ask the question "What is the source of nitrogen for plant growth?" Some, including the great German chemist

Liebig,[13] thought that the nitrogen of the atmosphere could be directly assimilated by plants, the value of manure residing wholly in the mineral content of its ash. Lawes disagreed with this view, as did Boussingault, whose opinion was that a manure's value lies mostly in the "azotized" portion.

Boussingault

Boussingault was born in Paris in 1802; his father was a shopkeeper. His education was rudimentary but from an early age he had a strong interest in chemistry which his mother encouraged by buying him Thenard's four volume work on theoretical and practical chemistry—a considerable drain on family finances. He joined a school for mining artisans, graduated to become a demonstrator and soon distinguished himself as a serious and resourceful chemist. He was anxious to travel and, through the influence of Alexander von Humbolt, became Professor of Chemistry at Bogota, Columbia. His most unusual assignment there was an order to cast a statue of Simon Bolivar in platinum. This was not possible because platinum could not then be cast and, moreover, the total quantity of this metal available in Columbia was only a few kilograms. When Boussingault returned to Europe with an established reputation, he first became Professor at Lyons and then was appointed to the chair of agriculture at the Conservatoire des Arts et Métiers in Paris.

In 1837 he started a series of experiments on manuring and crop rotation at his father-in-law's farm at Bechelbronn. These soon established that the masses of carbon, oxygen, hydrogen and nitrogen in the crop were greater than the masses of these elements provided in the manure. He thought that the hydrogen and oxygen came from water and the carbon from carbon dioxide. Only the source of nitrogen presented a problem; did it come from the atmosphere? In a four-course rotation of potatoes, wheat, turnip and oats he measured a total gain of 47.5 kg N per hectare, the analyses showing that this gain came from the clover. This was followed up with experiments on oats, wheat, clover and peas growing in pots of calcined soil; only the clover and peas gained in nitrogen. Mindful of Liebig's criticisms (see below), he repeated these experiments growing the plants in large glass containers supplied with washed air and added carbon dioxide. These experiments provided no evidence of nitrogen assimilation but again drew fire from the formidable Liebig who promulgated the view with almost religious fervor, that manures only needed to provide the elements in the plant's ash, to which Boussingault's rejoinder was that if so, farmers should burn their manure to avoid the cost of cartage! (Boussingault, 1838–1856).

Lawes and Gilbert

Unlike the other great men in our Pantheon, Lawes was a scion of the landed gentry, growing up in a sheltered country environment. He was educated at Eton and Oxford without taking a degree, though while at Brasenose he attended lectures in chemistry by Professor Daubeny. Lawes was only eight when his father died, but his mother, like Boussingault's, strongly encouraged his interest in chemistry, even allowing the conversion of a bedroom at Rothamsted Manor into a laboratory. His first unsuccessful experiments were to extract the active principles of drugs such as belladonna. This was followed by pioneering work on treating insoluble sources of phosphorus, such as animal charcoal, a waste product, with sulfuric acid to render them more available as manures. This became a successful business that financed his experiments and later provided a Trust Fund to continue the work after his death (Dyke, 1993).

Lawes' long-term experiment on the effect of fertilizers on the growth of wheat on Broadbalk field at Rothamsted, which still continues, had already been going for some years. It showed that yields increased with applications of sulfate of ammonia but not with minerals (P, K, Na and Mg) alone (Table 1). So it was with good reason that Lawes later wrote "the Broadbalk results are the best stick to beat Liebig." Earlier, but without rancour, Lawes had disagreed with Liebig's opinion that animal fat originated wholly in the carbohydrate content of its food. The dispute about the sources of nitrogen in plants was an altogether more serious matter. Liebig scorned the relevance of field experiments in their disagreements and was infuriated when Lawes pointed out inconsistencies in his opponent's case. Otherwise Lawes and Gilbert conducted their side of the argument by soberly reported experiments, as in their "Reply to Baron Liebig's Principles of Agricultural Chemistry" published by the Royal Agricultural Society of England in 1855.

Here was Lawes, a young man still only 31 and without academic qualifications, challenging the most eminent chemist of the age. Liebig was increasingly vituperative. He wrote in good colloquial English and the following letter to Mr. Mechi epitomizes the flavor of the dispute:

> "How ignorant stupid and devoid of all good sense must be the great mass of agricultural people to allow such a set of swindlers to lead them into all these questions. If you ask any scientific

Table 1. Grain Yields of Wheat on Broadbalk Field, Rothamsted (Lawes and Gilbert, 1895)

	No Fertilizer	*Minerals only P, K, Na ,Mg*	*Minerals +86 lb N/acre*	*Minerals +129 lb N/acre*
1852	14	17	27	28
1853	6	10	24	23
1854	21	24	45	49
1855	17	18	33	31
1856	15	19	28	39

Notes: N supplied as ammonium sulfate.
Yields expressed as bushels/acre.

man about the content and practical value of their paper . . . it is all humbug, most impertinent humbug. Lawes and Gilbert latch onto me like vermin and I must get rid of them by all means. There is cowardice among scientific men of England which I am unable to understand and it is an offence against public welfare that they have not the courage to take a public stand in that most degrading controversy between science and ignorance."

Decades later the rift was healed by the award in 1893 of the Liebig Silver Medal to Lawes and Gilbert for their work in agriculture.

Notwithstanding this quarrel, Liebig's reputation as the foremost chemist of the time was not in dispute. Almost single handedly he established the science of organic chemistry and in agriculture formulated the "law of the minimum" for growth and was among the first to warn of the dangers of taking more from the soil than was added in manure. He founded Germany's first agricultural research institute at Mockein and the journal named after him, "Liebig's Annalen der Chemie" is still published.

In sharp contrast Boussingault, whose extensive field work on fertilizers put him in a stronger position than Liebig to criticize Lawes, did so in a courteous and civilized manner as the following letter to Gilbert shows, translated from elegant French.

Liebfrauenberg van Woerth
Bas Rhin,
12 July, 1857

My dear Sir,

I regret very much not to be in a position to avail myself of the kind invitation you have extended to me on behalf of yourself and Mr. Lawes. However, I have just returned from a long journey to the southern part of the country and my belated arrival compels me to stay at the farm for the entire season. Moreover, I am involved in an experiment on tobacco which requires regular attendance and a good deal of analysis.

The experiment you are carrying out at the moment certainly interests me, although I am entirely convinced that plants do not fix gaseous nitrogen directly from the air. It would be too fortunate if this direct absorption took place; we would not have much difficulty in procuring fertilizer. I persist in my belief that in order for assimilation to occur it is necessary for the free nitrogen to contain ammonia or nitric acid.

Nevertheless, if in your experiments you should succeed in fixing in plants an appreciable proportion of the nitrogen of the air, I have so much confidence in the results from Rothamsted that I would be greatly inclined to change my mind. Between now and the end of the year I expect I will be in a position to publish some research which in its totality will serve to elucidate the question.

Please inform Mr. Lawes that I have a fond wish to make his acquaintenance and let him know of the extent of my regard for him.

I am, dear Sir,
Yours very truly,

Boussingault.

Between the writing of this letter and 1886, a very large body of work was published on plant nutrition in relation to nitrogen fixation. Schneider's review of 1893 comprised 275 papers, though by a curious oversight this did not contain any of Boussingault's. After 1886 much of this work soon sank into oblivion.

Hellriegel and Wilfarth

It is fairly well known that biological nitrogen fixation was finally proven in Germany in 1886 but the manner of this discovery is less familiar. It was announced as a brief report (tageblatt) at the 59th Congress in Berlin of German Scientists and Physicians entitled "Welche Stickstoffquellen stehen der Pflanze Gebote" (Hellriegel, 1886). It is, however, of interest to know why Hellriegel[15] and Wilfarth[14] succeeded where others, particularly the Rothamsted work of 1857–1858, failed.

In the 1857–1858 series of experiments Lawes, Gilbert and Pugh (1862) grew beans, cereals, etc., under bell jars provided with a supply of air scrubbed first in sulphuric acid and then in bicarbonate of soda. The plants were given ashed plant or soil material to provide the mineral complement, as well as various preparations containing mineral fertilizer. The bell jars were made gas-tight by standing in grooves filled with mercury in a slate slab. Plate 1 shows the experimental setup. The incoming gas stream was provided at above atmospheric pressure to reduce still further the possible ingress of outside air containing ammonia. Analysis of the harvested plants showed very small losses and gains of nitrogen, within the limits of experimental error, and they concluded that their experiments showed no evidence of nitrogen fixation. However, they commented on the very poor growth of the plants, and suggested if this could in some way be improved then nitrogen may possibly be fixed!

Had their experimental constraints been less rigorous and any of the bean plants become contaminated with rhizobia it is beyond doubt that they would have associated the improvement in growth with nodulation and would have concluded as did Hellriegel and Wilfarth thirty years later, that nodulated legumes fix nitrogen. Anyone who has tried to keep control pots of plants free from contamination knows how difficult this can be. Earlier claims by Ville and others may have been due to sporadic contamination by rhizobia or ammonia.

George Ville, the natural son of Louis Napoleon, was a colorful character, flamboyant and easily offended. A slapdash experimenter who claimed to be able to get a wide range of plants, mostly non-legumes, to assimilate atmospheric nitrogen. This led to a dispute with Boussingault and confrontation at the Academie des Sciences with Ville demanding that a committee of the Academie should examine his claim. Now everything went wrong! The committee's plants were affected by fumes from nearby paintwork, water for the experiment became contaminated by ammonia, and the chemist in charge had to decamp for urgent family reasons. Unsurprisingly the committee's report was inconclusive.

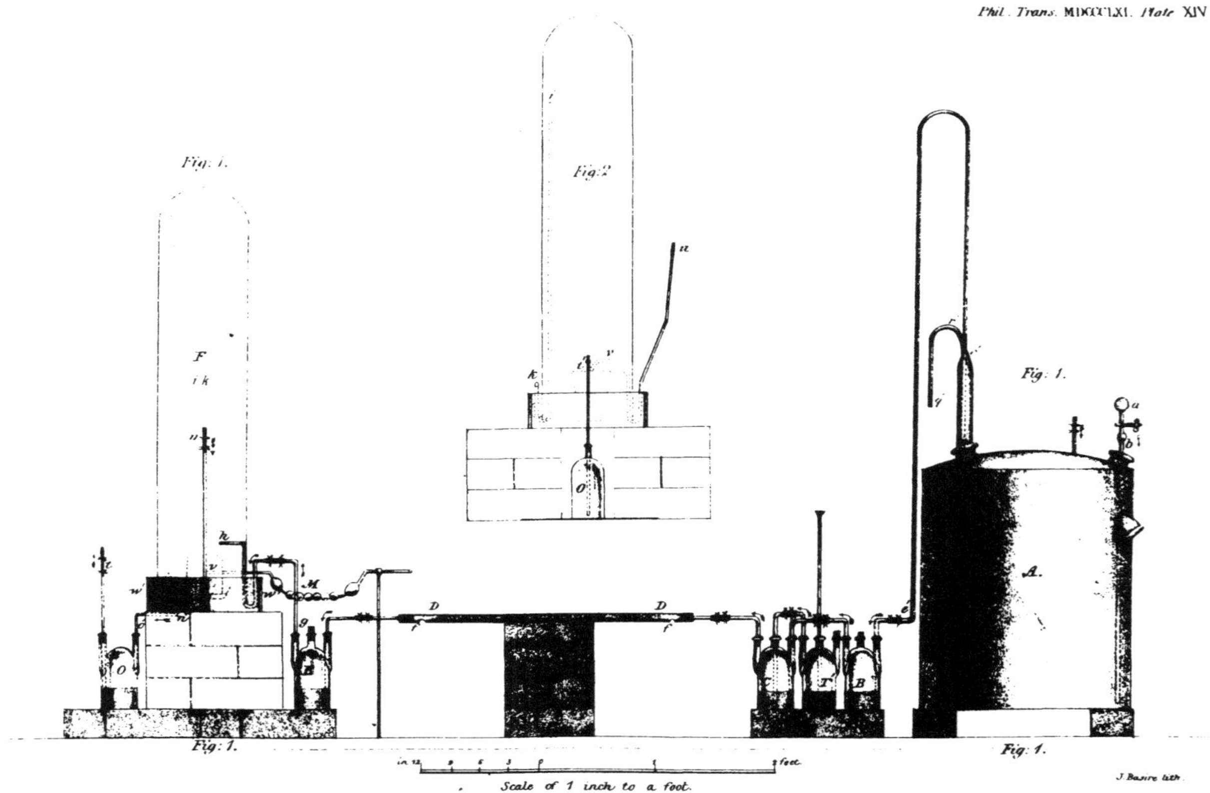

Plate 1. Apparatus used in 1858 to determine whether plants assimilate free nitrogen (Lawes, Gilbert and Pugh, 1862; reproduced with permission). A full description of the apparatus is given in the text.

Ville had been in touch with Lawes and had sent him a glass cabinet for his experiments. Pugh's opinion of Ville was not flattering. "I have told Mr. Lawes and Dr. Gilbert that such is my faith in Ville's dishonesty that I would not trust plants in his presence when salts of ammonia were within his reach" (Dyke, 1991).

Lawes and Gilbert worked with the fertile clay soils of Rothamsted and were cognizant of the problem of determining small gains in soil nitrogen whereas Hellriegel and Wilfarth were familiar with the impoverished light soils of North Germany where the contrasted responses of cereals and legumes to nitrogenous fertilizers were strikingly evident. This predisposed them to associate good growth with nodulation. Their inspiration to use an inoculum of soil from fields which had grown good legumes clinched the argument; irrefutably fixation could be attributed to a nodule-inducing ferment.

Not that botantists were unaware of nodules which were thought to be proteinaceous storage organs. However, Boussingault and Lachman as early as 1858 had speculated that fixation might be of "mycodermic" origin; a pregnant notion not followed up by experiment.

Hellriegel and Wilfarth's extensive and meticulous research on the effects of chemicals and soil "extracts" on crop yield (both legume and non-legume) carried out over many years at Dahme and Bernburg, showed beyond doubt that nodulated legumes could use the nitrogen of the atmosphere. This property was mediated by a "nodule inducing ferment" that could be transferred from plant to plant. The ferment only affected legumes and was to some extent specific as between one legume species and another. It was destroyed by heat and its effect was inhibited by the application of mineral N-fertilizer.

The pot cultures that established these results were placed on trollies so that when rain, possibly contaminated with ammonia, threatened they could be wheeled into the shelter of a glasshouse. All these precautions contributed to their success and it is sufficient to quote the essential results from one of their experiments to show the clear and unambiguous nature of their discovery (Table 2). Without inoculation or with a heat sterilized inoculum, the plants at harvest contained less nitrogen than was present in the seed sown, but with an inoculum the amounts of nitrogen fixed ranged from 356 to 1078 mg, depending on species. In other experiments, they observed that some of the control plants recovered, the indications of poor early growth disappeared, and the recovered plants bore nodules having "broad juicy organs bursting with nitrogen excess."

To control contamination more effectively, and building on the experiences of Lawes, Gilbert, and Boussingault, Hellriegel and Wilfarth also did experiments in which their plants were enclosed under bell jars or in carboys and provided with a stream of scrubbed air. Hellriegel and Wilfarth's plants with active "ferments" again showed fixation.

Although the full results were not published until 1888 in a 234-page paper, the presentation at Berlin was so impressive that it was accepted with acclaim by the distinguished audience with only one dissenting voice—that of Professor B. Frank.

Adolph Meyer described it as a sensational discovery reported with modesty. He could not recall a greater impression being made at a scientific meeting. Sir Henry Gilbert was the chairman at this momentous occasion. His feelings must have been mixed in view of his and Lawes' unsuccessful but equally careful experimentation over the previous several decades, already discussed.

Reading Hellriegel's papers, one has the impression that he was the leading partner in the enterprise. His other work also testified to his experimental prowess, especially his pioneering work on developing methods for plant culture in sand and water described in "Bieträge zu den Naturwissenschaftenlichen Grundlagen des Ackerbaus mit der besonderen Berücksichtigung der agrikulturchemischen Methode der Sandkultur, 1883." His biographer, von Glathe (1970), recorded he was a kindly man, respected by his colleagues, an excellent lecturer with a sense of humor.

Hellriegel was the son of a farmer. He studied agriculture and forestry at Tharandt Academy, qualifying as a chemist. He was much influenced by Liebig but disagreed with his contention that the value of manures lay only in their mineral content. Hellriegel was honored by German academic and agricultural institutions and his membership of the Royal Agricultural Society of England was proposed by Sir Henry Gilbert in 1891.

Hermann Wilfarth graduated from Rostock University as a chemist, later studying agriculture with Kühn at Halle. From there he moved to Dahme to start a lifelong association with Hellriegel. Wilfarth succeeded Hellriegel as director of the Bernburg Institute in 1895.

That the seminal discovery in nitrogen fixation was at last made at Bernburg owed something to the influence of Emperor Frederick of Prussia who was persuaded by

Table 2. Nitrogen Balances of Plants Given No Fertilizer Nitrogen in Hellriegel and Wilfarth's 1887 Experiment

	mg N Supplied in Seed and Inoculum	*Total N Yield mg*	*N Lost(–) or Fixed(+)*
Serradella			
No inoculum	23	1	–21
Inoculated	23	377	+354
Sterilized inoculum	23	1	–22
Lupin			
No inoculum	64	1	–52
Inoculated	64	1147	+1078
Pea			
No inoculum	38	14	–23
Inoculated	38	424	+386

Liebig to promote education and science. In Prussia by 1886 there were four higher colleges of agriculture staffed by 80 professors, 41 agriculture schools of lower status, and experimental farms specializing in many fields of agriculture, horticulture, fishery, etc.

In 1888, Beijerinck published a series of reports entitled "Die bakterien der Paplionacen-Knöllchen" which he called *Bacillus radicicola*. At about the same time (1885–1891) Bertholet, Warington and Winogradsky had isolated the main microorganisms from soil and water responsible for ammonification, nitrification and denitrification.

After 1888

Confirmation of Hellriegel and Wilfarth's discovery soon followed, the Rothamsted studies in 1890 being the most thorough. Using the "microbe seeding" method, fixation was demonstrated for nodulated peas, vetch, clover, lucerne, sainfoin and lupins, Hellriegel advising Lawes that lupins only flourish in light soils. The amounts of nitrogen fixed were determined and interestingly, in view of later controversy, no increase in nitrogen was detected in the medium in which the plants were grown. Lawes commented that soil-grown plants still obtained much of their nitrogen from the soil.

Miller and Willis were involved in these experiments and Edwin Grey (1922) in his "Reminiscences of Rothamsted" gives a fascinating account of their daily visits to these experiments which were set up in a greenhouse on the field originally used by allotment holders. The greenhouse was screened to prevent insect entry, provided with benches and blinds and locked against intruders, the key being hung on a nail beside Grey's famous balance. Each morning Sir Henry, with Willis and Grey in attendance, would take the key and proceed to the allotment greenhouse to water the plants and take notes. Grey recorded "as soon as they grew we saw a remarkable difference between those in the microbe-seeded pots and those in the pots where no extract had been applied . . ." Lawes had set aside the allotment field in 1857 for the villagers to grow their own vegetables. He also built a handsome clubhouse for their use. In an article he stated, "My idea of a clubhouse at that time was a large room where the members could have their beer and tobacco and a small room for a library." The clubhouse was built in about 1857; it had a thatched roof, later replaced by tiles, and a verandah going all the way round the building providing seats for the summer.

Charles Dickens visited the clubhouse and wrote an article about it entitled "The Poor Man and his Beer" for the April 1859 issue of the magazine *All the Year Round*. We do not know whether Dickens was shown the allotment greenhouse where such interesting work was later to be done.

In an uncharacteristic act of vandalism, the clubhouse was demolished after the second World War to make room for a parking lot and laboratories. During that war, the present writer cultivated one half of one of the surviving allotments, 20 poles in area (110 sq. yds), double that of most such plots today.

BIOLOGICAL NITROGEN FIXATION IN THE 1900s

Hellriegel died in 1895 and Wilfarth in 1905. They had done little to extend their work, except to update their reports to counter Frank's criticisms.

Research on BNF almost ceased during the first World War and, even before 1914, its focus had shifted from Europe to America. All movements have their scriptures: for BNF the Old Testament was Fred, Baldwin and McCoy's "The Root Nodule Bacteria and Leguminous Plants" (1932), and the New Testament, P. W. Wilson's "The Biochemistry of Symbiotic Nitrogen Fixation" (1940) both published by the University of Wisconsin Press in Madison.

The interwar years saw much work on specificity, nodule structure and function, physiology, biochemistry and legume agronomy. A controversy at the time was whether fixed nitrogen was secreted from the nodule into the soil, as maintained by Virtanen at Helsinki, or alternatively only became available to succeeding crops after the decay of the nodulated roots, as held by the Wisconsin school.

To resolve the question, joint experiments were conducted at Helsinki and Wisconsin in which varieties of pea, samples of soil, strains of rhizobia and experimenters were exchanged. There seemed to be some slight evidence that nitrogen might be excreted in the very long days of the Scandanavian summer. Eventually the consensus was that if secretion occurred it was small, sporadic and of no practical significance. Like the Liebig dispute it was somewhat ascerbic; Virtanen was known for his strong opinions.

The rise in the price of oil in the 1970s greatly increased the cost of fertilizer. Approximately 1.5 kg fuel oil is consumed to make 1 kg nitrogen in fertilizer by the Haber catalytic process. This promoted internationally-funded research to improve biological nitrogen fixation, especially of food pulses in the tropics. Each center was mandated to do research on particular crops, for example on *Cajanus cajan* (pigeon pea), *Cicer arietinum* (chick pea) and *Arachis hypogea* (ground nut) at the International Crop Research Institute for the Semi-Arid Tropics (I.C.R.I.S.A.T), Patancheru, India. Six other centers dealt with other pulses, each maintaining gene banks of cultivars and land races and culture collections of *Rhizobium*. Nevertheless, the International Board of Plant Genetic Resources has pointed out the serious genetic erosion of pulse and forage legumes worldwide.

No time was lost in putting Hellriegel and Wilfarth's work to use. The first direct application was by Salfeld in 1887 who used a soil transfer method on a range of legumes. He had, in fact, obtained positive effects of soil transfer several years earlier without knowing the reason behind his success. Indeed Hellriegel and Wilfarth candidly acknowledged that their discovery was accidental and at first even unwelcome (zunächst unerwüscht). In parenthesis, it is very probable that nomadic tribes who carried soil as talismans from place to place, were the first to practice inoculation—inadvertently. After 1886, many others, mainly in Germany, experimented with soil transfer, treating fields or seeds with soil, with or without liquid or other additions such as sugar or glue.

In 1890, Nobbe was the first to inoculate with pure cultures of *Rhizobium*, though the work was not published until 1896. The first company to market inoculants under the name "Nitragin" was Meister Lucius und Brüning, Hochst an Main, Germany, using Nobbe and Hiltner's patented culture. The successor of this company continues to produce, to the present day, a high quality inoculant, under the same name, in Wisconsin. At first it was provided as agar or liquid cultures, but later was supplanted by a peat-based product that was more convenient to use.

In the UK, J. F. Mason, MP for Windsor, who collaborated with Lawes in field experiments with legumes on his farm at Eynsham, Oxfordshire, provided funds to build the Mason Bacteriological Laboratory at Rothamsted where early work was done on the bacteriology of *Rhizobium* and on commercial-scale field inoculations. Under licence from Rothamsted an agar culture was made by Allen and Hanbury's (a milk product firm) from mother cultures provided by Rothamsted where every thousandth culture was tested for purity and effectiveness. Agar cultures, though messier to use, were easier to control because contaminants, usually fungi and actinomycetes, showed up more easily. Inoculants based on peat, soil, lignite, bagasse, etc, as carriers, have now replaced agar and liquid products.

When Allen and Hanbury's were taken over by Glaxo they wished to import a peat inoculum from Belgium. It was called "Nodosit." When examined, it was found to contain no rhizobia and smelled strongly of actinomycetes! The late director of Rothamsted, Sir Fred Bawden, wryly commented that it should be called "No does it!" When we reported to the producer the response was "We cannot withold this inoculant from the market."

Control laboratories have now been set up in many countries, such as the highly regarded control laboratory in Australia, and under the auspices of the International Biological Programme, a world list of *Rhizobium* Culture Collection's has been published (Skinner et al., 1983), the strains listed being freely available to bona fide enquirers.

Numbers of rhizobia in a culture are estimated by the "most probable number" method whereby dilutions of the inoculant are used to infect sterile-grown test plants. The nodulated test plants, because they are cultured without combined nitrogen, show by their growth whether the inoculant strain fixes nitrogen.

The best practice today uses sterile pochettes of a neutral peat and bacteriological medium which are injected with the culture. This continues to multiply for a period before the inoculum is used. The best inoculants specify a high viable count in excess of 10^9/g that matches strain to host species, or even cultivar. Regrettably, poor and even worthless inoculants continue to be produced.

TECHNICAL ADVANCES SINCE WORLD WAR II

Work on the biochemistry of BNF was for a long time hampered by the failure to obtain an active enzyme preparation from any source. This was largely because of the enzyme's extreme sensitivity to inactivation by oxygen. At the same time the nitrogen fixing enzyme, nitrogenase, has an absolute requirement for oxygen, later shown to be met, in the nodule, by a specific hemoglobin, leghemoglobin, which allows the transfer of oxygen at low partial pressure (Figure 1).

Post World War II, the most influential developments were (Chatt, 1980; Bergersen & Postgate, 1980):

1. The use of the stable isotope of nitrogen, ^{15}N, to confirm and measure fixation;
2. The advance in our knowledge of the coordinate chemistry of the transition elements and of recombinant DNA technology, with their bearing on nitrogenase action; and
3. The development of the acetylene reduction assay to measure nitrogenase activity.

This new assay technique was developed independently by American and Australian workers who first established the connection between nitrogenase and hydrogenase, and showed the former could reduce a number of substrates other than nitrogen, for example nitrous oxide, azide, cyanide and, most usefully, acetylene. The latter is reduced to ethylene and this forms the basis of the acetylene reduction test. The material to be examined is briefly gassed with acetylene and the ethylene formed measured by gas chromatography. While this technique was being refined,

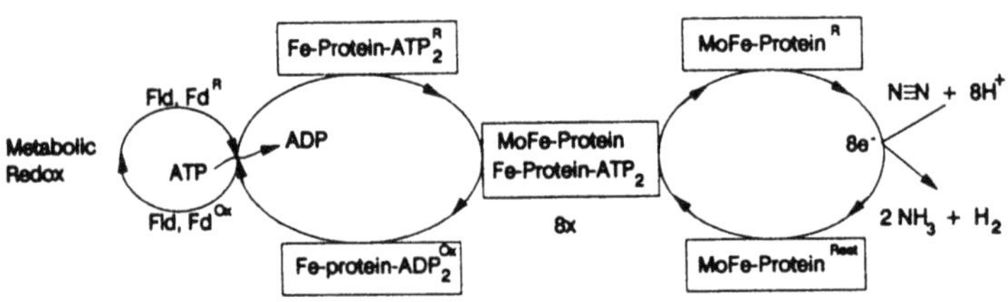

Figure 1. The Nitrogenase Reaction. The electron transfer proteins ferredoxin (Fd) and flavodoxin (Fld) serve to couple the nitrogenase reaction to metabolically generated reducing equivalents. Ammonia synthesis requires 8 electrons: 6 for the reduction of dinitrogen and 2 for the coupled, obligatory synthesis of H_2. These reactions are catalyzed by the terminal component in the complex, the MoFe-protein. The electrons are transferred to the MoFe-protein from the Fe-protein in a process coupled to the hydrolysis of 2ATP/electron (Howard and Rees, 1994, 1996).

the two groups reported to each other, and agreed to delay publication until the results were incontrovertible. Bob Burris gives a full account, with correspondence, of this happy collaboration (Burris, 1975).

The value of this assay is evident not least in the impressive research mainly in the United States, France and the Agricultural Research Unit on Nitrogen Fixation at Sussex University, UK. Nitrogenase synthesis and activity is directed by at least 20 plasmid-located cistrons. The enzyme consists of two proteins. The larger, (c. 220 kDa) contains Fe and Mo and acid-labile sulfite and is the center where nitrogen is reduced. The smaller protein, 68 kDa, also contains Fe and functions as an electron carrier. The tertiary structures of dinitrogenase and dinitrogenase reductase have been determined (Kim and Rees, 1992; Georgiadis et al., 1992).

The burgeoning of papers up to 1940 was the subject of a study by Wilson and Fred (1935) who predicted a levelling off to about 100 papers each year by the 1970–1980 decade. In the event this was exceeded as new discoveries and methodologies triggered spurts of exponential growth, accentuated at the end of this period by stimulation from the IBP and its associated funding.

Whereas earlier studies were mostly by individuals or pairs of workers like Hellriegel and Wilfarth and Lawes and Gilbert, newer studies were increasingly by teams, each member bringing a specialty to bear on the problem (see also Chapter 9). This was particularly so in the studies on the coordinate chemistry of dinitrogen complexes with transition elements and in the enzymology and genetics of the nitrogenases. Such collaborations often extended to laboratories across the world, as shown by shared authorship of papers.

Already by 1985, BNF research had peaked and its decline foreshadowed. At the same time, support of all kinds—especially from governmental sources—had lessened with the retirement or loss of some of the group leaders. Even Hellriegel had had his problems. In 1897 he wrote an article "Eine Plauderie über Forchungs Methoden" in which he remarks on difficulties in obtaining support for his work.

At present, as in other fields, support is for shorter term research with practical pay-offs in view. Among these are projects to introduce nitrogenase into cereals and other non-leguminous crops. In my view, this may be misdirected. Even where an initial introduction into an unaccustomed host may be successful, benefits are unlikely to accrue even in the long term. For the introduced nitrogenase to function effectively, many more changes would certainly be needed in both partners for fixation to be adjusted to the host's requirements. It is already known that in the legume this involves many components, some inherited polygenetically (Nutman, 1967; 1980; 1987). An equally valid (or even preferred) option may be to extend and improve upon the symbioses that nature has provided, reversing the current trend in agriculture from using fewer to using more species. Today in the UK nearly all the clover-grass mixtures in leys are based upon a few cultivars of one legume species only, *Trifolium repens*. It was not always so. Seed merchants used to stock *T. hybridum, T. glomeratum, T. incarnatum, T. frageratum* in addition to red clover, wild white, Dutch and Kent white clovers.

Allen and Allen (1981) have described the little that is known about the introduction of legume species into agriculture and the role of inoculation in this process. For example, alfalfa (the hay of the Medes), *Medicago sativa*, was taken by the Spanish from the Middle east into South and Central America, from which point it spread to the north without, one presumes, deliberate use of soil inoculation. None of its cross inoculation group (species of *Medicago, Melilotus* or *Trigonella*) occur in the New World. Soyabean was introduced into America from Asia much later, and today both crops are routinely inoculated.

The case history of sub-clover (*Trifolium subterranean*) as an important pasture legume is of special interest, as an involuntary migrant from the Northern to the Southern hemisphere, the phenomenal success of which depended on inoculation.

This small clover is native to the Mediterranean and Atlantic littoral regions and was unknown to pastoralists until the early years of this century. A few plants of this species were first noticed in about 1905 by Amos Howard at Mount Barker in South Australia. They grew luxuriantly and must therefore have been nodulated by rhizobia introduced accidently with the host in seed from Europe, since no *Trifolium* species or their rhizobia occur naturally in Australia. This plant, like the ground nut, buries its flower head when set so that the seed, formed within a spiny burr, is buried in the soil and naturally carries a high burden of contaminating soil microorganisms. Howard saved and multiplied the seed which he cleverly collected on a roller covered with sheepskin to which the clover burrs attached themselves. The use of sub-clover in sown pastures was started commercially by Howard in 1907, but development was slow until it was realized that an inoculum of rhizobia had also to be provided. Today there are many millions of hectares of sub-clover pastures in Australia, the number of named cultivars being more than 400. The resulting rise in fertility has allowed the wheat-growing areas of Australia to be greatly expanded. Subterranean clover has now been established as an important pasture species in Southern Africa and the Americas by introduction from Australia.

Howard's vision may possibly have owed something to his father's history. Howard senior was a gardener who lived near Watford, within a short distance of Rothamsted, and who emigrated to Australia in 1876. It is just possible that he may have heard of the goings-on at Rothamsted and of their experiments on the non-responsiveness of legumes to nitrogenous fertilizers and later to the well-publicized discovery of BNF by Hellriegel and Wilfarth. Anything Howard jr. may have heard about the special value of clover would have inclined him to take a closer look at this alien species growing at Mount Barker.

Even had the work at Bernburg not been successful, the discovery of biological nitrogen fixation could not have been long delayed; the time for it was over-ripe. Indeed, as already speculated, the discovery could have been made by Boussingault at Bechelbronn or Lawes at Rothamsted, or by other botanists or agriculturalists interested in proving or disproving that plants assimilate atmospheric nitrogen. Had the discovery been made at Rothamsted in the middle years of the last century it may not have been to the benefit of agricultural science. Instead of continuing for the next 50 years with their epoch-making research on fertilizers, Lawes and Gilbert might have been diverted with less profit into studying soil "ferments." At that time, microbiology was far from developing its own techniques and philosophy. It was not until 1864 that Pasteur laid the foundations of microbiology by showing that fermentation was caused, without doubt, by minute organisms that were not spontaneously generated, but came from similar forms of life present in the air, soil and water. Research on BNF following Hellriegel and Wilfarth's historical discovery reached its apogee in about 1984 and no doubt after a quieter period of consolidation (sic!) new growing points will arise, but where and of what kind it would be hazardous to guess. They will, however, owe as much to novel ideas and techniques as did the discoveries of the past.

REFERENCES

Allen, O.N. & Allen, E.K. (1981). The Leguminosae. A Source Book of Characteristics, Uses and Nodulation. Univ. Wisconsin Press, Madison.

Awramik, S.M., Schopf, J.W., & Walter, M.R. (1988). Carbonaceous filaments from North Pole, Western Australia. Are they fossil bacteria or ancient stromatolites? A discussion. Pre-Cambrian Research, 39, 303–309.

Beijerinck, M.W. (1888). Die Bakterien der Papilionaceen-Knöllchen. Bot. Ztg. 46, 725–735; 741–750; 757–771; 781–790; 797–804.

Bergersen, F.J. & Postgate, J.R. (1987). A century of nitrogen fixation research: present status and future prospects. Phil. Trans. Roy. Soc. Lond. B 317, 65–297.

Bertholet, M. (1885). Fixation directe de l'azote atmospherique libre par certaines terrains argileux. C. r. hebd. Seanc. Acad. Sci. Paris 102, 775–784.

Boussingault, J.B. (1838–1856). Recherche chemiques sur la vegetation, etc. Annls. Chim. Phys. ser. 2, 69, 353; ser. 3, 46, 5–41.

Burris, R.H. (1975). The acetylene-reduction technique. Nitrogen fixation by free-living microorganisms. The International Biological Programme 6, 249–257.

Chatt, J. (1980). Chemistry relevant to the biological fixation of nitrogen. Proc. Phytochem. Soc. Europe Symph. Ser. 18, 1–18.

Day, A. & Poulton, J. (1996). Extranuclear DNA. Foundations of Modern Biochemistry, Vol. 2, pp. 59–97. JAI Press, Greenwich, CT.

Dyke, G.V. (1991). John Bennet Lawes. The Record of His Genius. Research Studies Press Ltd., Taunton, Somerset, UK.

Dyke, G.V. (1993). John Lawes of Rothamsted, Pioneer of Science, Farming and Industry. Hoos Press, Harpenden, UK.

Fred, E.B., Baldwin, I.L., & McCoy, E.F. (1932). Root Nodule Bacteria and Leguminous Plants. Univ. Wisconsin Press, Madison, WI.

Georgiadis, M.M., Komiya, H., Chakrabarti, P., Woo, D., Kornuc, J.J., & Rees, D.C. (1992). Crystallographic structure of the nitrogenase iron protein from *Azotobacter Vinelandii*. Science 257, 1653–1659.

Glathe, H. von (1970). Hermann Hellriegel (1831–1895). In Grosse Landwirte, pp. 245–257. D.L.G. Verlag, Frankfurt, Germany.

Grey, E. (1922). Reminiscences, tales and anecdotes of the laboratories, staff and experimental fields, 1872–1922. Rothamsted Exptl. Station.

Hellriegel, H. (1883). Beiträge zu den naturwisshenschaftlichen grundlagen des Akerbaus mit besonderer Berücksichtigung der agrikulturchemischen Methode der Sandkultur. Braunschweig: Vieweg und sohn.

Hellriegel, H. (1886). Welche Stickstoffquellen stehen der Planze zu Gebote? Tageblatt der 59. Verslammlung duetscher Naturforscher und Aerzte, Berlin, 18–24 Sept. p. 290.

Hellriegel, H. (1897). Eine Plauderei über Forschungs-Methoden. Arb. dt. LandwGes. 24, 1–19.

Hellriegel, H. & Wilfarth, H. (1888). Unterschungen über die Stickstoff—ernährung der Gramineen und Leguminosen. Beil. Z. Ver. dtZuck Ind.

Howard, J.B. & Rees, D.C. (1994). Nitrogenase: a nucleotide-dependent molecular switch. Annu. Rev. Biochem. 63, 235–264.

Howard, J.B. & Rees, D.C. (1996). Structural basis of biological nitrogen fixation. Chem. Rev. 96, 2965–2982.

Kennedy, I.R. (1997). Biological nitrogen fixation: the global challenge and future needs pp. 6–12. The Rockfeller Foundation Bellagio Conference Center, Lake Como, Italy.

Kim, J. & Rees, D.C. (1992). Crystallographic structure and functional implications of the nitrogenase molybdenum-iron protein from *Azotobacter vinelandii*. Nature (London) 360, 553–560.

Lawes, J.B. & Gilbert, J.H. (1855). Reply to Baron Liebig's "Principles of Agricultural Chemistry." Roy. Agric. Soc. 16, 1–90.

Lawes, J.B. & Gilbert, J.H. (1895). The Rothamsted experiments over 50 years. Blackwood & Sons, Edinburgh & London, UK.

Lawes, J.B., Gilbert, J.H., & Pugh, E. (1862). On the sources of nitrogen of vegetation; with special reference to the question whether plants assimilate free or uncombined nitrogen. Phil. Trans. Roy. Soc. Lond. 151, 431–577.

Mojzsis, S., Arrhenius, G., McKeegan, K.D., Harrison, T.M., Nutman, A.P., & Friend, C.R.L. (1996). Evidence for life on earth before 3800 Million years ago. Nature (London) 385, 55–59.

Nobbe, F. (1896). Einige neue Beobachtungen betreffend die Bodenimpfung mit rein kultivierten Knöllchenbakterien für die Leguminosenkultur. Bot. Zbl. 63, 171–173.

Nutman, A.P., Mojzsis, S.J., & Friend, C.R.L. (1997). Recognition of ≥ 3850 Ma water-lain sediments in West Greenland and their significance for the early Archaean earth. Geochimica et Cosmochimica Acta 61, 2475–2480.

Nutman, P.S. (1967). Varietal differences in the nodulation of subterranean clover. Aust. J. Agric. Res. 18, 381–425.

Nutman, P.S. (1980). Adaptation. Proc. Phytochem. Soc. Europe Symph. Ser. 18, 335–354.

Nutman, P.S. (1987). Centenary Lecture, Royal Society Lond. Phil. Trans R. Soc. Lond. B 317, 69–106.

Pidgeon, R.T. (1978). 3540 M.y. old volcanics in the Archaean layered greenstone succession of the Pilbara block, Western Australia. Earth Planet Sci. Let. 37, 421–428.

Schneider, A. (1893). A new factor in economic agriculture. Ill. Agric. Expt. Stn. 29, 301–319.

Salfeld, A. (1896). Die Boden-Impfung zu den Pflanzen mit Schmetterlingsbluten im landwirtschlaftlichen Betriebe. M. Hensius, Bremen, 100 pages.

Schopf, J.W. (1993). Microfossils of the early Archaean Apex chert. New evidence for the antiquity of life. Science 260, 640–646.

Skinner, F.A., Hamatova, E., & McGowan, V.F. (1983). World catalogue of Rhizobium collections, 2nd ed. UNESCO/UNEP, World Data Center for Microorganisms.

Warington, R. (1891). On nitrification. J. Chem. Soc. 59, 484.

Wilson, P.W. (1940). The Biochemistry of Symbiotic Nitrogen Fixation. Univ. Wisconsin Press, Madison.

Wilson, P.W. and Fred, E.B. (1935). The growth curve of a scientific literature. Scient. Mon, 41, 240–250.

Winogradsky, S. (1890). Sur les organismes de la nitrification. Annls. Inst. Pasteur, Paris, 4, 213–231; 257–275; 760–771.

Chapter 6

PLANT HORMONES:
A HISTORY OF DISCOVERY AND SCIENTIFIC FASHION

Daphne J. Osborne

DISCOVERIES

Recently a colleague said "*Nature* won't publish it unless you have cloned it or it cures cancer." Of course, that's not true but it does illustrate a point. The point is fashion; there are fashions in science as in other areas of life. Over a long scientific career in which hormones have always played a part, I can look back to waves of fashion starting from the days when each individual hormone was discovered to the present day when plant scientists strive to learn how each hormone may function at the molecular level of the gene.

Foundations of Modern Biochemistry,
Volume 4, pages 217–243.

ISBN: 0-7623-0351-4

The word "hormone"—to arouse to activity—was first used by Starling (1905) in describing the function of *secretin* in animals but it was not until 1910 that Fitting used it for the substance present in orchid pollen that caused the gynostemium of the orchid flower to swell. Long before the landmark discovery of auxin by F.W. Went[19] in 1928, it was evident that plants produced "substances" that could modify growth. Darwin[6] had reported in his book *The Power of Movement in Plants* that, when the coleoptile of *Phalaris canariensis* (the canary grass) was illuminated from one side only, there was a subsequent strong curvature towards the light source but if the tip of the coleoptile was covered or removed the coleoptile usually failed to bend at all. Darwin concluded (1881, p. 474) "Some influence is transmitted from the upper to the lower part, causing the latter to bend." By 1911, through the diligent efforts of mainly Danish, Dutch and German botanists, it was clear that unilateral light caused the unequal distribution of a substance between the light and dark sides. This substance *migrated* (not diffused) down the dark side causing there an acceleration of growth and hence a bending towards the light (Boysen Jensen, 1911).

The First Hormone: Auxin

The bending of a coleoptile, and the enhanced growth that it represented, became the fashionable test object for plant scientists until World War II. Agar blocks containing innumerable individual chemicals, or diffusates from specific plant parts were applied unilaterally to detopped coleoptiles in the dark and some remarkable discoveries were made. The presence of a growth enhancing substance in Fitting's pollen was fully vindicated and the substance was found to be similar to that in coleoptile tips (Laibach and Kornmann, 1933), but more interesting in the light of work that followed was the demonstration by Seubert (1925) that human saliva was highly active. However, it was the Dutchman Frits W. Went, working in his father's laboratory in Utrecht, who made the isolation breakthrough (Figure 1). Collecting tips of *Avena* coleoptiles, he placed them cut side down onto tiny blocks of agar for 2 h, then placed single agar blocks asymmetrically onto the cut surface of coleoptiles of *Avena* seedlings decapitated some 40 min. earlier. After leaving them for about 120 min. he showed a curvature in those coleoptiles of up to 20°—the extent of the angle depending upon the concentration of the collected substance that the block contained (Went, 1928). This became a standard assay for growth substances and was carried out, as in Utrecht, in dark rooms at high humidity using dim red light for observation.

The chemical identification of one active substance was eventually achieved by Kögl and his associates in Holland in the early 1930s. A major source of starting material was, surprisingly, human urine, but by then it was already known that the sources of a substance active in the *Avena* assay were not restricted to higher plants. Fungal cultures (*Aspergillus*) and urine (like the saliva recorded by Seubert in 1925), were all rich in activity, as were maize oil, malt and yeast. The substance

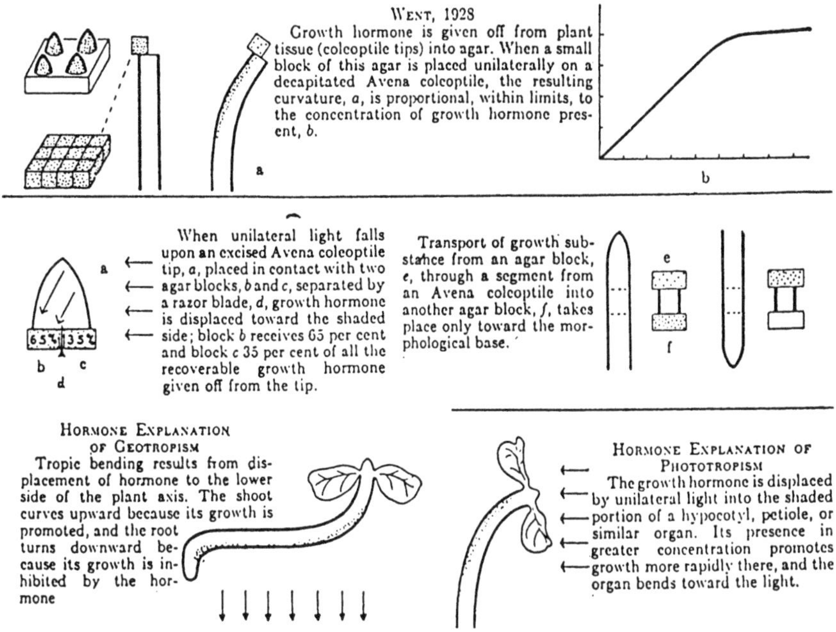

Figure 1. Critical experiments that are the foundation of hormonal control of plant growth. (Figure modified from Boysen Jensen et al., 1936)

isolated was an indole compound—indole-3-acetic acid (see Figure 2a)—and was apparently present in all parts of plants to different extents (Kögl et al., 1933). They called it auxin and it was meant as an inclusive term for growth substances that bring about cell enlargement. The ubiquity of this active material led Frits Went to write "Ohne Wuchsstoff, kein Wachstum"—no auxin, no growth!

The molecular characterization of auxin was not easy and mistakes were made. While Kögl's identification of indole-3-acetic acid was correct, other substances reported—auxin a (auxintriolic acid) and auxin b (auxinolonic acid) (Kögl et al.,1934) could not later be substantiated despite the efforts of many. Bennet-Clark, Tambiah and Kefford (1952), working in London during the post-Second World War years, used urine from high-fat intake volunteers (including Dutchmen) to try to resolve the different auxins by using the then newly developed paper chromatographic techniques. Their test room for *Avena* curvature assays lay under the Strand with a manhole cover in the ceiling; each time a pedestrian stepped on this the rattle disturbed the workers below. The presence of indole-3-acetic acid was readily confirmed, but bacteria in the gut and the mouth were established as contributors to the auxin production. The intake of oils and fats in the diet led to considerably higher levels of auxin later being excreted. When it was finally fully appreciated that much of Kögl's original urine samples had come from a hospital,

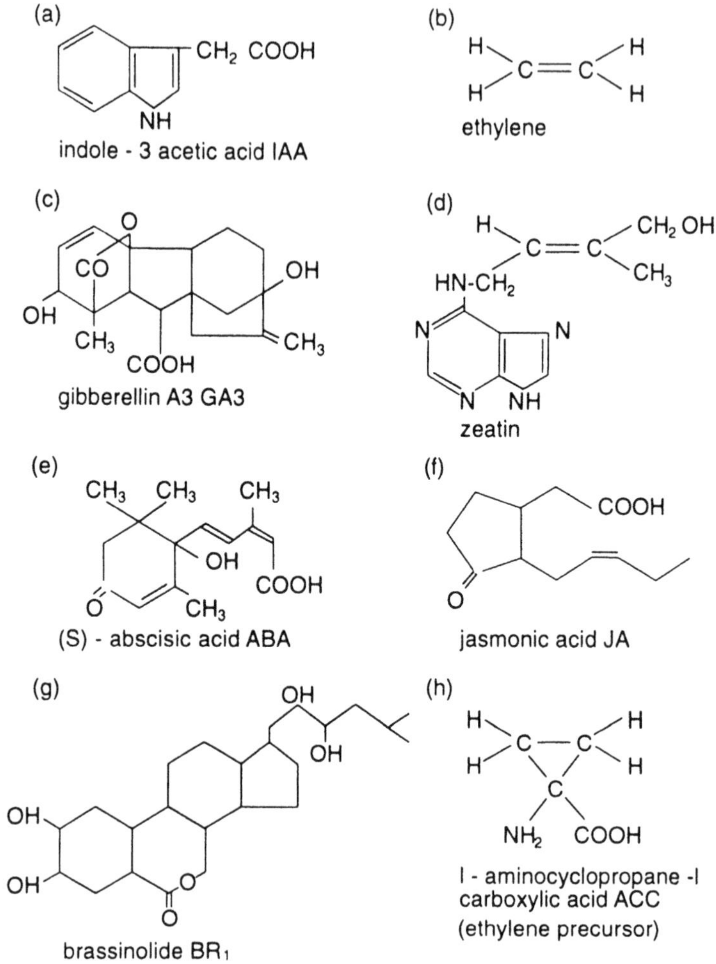

Figure 2. The various chemical structures of native plant hormones.

the quest for auxin a and auxin b was finally abandoned. So it was that in 1940, when I was at school, only one plant hormone was known—indole-3-acetic acid, called auxin.

The effects of auxin upon the growth of plants were truly remarkable. As a result of the concerted effort and interest that plant scientists devoted to this highly fashionable and exciting field of research, much of what we know of the control of plant growth and development by auxin was discovered in these highly productive years. The hormone was derived from the aromatic amino acid tryptophan and was relatively abundant in the rapidly growing meristematic parts of plants such as apical buds and root tips. Inactive bound forms existed, as in seeds, where hydrolysis led to the liberation of "free" auxin. Auxin enhanced the rate of growth and extensibility in immature cells of shoots, but similar concentrations depressed that of roots. It delayed shedding of defoliated petioles or decapitated fruit stalks and flowers, enhanced cell divisions, suppressed lateral bud development but induced the rooting of stem cuttings and gave rise to the maturation of seedless (unfertilized) fruits of many species.

The phototropic responses of shoots first recorded by Darwin, and the equally fast, gravity-controlled, bending responses when plants were displaced from the vertical position, were all attributed to the greater elongation growth that a redistributed auxin would induce on the darkened side of a unilaterally illuminated shoot or on the lower side of a horizontally placed seedling or shoot. Roots curved in an opposite manner, in accordance with the finding that auxin depressed root extension on the lower side. Any disturbance of the normal distribution of auxin was shown to lead to a bending response. Temporary gross distortions in the orientation and twisting of stems and petioles, referred to as "epinasty," occurred when plants were sprayed overall with an auxin solution.

Many related and unrelated chemical compounds were tested for their ability to cause similar growth reactions to those of auxin and the early treatises describing this work make fascinating reading today. For the history of auxin exploration Boysen Jensen, Avery and Burkholder's (1936) "Growth Hormones in Plants" is one of the best and that of Went and Thimann[24] (1937) one of the most personalized. The contributions of the Boyce Thompson Institute of Plant Research (up to 1940) provide an ongoing account of plant responses to auxin and to the many other synthetic growth regulating compounds that were studied by Hitchcock, Zimmerman and colleagues in Yonkers, New York. The growth effects of many of these synthetic substances were highly persistent, unlike the short-term responses of a single application of indole-3-acetic acid. It is now known that auxin itself is quite rapidly (within a few hours) metabolized in the plant to other compounds (mainly glycan or amino acid conjugates) that are non-growth regulatory, but many of the synthetics such as chloro-substituted phenoxy acetic acids, of which perhaps the 2,4-dichlorophenoxy acetic acid (2,4-D) is the best known, remain for much longer. These synthetic growth controlling substances later revolutionized the agricultural industry throughout the world. This was essentially because the effects of synthetic auxins were long-term. Dicotyledonous plants in particular were killed due to their over-response while monocotyledons (the grasses and wheats and barleys) were little affected. With these findings, the era of the *selective herbicide* had been initiated, but other synthetic compounds too were shown to improve on the auxin-induced responses. The swelling of the orchid gynoecium when the active substance from pollen was applied in Fitting's classic experiment of 1915 was, in fact, the forerunner for the horticultural industry's development of naphthoxy and naphthalene derivatives for the commercial improvement of fruit set and the parthenocarpic production of tomatoes, apples and pears. Synthetic preparations to enhance rooting of cuttings or to retard the shedding of plant parts, all in common use today, are derived from the findings of those early pioneering experiments with the "active substance" auxin, the first plant hormone to be isolated and characterized.

Two of the very early properties discovered for auxin, or IAA as it is often known, remain mechanistically unresolved and are still a focus for debate and research. One is polar transport. Frits Went (1928) showed that auxin is actively moved in living plant tissues in a morphologically basipetal direction at a rate of approximately 1 cm per hour. The acropetal upward movement in a tissue is much less and is attributable to diffusion. One of Went's clever experiments was to place an agar block containing auxin on one cut surface of an excised segment of coleoptile and a plain agar block on the other. In one set, the auxin block was placed on the morphologically top cut surface and in the other, the auxin block was placed on the basal end cut surface. When both sets of plain agar blocks were later removed and analyzed for their auxin content, only the receivers from the apically donated auxin contained auxin, the receivers from the basally applied auxin contained barely detectable levels (Figure 1). By this simple means, Went had shown that the movement of auxin takes place preferentially in the basipetal direction in living plant tissues. The precise mechanism by which this is achieved is still not fully clear today (see later). Synthetic auxins too, exhibit a degree of polar transport but less intensively than the movement of natural auxin. Plants alone possess this strictly directional cell to cell transport of the auxin molecule and no other plant hormone is moved in this way. Not only did Went show the polarity of transport of auxin, he also demonstrated that plant tissues possess an overall inherent polarity. A slice cut from a dandelion root for example, will eventually produce new shoots at the apical cut edge and roots at the basal edge. Inserting another slice between the two cut surfaces prevents the orderly production of new roots and shoots only if it is inserted in the upside-down orientation. In other words, the polarity of the segment as a whole is disrupted by such discontinuity of axiality. These findings are of central importance to our current understanding and research into how plant cells communicate and how chemical signals are transmitted.

The other seminal discovery, that auxin causes an immature plant cell to increase its rate of elongation, has been the subject of considerable fluctuations in scientific fashion. The decapitation of coleoptiles and the application of tiny agar blocks, followed by measurements of the angles of bending achieved, is a laborious and quite tedious process. Many workers began, instead, to cut sections from below the coleoptile tip and float them on a solution of the substance under test. Instead of measuring angles, they measured the increment in length achieved while in the solution. The method was quantitatively reliable, easy to handle and opened up new possibilities to explain auxin action. In 1934, Strugger immersed segments of *Helianthus* hypocotyls in buffer solutions at different pH. He found that lowering the pH increased the growth rate and the extensibility of the cell walls. This was the start of the "acid growth" theory of cell extension. Auxin was almost inactive at pH 7.0 but increasingly active as the pH was lowered to 4.0. Bonner[18] (1934) confirmed this result with coleoptiles, but it remained unclear whether it was due to the enhanced uptake of undissociated molecules as the pK of auxin was reached or whether some other pH related events were the cause. Peter Mitchell's[1] Nobel Prize-winning researches into proton pumping ATPases and the electrochemically induced changes in the potentials across membranes (see Ferguson, 1997) heralded a renewed wave of interest in the "acid growth" theory of plant cell extension. From a decline of interest there then arose in the sixties new and vigorously active research that continues today (see later).

Ethylene—Is it a Hormone?

Since 1864 there had, however, been another potential plant hormone waiting in the wings to be "discovered." This was the volatile olefine ethylene (Figure 2b). The source, however, was not from plants. The first publication came from Germany and indicated that a volatile substance emanating from faulty gas mains could have caused the leaves of nearby trees to shed prematurely (Girardin, 1864). Some 20 years later Molisch (1884) noted that abnormal geotropic responses and horizontal growth of seedlings could be related to the presence in the air of traces of illuminating gas or smokes. Later, flower growers observed that traces of the gas escaping in their heated greenhouses led to early closure of blossoms (Knight and Crocker, 1913) and the Californian citrus growers discovered that fumes from their kerosene anti-frost pots could be used to ripen green lemon fruits (Sievers and True, 1912). Although we now know that all these responses are attributable to traces of ethylene in the air there was no indication at that time that the volatile material could be of plant origin. Even when Neljubov[16] (1901), working in St. Petersburg, showed that it was the ethylene component of illuminating gas that caused the stunting and stem swelling of his pea seedlings, there was no reason to consider that this remarkably active substance (active at parts per million in the air) could be a hormone. Interest in the possibility was however seriously aroused when Elmer (1932) reported that potatoes stored in containers with ripe apples or pears showed an inhibition of shoot sprout growth, and Botjes (1933) showed that volatiles from ripe apples caused epinasty, similar to that produced by auxin, in tomato plants enclosed with them. The next year Gane (1934) provided the indisputable proof of ethylene as a natural product which could function as a plant hormone. For four weeks he passed an air stream over Pearmain apples into bromine water and treated the brominated products (which then contained ethylene dibromide) with aniline so producing N,N′-diphenylethylenediamine, thereby showing that ethylene had been produced and released by the apples into the collected air stream. Ethylene then became known as the "*ripening*" hormone. Now it is a recognized product of all plant parts—leaves, stems, flowers and roots, though the highest levels of production recorded are still from ripening fruits.

The demonstration that ethylene was a natural product, together with intensive research (much at the Boyce Thompson Institute by Crocker, Zimmerman and

Hitchcock, 1935) which showed that ethylene gas would induce many of the growth responses induced by auxin did not, at first, establish the importance of ethylene other than as a rather specialized volatile substance associated with ripening. Indeed, in the commonly used biology textbook when I was at school—Brimble's (1939) *Intermediate Botany*—there was no mention of ethylene. Scientists took sides as to whether or not the effects it produced were all due to auxin changes or whether it was just a "waste" product of metabolism like respiratory CO_2. Thimann felt strongly that auxin was the master regulator which controlled all the plant responses. Those working on the many synthetic auxins studied until 1940, in particular Crocker, Zimmerman and Hitchcock (summarized in Crocker, 1948), doubted this and considered ethylene to be a "phytohormone." The fact that auxin caused *retardation* of abscission of debladed leaf petioles (La Rue, 1936) while illuminating gas (or ethylene) *induced* leaf shedding was difficult to explain by the master hormone concept. As late as 1979 there were still a few like W.P. Jacobs who wrote (in the Preface to his book *Plant Hormones and Plant Development*) "The gas ethylene does not seem to be a hormone in the strict sense because of lack of evidence that it moves within the plant from site of production to site of action. However, it acts at hormonal levels in various developmental processes. . . ."

The Hormone from Japan—Gibberellin

After the horrific destruction of World War II, one thing for certain was improved—communication. With the defeat of the Japanese, their science became open to the interest and scrutiny of the West. The West then received a shock. A natural growth regulator, that was not auxin, had been discovered in Japan in 1926 by Eishii Kurosawa.[20] He was investigating why rice seedlings infected with the *Fusarium* fungus (*Gibberella fujikuroi*) showed excessive elongation of their stems, standing high above their uninfected neighbors. He showed that culture broths, from which the fungus had been removed, would cause this enhanced elongation growth when applied to rice plants. The substance was named gibberellin. Unlike auxin, gibberellin did not cause epinasty, nor did it appear to be involved in tropisms or moved in a polar way. Twelve years after its discovery, Yabuta and Sumiki at the University of Tokyo succeeded in crystallizing active material from the broth medium but did not identify the mixture. In their half page publication in Japanese they called their two active isolates gibberellin A and gibberellin B (Yabuta and Sumiki, 1938). Sadly, all this work ceased during World War II.

In the great wave of fashion for the chemical control of plant growth that swept the forties and the fifties, Western physiologists took gibberellin to their hearts. In the United Kingdom, ICI realized the potential importance of this new growth regulator and funded P. W. Brian to explore its function and isolate a purified product. His group achieved both objectives publishing the paper in 1955 (Brian et al., 1955). The chemical identification of the substances was eventually achieved by Cross et al. (1961) from fungal cultures, and from bean seeds (MacMillan et al.,

1961). The structure was quite unlike auxin. It was a sesquiterpene with three interlocking rings each with a number of possible ring substitutions. This provided some five different molecules each of which possessed activity. Gibberellins turned out to be a multi-family of closely similar molecules of which over 90 are now known. When gibberellic acid (gibberellin A_3, or GA_3, see Figure 2c) from the fungal source became generally available to plant physiologists in the late fifties, a number of unusual regulatory properties became apparent that were quite unlike those of auxin. In plants with single sex flowers, such as hops and cucurbits, treatment of the developing flowers led to the formation of male, staminate forms (Galun, 1959); whereas auxins induced the female form. Dwarf lines of French beans, maize (corn) and peas were all normalized to tall types when treated with a gibberellin. The elongation response for the first leaf base of a single-gene dwarf mutant became a standard quantitative bioassay for gibberellins (Phinney, 1957).

One of the most intriguing responses revealed was the induction of flowering by gibberellin in "long day" rosette plants while they were held under non-flowering "short day" conditions. The experiments of Anton Lang[17] (1957) with *Hyoscyamus niger* are classics in this respect. Later he showed that flowering under long days was normally associated with an increase in the level of endogenous gibberellins. For a while, gibberellin took on the mantle of the "flowering hormone."

In the germination of cereal seeds, it was long known by brewers that if the embryo was excised (or dead) the endosperm would not be hydrolyzed and sugars would not be released. In 1960, Paleg showed that amylolytic activity in the embryo-less half could be fully restored in the presence of gibberellin. In other words, the substance that passed from the embryo to the endosperm (or rather, to the living cells of the aleurone layer that encloses the dead endosperm) induces there the synthesis of α-amylase which is responsible for hydrolysis of the stored starch reserves held in the endosperm. The extent to which the α-amylase was induced became another bioassay for gibberellin.

Despite the very clear evidence that the different gibberellins which had been isolated could cause effects in plants that were not attributable to auxin, the sceptics still remained and it was difficult to rebut those who argued that gibberellins functioned by altering levels or rates of metabolism or synthesis of auxin. In fact, all of these events could be demonstrated. There was great loyalty to the early concept that auxin was the "master hormone" (Nitsch and Nitsch, 1959). Perhaps the evidence that best indicated the independent hormonal status of gibberellin came from the experiments of Varner, Ram Chandra and Chrispeels (1965) showing that isolated aleurone tissue from barley would start to produce a newly synthesized α-amylase 3–4 hours after being treated with gibberellin and that this was dependent upon the formation of new mRNA. If the latter synthesis was blocked by the presence of actinomycin D then no new amylase was produced. Auxin would not substitute for gibberellin. At this time too, Horton and Osborne (1967) showed ethylene would induce in separating abscission zones an enzyme (cellulase) that degraded carboxymethyl cellulose. This enzyme activity was also associated with an increase in both RNA and protein synthesis at the zone prior to cell separation as judged by autoradiographic analysis of radio-labelled precursors (Osborne, 1968). Any remaining doubts whether either gibberellin or ethylene were truly hormones were now fast disappearing. Inhibitors of protein (puromycin) and RNA (actinomycin D) synthesis had by this time also been shown to block auxin-induced elongation growth. The era of hormone regulated gene activity in plants was dawning (see later).

It is doubtful if Frits Went ever thought of plants as possessing a single regulatory hormone. He developed many concepts during the late twenties and thirties including a multiplicity of signal molecules that could coordinate and integrate the organized growth of the whole plant. He proposed a phyllocaline from leaves, a rhizocaline from roots in addition to a caline from buds (see Boysen Jensen et al., 1936). Auxin was certainly a bud caline for it had become very clear that auxin either from a terminal bud, or applied to a de-topped shoot, would inhibit the outgrowth of the laterals below. This concept of *apical dominance* has essentially stood the test of time, though there is the persistent question of what permits the onset of cell division in the bud on release of the repression.

Cell Division Hormones—Cytokinins

There had been good evidence for the existence of a cell division hormone (Figure 2d) since 1895 when Haberlandt first showed that slices of kohlrabi or potato tubers would, within a few days, form a new corky layer on the cut surface which was preceded by active cell division of several cell layers below the damaged cells. However, if the slices were well washed after the initial cutting, so that all traces of the damaged cells were removed, the divisions and corky layer failed to form—but would do so if crushed cells were re-applied (for summary see Haberlandt, 1921). This was not one of the properties of auxin.

For very many years unsuccessful attempts were made to induce plant cells in to multiply in culture, a basic problem being the failure to maintain cell division. In 1941, Van Overbeek et al. were investigating the age at which embryos could be removed from developing seeds, survive in culture and eventually grow. Premature embryos of *Datura* would not survive on normal culture medium, but (perhaps not surprisingly) would do so if supplied back with extracts of *Datura* seed tissues. In order to obtain a larger supply of active substance (200 ml at least), they tested the liquid endosperm "milk" of coconuts that nourishes the coconut embryo. When the "milk" was included in the culture medium, the *Datura* embryos flourished. Another tremendous fashion arose, to hunt for the cell division hormone or the "coconut milk factor." One of the best financed groups was at Cornell University where F. C. Steward[22] and his team were studying the growth and differentiation of carrot root tissues in culture. As with auxin, it was clear that there were many sources of the cell division factor—wheat germ, yeast extract, the milky stage of maize kernels and young carrot leaves. The active material from coconut was heat stable and water soluble and a crystalline isolate was active at 0.25 ppm (Schantz and Steward, 1952).

Identification of the active cell division substance took several more years and was finally achieved by Folke Skoog and Carlos Miller at Wisconsin, but not from coconut (Skoog and Miller, 1957). Instead they first used yeast, but as it seemed that the compound they sought had purine-like properties they actually tried an old bottle of DNA that had been sitting for some years on the laboratory shelf. The sample was amazingly potent and on a weight basis they recovered as much active material as they had from four years work with coconuts. It must be some product like DNA! But on extracting fresh DNA the material obtained was inactive. It was a product of *old* DNA! They soon discovered that fresh DNA could be converted to "old" DNA by autoclaving at pH 4.0 for an hour at 120°C, and from this source, a crystalline active material was obtained in December 1954. By March 1955, the structure was known. It was a 6-substituted amino purine, and synthesis confirmed the 6-substitution as a furfurylamino group. The substance was called "kinetin," since the biological tests used in its isolation had been directed to cytokinesis in tissue cultures. Plant physiologists now had a cell division factor to add to their list. But was it a hormone, did it occur naturally and was it the stuff of coconut milk?

In fact, kinetin was not the natural cell division hormone, although as an adenine derivative it is chemically closely related to the native kinins that have been isolated from plant sources. It was D. S. Letham in New Zealand (1967) who first succeeded with a crystalline cytokinin isolated from the milky stage of immature maize (*Zea mays*) kernels. It was 6-substituted *iso*-pentenylamino purine (N^6-(Δ^2-isopentenyl)) adenosine (Figure 2d). He called the substance zeatin, and zeatin riboside was, in fact, the major active constituent of coconut milk (Letham, 1974).

A large number of different substitutions were found in plant tissues, either as the base, the riboside or the ribotide. Their test for cytokinin activity was to induce mitotic events in cultures of plant cells (usually those of tobacco pith) and with auxin also in the medium, to determine whether roots or buds developed on callus cultures. But other important attributes were soon discovered. These were cell enlargement in leaves (a condition not achieved with auxin), the breaking of apical dominance when the substances were applied to lateral buds suppressed by auxin, and very importantly, the delaying of senescence in cytokinin treated herbaceous leaves, first demonstrated in *Xanthium* by Richmond and Lang (1957). The treated regions of the leaves stayed green, photosynthetically active, and acted as a locus for the accumulation of metabolites from neighboring cells.

But perhaps the discovery that caused the greatest surprise was the presence of these substituted adenine derivatives in the anti-codon loops of several transfer RNAs—not only in plant tRNAs but in those for serine and tyrosine in yeast (Bergquist and Matthews, 1962) in *E. coli* and probably in all other organisms.

These findings started a tremendous wave of research interest (a fashion) into the potential role of cytokinins in nucleic acid synthesis (and hence the control of protein synthesis) in plants. Looking back on those years of the early 1960s, the burning questions such as, are cytokinins derived from the degradation of adenosine residues alkylated *in situ* in RNA or are they synthesized *de novo*? have faded almost completely, for we now have evidence that they are produced by the addition of dimethylallyl pyrophosphates to the N^6-position of AMP by isopentenyl transferases to yield the ribotide. From this, the ribosides and free bases are derived (see Binns, 1995). Certainly, the amounts that could arise from tRNA degradation alone appear much too low to account for the levels present in plant tissues.

It became clear, however, that in the whole plant the major site of cytokinin synthesis was in the roots from whence it was transported to the rest of the plant via the xylem transpiration stream. Perhaps kinins were a kind of rhizocaline?

Growth Inhibitors—Dormin: Abscisin

By the late fifties and early sixties, attention and fashion was directed not only to substances that enhanced the growth of cells and tissues, but also to those substances which might impose quiescence (Figure 2e). The ideas were not new, for Hemberg (1949) had already shown that potato peel contained substances that inhibited growth of *Avena* coleoptile sections and that the extracts became less inhibitory as the tubers broke dormancy. Similar inhibitors were found from *Fraxinus* buds which also depended upon the depth of their dormancy. In their fractionations of plant extracts by paper chromatography Bennet-Clark et al. (1952) had demonstrated in their *Avena* assays a growth inhibitory region at Rf 6-7: They named the material it contained "inhibitor β." Some years later in Aberystwyth, Wales, Phillips and Wareing[21] (1958) studying the reasons for the onset of dormancy in autumn in the shoots of sycamore (*Acer*), showed that inhibitor levels in the terminal buds were high in October, fell during winter and were at a minimum in June. When crude extracts were applied to the leaves of actively growing sycamore shoots, growth was inhibited and the terminal bud became dormant. They called their inhibitor "dormin." They tried hard to isolate and characterize the chemical substance responsible, but resolution came via California and the Shell Company's Milstead Laboratory in Kent—in a quite remarkable way.

All research scientists have goals and for the biologist, the demonstration, identification, the chemical characterization and synthesis of a regulatory molecule that determines the way living cells grow and differentiate is one of the most highly prized objectives. As early as 1951, F. T. Addicott and his colleagues were trying to isolate from shed cotton bolls a factor which when applied back to debladed cotyledonary "explants" excised from young cotton seedlings, would induce premature abscission of the cotyledon petiole stump. I well remember visiting his laboratory in Los Angeles and being impressed by the arrays of fractionation columns, yellowing cotton bolls and large numbers of dishes of explants in different stages of petiole separation from their stems. It was another ten years before they published their first paper describing the chemical nature of "abscisin II" (Okhuma et al., 1963). They obtained 9 mg of this substance from 225 kg of immature, shed, cotton fruits and published the planar structure in 1965 (Okhuma et al.).

P. F. Wareing and his "dormin" group had, meanwhile, joined with Cornforth's team in the Shell Research Laboratories and they too succeeded in isolating the active compound for dormancy in shoots of sycamore (Cornforth et al., 1965). It had the same structure as Addicott's abscisin II! Confirmation that both groups had isolated the identical substance came in the same year with Cornforth, Milborrow and Ryback's (1965) chemical synthesis of abscisin II. It was a sesquiterpene, the stereoisomer (S)-conformation being native (Figure 2e). Two pathways of synthesis seemed operative, *de novo* via mevalonate and the isoprene route, or as a degradation product of carotenoids.

Not only were abscisin II and dormin the same, but Hemberg's potato peel inhibitor and Bennet-Clark's inhibitor β all turned out to be similar. Finally, it was agreed that this new natural growth regulator (hormone) should be called abscisic acid (ABA for short) in recognition of its isolation as an abscission accelerator. Interestingly, over the years, it has become evident that abscission acceleration is one of its least important physiological properties. The regulation of dormancy in buds and particularly seeds and the lack of dormancy in mutants lacking an ABA-synthesis gene have become dominant aspects of ABA research. Perhaps its most intriguing activity is that of inducing stomatal closure and the forty-fold increase in internal leaf concentrations (from 10–400 μg/kg fresh weight) that can occur within the hour if a plant (or leaf) is subjected to wilting (Wright and Hiron, 1969).

The Jasmonates

In this flurry of excitement over the discoveries of the major plant hormone other smaller searches were in progress. The hunt for the substance present in senescent leaf material which accelerates abscission but is neither ethylene nor abscisic acid is still not resolved (Osborne et al., 1972), though the demonstration that some factor from the stelar tissue of senescing bean pulvini is essential for abscission in the zone below has re-confirmed its existence (Thompson and Osborne, 1994). Another group of substances functioning as cell growth inhibitors and as promoters of leaf senescence are the jasmonates (cyclopentanones) first known as the fragrant components of essential oils. The methyl ester of jasmonic acid and the free acid (Figure 2f) are ubiquitous in both higher and lower plants and interestingly are active as abscission accelerators (for review see Parthier et al., 1991). The aromatic fragrance of the jasmonates both B. V. Milborrow and I recognize as not unlike our own active purified senescent factor samples. NMR analyses indicate that either our sample contains a jasmonate, perhaps co-fractionating, or is a substance with certain similar groupings.

The Brassins

The enormous cost, time and effort involved between the initial discovery of an active substance with properties different from those already attributable to the major hormones in plants, and to the isolation and chemical characterization of the new active principle is well illustrated by the hunt for the "brassins" (Figure 2g). In 1970, Mitchell and his group in the United States (Mitchell et al., 1970) found a novel response of elongation, bending, stem swelling and splitting in a bean stem assay treated with lipid-containing extracts from rape pollen (*Brassica napus*). 227 kg bee-collected rape pollen yielded 10 mg of brassin (called brassinolide). After 10 years of work and a more than $1 million investment by the United States Department of Agriculture, "brassin" structures were resolved—a family of naturally occurring polyhydroxy steroids (Figure 2g) characterized as 5-α-cholestane derivatives (for review see Mandava, 1988).

More about Ethylene and its Precursor

Although ethylene was the first plant growth regulator to be identified with certainty (though not then as an endogenous hormone) the immediate precursor took many more years to unravel. Since ethylene is produced within the plant cell it is, of course, in solution and ethylene is soluble in water (depending on temperature and pressure) to several hundred parts per million—well in excess of its threshold activity at parts per million. Studies of the biosynthesis of this volatile hormone (incidentally, it is a plant pheromone, since it will influence the behavior of plants close-by, as well as the producer) were quite difficult and generally depended upon gas chromatographic identification of the product from radiolabelled potential precursors. Working with slices of apple fruit Lieberman and colleagues (1966) showed that methionine was the ethylene precursor. The process required oxygen and under nitrogen a non-protein amino acid was shown by S. F. Yang and his group in Davis, California (they too used mature apple fruit tissue) to accumulate. There was a burst of ethylene production when oxygen was re-admitted to the system (Adams and Yang, 1979). They identified this amino acid as 1-aminocyclopropane-1-carboxylic acid (ACC; Figure 2h; see review by Yang and Hoffman, 1984). Fact is often stranger than fiction and, looking back into the literature, the Yang group found that L. F. Burroughs, working in the Cider Research group at Long Ashton Research Station in England, had first isolated this unusual amino acid from ripe perry pears. Burroughs (1957) was convinced it had some role in ripening because it then increased to very high levels, but at that time there was no reason to link the rise in ACC to the high rates of ethylene production associated with fruit ripening. We now know it to be the immediate precursor for ethylene formation in higher plants.

Since almost all plant cells produce ethylene (animal cells do not) how ACC production is regulated has become of tremendous importance in the understanding of the roles of ethylene in plant growth manipulation. Ethylene itself diffuses readily along a steep diffusion gradient from water to air and is therefore preferentially lost to the environment rather than being transported long distances and directly affecting remote parts of the plant (hence W. P. Jacobs' non-acceptance of it as a "true" hormone). In contrast, ACC is readily transported in the vascular tissues to all parts of the plant and is converted to ethylene by the cells that receive it.

The most recent discoveries in the methionine to ethylene pathway are the demonstration of S-adenosylmethionine as the intermediate and the existence of the multigene family of ACC synthases that convert S-adenosylmethionine to ACC (for review see Kende, 1993). The expression of the different genes in different tissues is determined by different stimuli such as ripening, tissue wounding or the status of cell growth responses. The isolation of the oxidase enzyme that converts ACC to liberate the free ethylene molecule *in vitro* (Ververidis and John, 1991) was another breakthrough, particularly because for many years it was thought that the enzymes concerned would operate only on an intact membrane system *in vivo*.

However, it should be pointed out here that not all plants follow the same synthetic pathway for ethylene. Lower plants (liverworts, mosses, ferns, lycopods) do not produce it from methionine, nor from ACC—an alternative ethylene pathway therefore exists (Osborne et al., 1996). In evolutionary terms this is very significant and it remains to be established whether any cells in higher plants still retain this earlier primitive route for ethylene synthesis as part of their metabolic repertoire.

OTHER HORMONES?

A glance at the chemical structures so far established for the different natural regulatory compounds in plants shows that both auxin and ethylene are derived from amino acids, the gibberellins via the isoprenoid pathway; abscisic acid and the jasmonates from carotenoid conversions; cytokinins from purines; and the brassins via steroids. Conspicuous by their absence as hormones, particularly when compared with animal cells, are the peptides and proteins, but this may just be because they have not yet been "discovered." We know already that a fungal infestation of plants (*Phytophthora megasperma* in tomato) or wounding can elicit the production of specific peptides which result in the systemic induction of the plants' proteinase inhibitors (Pearce et al., 1991). Whether peptides produced in response to pathogens or damage can be considered as "hormones" is doubtful today—but our views may change in the future.

There is however one very good candidate for a natural protein (peptide) hormone in plants. The flowering hormone or "florigen" certainly exists and one of the very earliest experiments of Zeevaart (1958) provides irrefutable evidence for a "flowering inducing principle" that can be transmitted from one plant to another plant by grafting. This classic experiment was carried out by grafting a leaf from a short day flowering *Perilla* plant onto a non-flowering vegetative receptor *Perilla* kept under non-flowering long day conditions. The receptor plant flowered, but only after protoplasmic cell–cell communication had been established in the graft union. The grafted leaf could then be removed and grafted onto other plants to induce flowering. So it was not a question of a simple diffusible substance, but rather the passage of a factor requiring effective cytoplasmic contact between the grafted leaf cells and those of the stock plant. Large molecules, like viruses, will move from cell to cell in this way. Florigen has been variously considered as a complex of molecules, a lipid or a peptide. Whichever it may be, the passage of the flowering stimulus from a single photo-induced leaf to the apical bud of the stock plant causes that apical bud to be converted from a vegetative to a flowering morphology. In the process, the factor must be amplified many times since a leaf can also be removed from the stock and successfully grafted onto another vegetative plant, causing it to flower. Current research is underway to seek a peptide florigen (B. V. Milborrow, personal communication). Can we now envisage a molecule—such as a peptide—being transmitted from a single photo-induced (short day or long day) leaf to the growing buds in the rest of the plant and there inducing the de-repression of specific flowering-linked genes in such a way that at each cell division of the meristem, this flowering expression is retained? Of course we can, because these are the facts of flowering behavior, but we await the isolation and chemical identification of the active principle that controls this change.

WHERE ARE THE FASHIONS NOW?

Whereas in the past, plant physiologists were very much committed to their favorite hormone and there were auxin physiologists and gibberellin physiologists and others dedicated to kinins, ethylene or abscisic acid. These demarcations slowly changed with time, and it is now accepted that all the hormones probably play a part in any process and the outcome will be influenced by the status of the others, particularly since all cells appear to possess the capacity to produce each hormone to a smaller or greater extent. Furthermore, the function or availability of each can also be regulated by one of the others. Seminal in this respect are the auxin-ethylene interactions. IAA will enhance ethylene production, through an induction of ACC-synthase expression (see Yang and Hoffman, 1984), while ethylene will inhibit the polar transport of IAA and hence its movement to different parts of the plant. In aleurone cells, abscisic acid will suppress the gene for α-amylase production and can overcome the induced expression of this gene by gibberellin. Whereas cytokinins will suppress proteolysis during leaf senescence and maintain protein synthesis, the jasmonates will reverse the effect of cytokinins. But jasmonates will enhance ethylene production and ethylene, too, enhances the rate of senescence in leaves that have already started to degrade their chlorophyll (see Sembdner and Parthier, 1993). All these interactions are far less easy to understand than the earlier efforts at hormone isolation, identification, and synthesis and their quantification *in vitro* and *in vivo* when great efforts were made to relate plant development to internal hormonal levels and to the effects induced when the compounds were added back to plants.

HORMONE FUNCTION

Then there is the question of how *do* hormones actually function. When the concepts of plant hormone regulation of gene expression were in their infancy, I remember the enthusiastic thrust of James Bonner's experiments in seeking the direct binding of gibberellins or auxin to chromatin isolated from pea shoot nuclei to explain their "effector" role in enhancing cell growth (Bonner, 1965). However, it was soon apparent that a direct interaction with chromatin was not how the hormones functioned. Instead, there were transduction pathways in the cell from the site of perception of the hormone (receptor proteins located in membranes) through a cascade of cytoplasmic "secondary messengers" that led eventually to activation of the promoters of the specific coding sequences of specific genes.

Hormone Receptors

During the 1970s and 1980s many investigators looked for receptor proteins and the literature abounds in fashionable citations for each of the five major hormones depicting Scatchard plots using fractions from density gradient sedimentations from tissues which showed large responses to the hormone in question. The most convincing studies were those on auxin binding proteins from maize coleoptiles as illustrated first by Venis (1977). Maize has continued to be a favored starting material for such studies, despite the low abundance of these proteins in plasma membrane or ER fractions. Photoaffinity labelling and photoaffinity chromatography have finally been successful in the identification of one protein (pm 23) from plasma membrane vesicles which is structurally similar to the auxin binding protein (Zm-ER abp 1) of the lumen of the endoplasmic reticulum. The primary structure of this 22 kDa glycoprotein has now been deduced from cDNAs isolated from maize coleoptiles (see Palme et al., 1993). Antibodies raised against Zm-ER abp 1 or from the cloned protein have been shown to specifically inhibit changes in potential differences elicited across the plasma membranes of isolated tobacco protoplasts (Barbier-Brygoo et al., 1991). In cultured maize coleoptile cells, the immediate depolarization by active auxin (IAA and synthetic auxins) was pH dependent, indicating an influx carrier co-transporting $2H^+$ with each auxin negative ion.

These results now link logically with the older and much studied questions relating to acid-induced cell growth, first reported by Strugger in 1934. Later it was found that auxin-induced elongation of tissue segments led to acidification of the medium as well as to acidification of the cell walls of the elongating material. In the wake of the Mitchell hypothesis of proton pumping in animal cells, the plant physiologists sought and identified auxin-activated ATP-ases in the plasma membrane which linked with acidification of the wall and medium and the active uptake of auxin molecules. At least this aspect of the first stages involved in auxin-induced cell growth have become more understandable (see review by Rayle and Cleland, 1992). The topology of auxin receptor-linked ATPases is also potentially important in resolving the polarity of auxin transport mechanism. In the 1980s a new fashion arose for following ion fluxes, ion channels and other ion transport proteins by the "patch-clamp" technique. Whole protoplasts, vacuoles or areas of plasma membrane were sealed to fine glass capillaries which functioned as tiny electrodes to follow current changes on administering test solutions to the protoplast, vacuole or membrane. There is no doubt of the immediate and relatively large changes in membrane potentials that can be elicited, but we still await the linkage to the transduction pathway and to new gene expressions. Also, until we can determine or mark the apical and basal ends of protoplasts we shall not resolve polar transport of auxin with assurance. We need this precise information for all the plant hormones and I predict the hunting and cloning of the receptors will continue until this is achieved.

Secondary Messengers

Plant cells, although they may look macroscopically very similar, encased as they are in their cell walls, perform very differently one from another. The plant is made up of a complex of "target cells" (Osborne, 1984), each expressing proteins that are markers of their target class, with predictable responses to hormones, e.g., abscission cells (McManus and Osborne, 1990). The two guard cells that border the stomatal pores of leaves and stems are an excellent example of their special target status with respect to abscisic acid, which induces stomatal closure, and cytokinins which induce opening. Both effects are dependent upon water fluxes and the swelling or contraction of these cells to open or close the pore. Patch-clamped protoplasts from these guard cells show Ca^{2+} dependent ion pumping enhanced by abscisic acid while microinjection procedures with a cytokinin, Ca^{2+}, abscisic acid or inositol triphosphates (IP_3) leads to stomatal opening or closure (for review see Mansfield et al., 1990). The fact that calcium-channel blockers also can prevent an inhibition of pore opening by abscisic acid seems to support a calcium mediated, K^+ ion flux-determined, hormonal control of these relatively rapid (minutes) and reversible cell volume changes. The hormonal control of ion-gating has therefore become a fashionable (and rewarding) area of research (for review see Blatt and Thiel, 1993).

But how far have plant physiologists resolved the chain of events that follow hormonal perception? Auxins, gibberellins, cytokinins and abscisic acid have all been shown to regulate free Ca^{2+} cytosolic levels in target tissues, by a hormonally regulated, calmodulin-mediated release from internal stores. Calcium has therefore received attention as a "second messenger" in plants. The membrane-localized presence of a guanine nucleotide binding protein (G-protein), the hydrolysis of phosphatidylinositol 4,5-bisphosphate to generate IP_3 and diacylglycerol with an ensuing phosphorylation of specific proteins by an activated Ca^{2+} modulated kinase, have all been demonstrated in one or other tissue in response to one or other of the major plant hormones (for a discussion, see Roberts et al., 1990). Despite these apparent similarities to the hormone induced events in animal cells (driven no doubt by the fashions and discoveries made first in animal cells) our knowledge of transduction pathways from perception to the gene response, is still fragmentary in plants and may turn out to differ in many respects. For example, an involvement of cyclic nucleotides has not been found in plants, despite efforts to discover one.*

THE CELL WALL—A NEW FRONTIER

Considerable advances have, however, been made on views of what was once considered to be a relatively inert part of the plant cell—the cell wall. Apart from the cellulose, hemicellulose, and pectins that make up the wall complex there is also a significant and variable protein complement with enzymic activity. Auxin-induced acidification of the wall via the activation of plasma membrane ATP-ases and the resultant proton secretion leads to lowering of the pH of the wall environment and a shift to enhanced activity of certain of these wall-associated enzymes. Local pH environments in cellular ecology have long been a subject for the metabolic biochemists, but the role of pH in the wall (the apoplast) is now seen as critical in many developmental events (Grignon and Sentenac, 1991). In the permanent cell wall extensions linked to cell growth, it was known from the days of Went (1928) that the wall became loosened, so accommodating the uptake of water and the increase in cytoplasmic volume associated with cell growth, but the mechanisms by which this is achieved are still a field of fashionable exploration. The protein complement of the wall, usually not more than 10%, changes during cell maturation and has been shown to contain ethylene enhanced non-enzymic, but crosslinking, hydroxyproline-rich proteins "extensins" linked to growth cessation (Kieliszewski and Lamport, 1994) and auxin promoted "expansins" that will catalyse the breakage of hydrogen bonds in the walls *in vitro* (McQueen-Mason and Cosgrove, 1994).

A new impetus to the study of the cell wall has come from the discovery of certain products of cell wall change associated with expansion growth, wall softening during fruit ripening, wall rigidification during xylogenesis and the separation of one cell wall from its neighbor during abscission. These are all associated with enzymically regulated changes in wall structure and include the liberation of small oligosaccharide fragments possessing what is nicely called "signal molecule activity." Hydrolases, both exo- and endo-, transform the molecular aggregates of cellulose, xyloglucans, and other mixed-linkage polysaccharides into smaller saccharide moities, while endo-transglycosylation in the absence of hydrolysis provides a means for molecular rearrangements *in muro* during growth and differentiation. These enzymically-induced changes in the cell wall during growth have been described by Fry (1995), while Albersheim and Darvill (1985) have explained how they envisage oligosaccharins liberated from pectins might function as what were once termed "morphogens."

The oligosaccharin products of microbial attacks on plant cell walls will elicit synthesis of phytoalexins (anti-microbial compounds of low molecular weight) in the host. The phytoalexins also interact with the plant's hormone responses. For example, a phytoalexin-inducing oligogalacturonide obtained from a *Fusarium* digest of polygalacturonic acid will inhibit auxin induced cell elongation in pea stems, auxin induced rooting in tobacco leaf explants, and will compete with auxin for auxin-binding sites in carrot membrane preparations (Filippini et al., 1992). Certain of the fungal enzymes themselves will directly change hormonal levels when they attack plant cells. An endo-xylanase from *Trichoderma*, for example, induces high levels of ethylene production by tobacco cells at the extremely low concentration of 10^{-9} M (Sharon et al., 1993).

It is then a simple step to consider cell wall oligosaccharide fragments from uninfected growing and differentiating cells as likely regulatory signal molecules that normally modify the effectiveness of the hormonal complement. Senescing cells, ripening fruit cells and separating (abscission) cells are good candidate sources for native, active oligosaccharins. There is no evidence, however, that their influence is any other than local, or extends to gene control. In this respect they appear to differ from hormones.

MOLECULAR BIOLOGY AND THE STUDY OF MUTANTS

It is significant that in 1988, the Annual Review of Plant Physiology added "and Plant Molecular Biology" to its title. Year by year the numbers of plant molecular biologists has risen as the identification and study of hormone regulated genes has become the fashion of the age. New techniques have made questions posable that were previously unanswerable.

What have we learned that is new? First, plant physiologists have a new test object in *Arabidopsis thaliana*—an insignificant little crucifer with a tiny genome. It has a generation turnover time of 40 days, it can be cultivated easily in growth cabinets or in the greenhouse and it readily produces mutants. In addition, it has low concentrations of those substances such as phenolics and other secondary products which interfere with the extraction of nucleic acids and other cytoplasmic constituents. In the hormone field, this little plant has facilitated important molecular research on the production of ethylene. There has been a tremendous impetus from the successful isolation, cloning and sequencing of genes which control different stages in ethylene biosynthesis, the binding and transduction of the ethylene signal, and the expression of genes controlled by different levels of ethylene production and by the extent of ethylene-response competence in different cell types. We learn from these studies that the organization of plant cells is even more like that of animal cells than we had ever supposed.

Back to Ethylene

Why has ethylene, the one-time doubtful candidate for hormonal status, now received so much attention from molecular biologists in the 1990s? I think because of the central role it now plays in large-scale food production in the regulation of harvesting, ripening, storage, and "improvement" of plant produce. Fashion has given way to investment activated by the food industry, with the funding this directs to plant research bringing great financial reward to agriculture and horticulture. By screening *Arabidopsis* seedlings for mutants that did not respond to ethylene by growth inhibition or stem swelling (tests originally devised by Neljubov in his ethylene studies of 1901), a single gene, ethylene receptor 1 (*etr*1) was identified via mRNA, cDNA, and cloning into vectors and the DNA then sequenced (Chang et al., 1993). *etr1* expression was absent from mutants that failed the ethylene response tests. When *etr1* was expressed in yeast, it was found to be (as in *Arabidopsis* itself) a membrane associated, di-sulphide dimer, with a COOH-terminal region having homology to histidine kinases of bacteria. Yeasts expressing the ETR1 protein *in vivo* have a high-affinity binding site for [^{14}C] ethylene, which is displaceable by competition with unlabelled ethylene. Wild type *Arabidopsis* reportedly contains 1 pmole of these ethylene binding sites per gram of tissue (Harpham et al., 1991).

Introducing the *etr1*-mutant gene (which interestingly is a dominant) into wild type *Arabidopsis* by *Agrobacterium* transformation, results in plants lacking the ethylene response. But in *Arabidopsis*, *etr1* functions with two other genes essential for ethylene signal transduction—*etr1* acts upstream of one of these, *ctr1*, a putative serine/threonine kinase (similar to the Raf tyrosine protein kinases of animals and *Drosophila*). Another receptor associated gene, *ers*, with a high level of identity to *etr1*, has now been cloned from *Arabidopsis*, suggesting either a redundancy in function, overlap, or a two component sensing of ethylene similar to that in bacteria (Schaller and Bleeker, 1995).

The use of mutants and the ability to isolate and study the regulation of genes responsible for mutant performance has reopened long known genetic phenomena for molecular interpretation. One of these concerns the never-ripe mutant of tomato (*Nr*) which fails to redden or soften its fruit with ethylene (Rick and Butler, 1956). It transpires that the amino acid sequence from the *etr1* gene in tomato is identical to that of the *Nr* locus and one amino acid change in the NR coding for the expressed NR protein is sufficient to render *Nr* incapable of responding to ethylene by normal ripening. Now we know that ETR of tomato is 70% identical with that of *Arabidopsis* and that tomato ETR mRNA is not produced by the *Nr* mutant, although it is abundantly produced during ripening or ethylene-induced ripening of the wild type (Wilkinson et al., 1995). So lack of response to ethylene by the *Nr* mutant is a reflection of lack of ethylene receptor. The extent of ETR/ERS expression may therefore be central also to distinguishing one cell type from another as in the differential responses of separation or non-separation across the tissue of an abscission zone. (For a general update on ethylene biosynthesis and perception see Fluhr and Mattoo, 1996.) Perhaps the most satisfying and non-food financed molecular research with the ethylene hormone and its receptor comes from an unexpectedly ecological quarter. The ethylene induced extension growth in cells of the petioles of semi-aquatic plants which brings their leaves to the surface again once submerged, results from a restriction of outward diffusion of ethylene from within the submerged tissues and the resultant increase in internal ethylene concentrations. When at the surface again the ethylene is dispersed and the normal slow aerial growth is resumed. This is in strict contrast to land plants (and *Arabidopsis*) where ethylene is a suppressor of extension growth as Neljubov first showed. A cDNA clone has just been isolated from the water dock (*Rumex palustris*) with similarities to the *ers* gene from *Arabidopsis* and when *R. palustris* is flooded (or exposed to ethylene) these submerged petiole cells up-regulate their production of RP-ERS1 mRNA leading to enhanced ethylene binding, signal transduction and extension growth. But on returning from the water to air, the RP-ERS1 mRNA transcripts decrease by 37% within the hour (Vriezen et al., 1997). This neatly demonstrates that the levels of ethylene *in vivo* alone are not enough to regulate plant performance; the whole transduction pathway must be operative too.

Molecular Biology and Auxin

At the present time, molecular research into ethylene outstrips that into other plant hormones, although auxin studies follow closely. All cells investigated thus far show some response to auxin so it may be concluded that Went was right with his "Ohne Wuchsstoff, kein Wachstum" and that all plant cells perceive auxin and transduce signals to the nucleus via secondary messages (see Napier and Venis, 1993). Although rapidly induced (or suppressed) auxin regulated proteins are well researched and documented (in particular those associated with cell growth) we currently do not understand how the promoter sites for these genes are actually controlled following signal transduction. There may be a multiplicity of mechanisms for nuclear and cytoplasmic interaction awaiting discovery in plants as, for example, in certain auxin resistant mutants of *Arabidopsis*. There, identification and cloning of the *axr1* gene showed it to be related to an ubiquitin-activating enzyme E1 (Leyser et al., 1993). The *axr1* mutant has a defective protein which may fail to function in ubiquitin attachment to proteins destined for subsequent rapid proteolysis. This can have profound effects on cell performance, particularly in senescence.

Whereas so many pieces of the jigsaw are now known for each hormone and all appear to reflect some similarities with regulating systems in animal cells, we still lack a proper understanding of how plant hormones coordinate their many different types of cells with such exquisite precision. I venture to predict that the molecular biology of ethylene and auxin will stay at the forefront of our investigations into hormones for a long time to come, but as we come to better understand these two, the way that other hormones and signal molecules function will doubtless fall into place too.

POSTSCRIPT

Sitting writing this in 1998, I am filled both with considerable excitement but not a little sadness. I was fortunate to be a visiting postdoctoral in many of the great hormone-centered laboratories of the world and knew most of the great discoverers well. Most of them are no longer with us, James Bonner, Anton Lang and Philip Wareing died last year, and the great Kenneth Thimann this January, but the legacy that they all have left of the urgency of the chase, the need for irrefutable evidence to back the speculations of the yet unknown and the anticipation as each new experiment yielded up its secrets, will never dim. And it will always be so! But as T.A. Bennet-Clark once warned "the devil always sends positive results first—repeat before you publish"!

ACKNOWLEDGMENT

I wish to thank Mrs. Vivian Reynolds for her unfailing help in preparing the manuscript.

NOTE

*Efforts may now have been successful! Whereas normal tobacco cells require auxin for division, sequence tagged (TDNA) lines encoding an adenylyl cyclase were obtained which were auxin-independent but cAMP-dependent. From one line (axi 141), a complementary DNA encoding adenylyl cyclase has been isolated with characteristic leucine repeats and similarity to yeast adenylyl cyclase (Ichikawa et al., 1997). The result seems not to be the expression of an alternative division pathway from the normal auxin-driven division since it is blocked by auxin inhibitors and is activated by cAMP and the cyclase activator forskolin. Perhaps a link to G-protein at the membrane will now bring plant growth regulation even closer to that of animals.

REFERENCES

Adams, D.O. & Yang, S.F. (1979). Ethylene biosynthesis: identification of 1-aminocyclopropane-1-carboxylic acid as an intermediate in the conversion of methionine to ethylene. Proc. Natl. Acad. Sci. USA 76, 170–174.

Albersheim, P. & Darvill, A.G. (1985). The oligosaccharins. Scientific American 253, 58–64.

Barbier-Brygoo, H., Ephritikhine, G., Klambt, D., Maurel, C., Palme, K., Schell, J., & Guern, J. (1991). Perception of the auxin signal at the membrane of tobacco mesophyll protoplasts. Plant Journal 1, 83–93.

Bennet-Clark, T.A., Tambiah, M.S., & Kefford, N.P. (1952). Estimation of plant growth substances by partition chromatography. Nature 169, 452–453.

Bergquist, P.L. & Matthews, R.E.F. (1962). Occurrence and distribution of methylated purines in the ribonucleic acids of subcellular fractions. Biochem. J. 85, 305–313.

Binns, A.N. (1994). Cytokinin accumulation and action: biochemical, genetic and molecular approaches. Ann. Rev. Plant Physiol. Plant Mol. Biol. 45, 173–196.

Blatt, M.R. & Thiel, G. (1993). Hormonal control of ion channel gating. Ann. Rev. Plant Physiol. Plant Mol. Biol. 44, 543–567.

Bonner, J. (1934). The relation of hydrogen ions to the growth rate of the *Avena* coleoptile. Protoplasma 21, 406–423.

Bonner, J. (1965). The Molecular Biology of Development. Clarendon Press, Oxford, UK.

Botjes, J.O. (1933). Aethyleen als vermoedelijke oozaak van de groeiremmende verking van rijpe appels. Tijdschr. Plantenziekten 39, 207–211.

Boysen Jensen, P. (1911). La transmission de l'irritation phototropique dans l'*Avena*. K. Danske Vidensk. Selskab. Forhandl, (Bull. L'Acad. Roy. Sci. Lettres, Danemark) 3, 1–24.

Boysen Jensen, P., Avery, G.S., & Burkholder, P.R. (1936). Growth Hormones in Plants. McGraw-Hill, New York.

Brian, P.W., Elson, G.W., Hemming, H.G., & Radley, M. (1955). The growth promoting properties of gibberellic acid, a metabolic product of the fungus *Gibberella fujikuroi*. J. Sci. Food Agr. 5, 602–612.

Brimble, L.J.F. (1939). Intermediate Botany. Macmillan, London.

Burroughs, L.F. (1957). 1-aminocyclopropane-1-carboxylic acid: a new amino acid in perry pears and cider apples. Nature (London) 179, 360–361.

Chang, C., Kwok, S.F., Bleeker, A.B., & Meyerowitz, E.M. (1993). *Arabidopsis* ethylene-response gene ETR1: similarity of product to two-component regulators. Science 262, 539–544.

Cornforth, J.W., Milborrow, B.V., Ryback, G., & Wareing, P.S. (1965). Identity of sycamore "dormin" with abscisin II. Nature (London) 205, 1269–1270.

Cornforth, J.W., Milborrow, B.V., & Ryback, G. (1965). Synthesis of (±)-abscisin II. Nature (London) 206, 715.

Crocker, W., Zimmerman, P.W., & Hitchcock, A.E. (1935). Similarities in the effects of ethylene and the plant auxins. Contrib. Boyce Thompson Inst. Plant Res. 7, 231–248.

Crocker, W. (1948). Growth of Plants. Reinhold, New York.

Cross, B.E., Grove, J.G., McCloskey, P., MacMillan, J., Moffatt, J.S., & Mulholland, T.P.C. (1961). The structure of fungal giberellins. Adv. Chem. series 28, 3–17.

Darwin, C. & Darwin, F. (1881). The Power of Movement in Plants. D. Appleton-Century, New York.

Elmer, O.H. (1932). Growth inhibition of potato sprouts by volatile products of apples. Science 75, 193.

Ferguson, S. (1997). Bioenergetics after 1960: from the chemiosmotic hypothesis to structure-based molecular mechanisms. Foundations of Modern Biochemistry, Vol. 3, pp. 3–22. JAI Press, Greenwich, CT.

Filippini, F., Lo Schiavo, F., Terzi, M., Branca, C., Bellincampi, D., Salvi, G., Desiderio, A., de Lorenzo, G., & Cervone, F. (1992). Phytoalexin elicitor-active α-1,4-D-oligogalacturonides reduce auxin perception by plant cells and tissues. Progress in Plant Growth Regulation (Karssen, C.M., Van Loon, L.C., & Vreugdenhil, D., Eds.), pp. 517–521. Kluwer Academic, The Netherlands.

Fitting, H. (1910). Weitere entwicklungs physiologische Untersuchungen an Orchideenblüten. Zeitschr. Bot. 2, 225–267.

Fluhr, R. & Mattoo, A.K. (1996). Ethylene—biosynthesis and perception. Current Revs. in Plant Sciences 15, 479–523.

Fry, S.C. (1995). Polysaccharide-modifying enzymes in the plant cell wall. Ann. Rev. Plant Physiol., Plant Mol. Biol. 46, 497–520.

Galun, E. (1959). Effects of gibberellic acid and naphthalene-acetic acid on sex expression and some morphological characters in the cucumber plant. Phyton (Buenos Aires) 13, 1–8.

Gane, R. (1934). Production of ethylene by some ripening fruit. Nature (London) 134, 1008.

Girardin, J.P.L. (1864). Einfluss des Leuchtgases auf die promenaden und strassen Baume. Jahresb. Agr. 7, 199–200.

Grignon, C. & Sentenac, H. (1991). pH and ionic conditions in the apoplast. Ann. Rev. Plant Physiol. Plant Mol. Biol. 42, 103–128.

Haberlandt, G. (1921). Wundhormone als Erreger von Zellteilungen. Beitr. Allgem. Botan. 2, 1–53.

Harpham, N.V.J. et al. (1991). The effects of ethylene on the growth and development of wild-type and mutant *Arabidopsis thaliana* (L) Heynh. Annals Bot. 68, 56–61.

Hemberg, T. (1949). Significance of growth-inhibiting substances and auxins for the rest-period of the potato tuber. Physiol. Plant. 2, 24–36.

Horton, R.F. & Osborne, D.J. (1967). Senescence, abscission and cellulase activity in *Phaseolus vulgaris*. Nature (London) 214, 1086–1088.

Ichikawa, T. et al. (1997). Identification and role of adenylyl cyclase in auxin signalling in higher plants. Nature (London) 390, 698–701.

Jacobs, W.P. (1979). Plant Hormones and Plant Development. Cambridge University Press, New York.

Kende, H. (1993). Ethylene biosynthesis. Ann. Rev. Plant Physiol. Plant Molec. Biol. 44, 283–307.

Kieliszewski, M.J. & Lamport, D.T.A. (1994). Extensin: repetitive motifs, functional sites, post-translational codes and phylogeny. Plant J. 5, 157–172.

Knight, L.I. & Crocker, W.C. (1913). Toxicity of smoke. Bot. Gaz. 55, 337–371.

Kögl, F., Haagen Smit, A.J., & Erxleben, H. (1933). IX Mitteilung. Über ein Phytohormon der Zellstreckung. Reindarstellung des Auxins aus menschlichem Harn. Hoppe-Seyl. Zeitschr. Phys. Chem. 214, 241–261.

Kögl, F., Erxleben, H., & Haagen Smit, A.J. (1934). X Mitteilung. Über die Isollerung der Auxine a und b aus pflanzlichen Materialien. Hoppe-Seyl. Zeitschr. Phys. Chem. 225, 215–229.

Kurosawa, E. (1926). Experimental studies in the secretion of *Fusarium heterosporum* in rice plants. J. Nat. Hist. Soc. Formosa. 16, 213–227.

Laibach, F. & Kornmann, P. (1933). Zur Methodik der Wuchsstoffversuche. Planta 19, 482–484.

Lang, A. (1957). Effect of gibberellin upon flower formation. Proc. Natl. Acad. Sci. USA 43, 709–717.

La Rue, C.D. (1936). Effect of auxin in abscission of petioles. Proc. Natl. Acad. Sci. USA 22, 254–259.

Letham, D.S. (1967). Chemistry and physiology of kinetin-like compounds. Ann. Rev. Plant Physiol. 18, 349–364.

Letham, D.S. (1974). The cytokinins of coconut milk. Physiol. Plant. 32, 66–??.

Leyser, O.H.M., Lincoln, C.A., Timpte, C., Lammer, D., Turner, J., & Estelle, M. (1993). *Arabidopsis* auxin-resistance gene AXR1 encodes a protein related to ubiquitin-activating enzyme E1. Nature (London) 364, 161–164.

Lieberman, M., Kunishi, A., Mapson, L.W., & Wardale, D.A. (1966). Stimulation of ethylene production in apple tissue slices by methionine. Plant Physiol. 41, 376–382.

MacMillan, J., Seaton, J.C., & Suter, P.J. (1961). Isolation and structures of gibberellins from higher plants. Adv. Chem. Ser. 28, 18–25.

Mandava, N.B. (1988). Plant growth-promoting brassinosteroids. Ann. Rev. Plant Physiol. and Plant Mol. Biol. 39, 23–52.

Mansfield, T.A., Hetherington, A.M., & Atkinson, C.J. (1990). Some current aspects of stomatal physiology. Ann. Rev. Plant Physiol. Plant Mol. Biol. 41, 55–75.

McManus, M.T. & Osborne, D.J. (1990). Identification of polypeptides specific to rachis abscission zone cells of *Sambucus nigra*. Physiologia Plant. 79, 471–478.

McQueen-Mason, S. & Cosgrove, D.J. (1994). Disruption of hydrogen bonding between plant cell wall polymers by proteins that induce plant wall extension. Proc. Natl. Acad. Sci. USA 91, 6574–6578.

Mitchell, J.W., Mandava, N.B., Worley, J.F., Plimmer, J.R., & Smith, M.V. (1970). Brassins: a new family of plant hormones from rape pollen. Nature (London) 225, 1065–1066.

Molisch, H. (1884). Uber die ablenkung der wurzeln von ihrer normalen wachsthumsrichtung durch gase (aërotropismus). Sitzungsber. Kais Acad. Wiss (Wein) 90, 111–196.

Napier, R.M. & Venis, M.A. (1993). Tansley Review No. 79. Auxin action and auxin binding proteins. New Phytol. 129, 167–201.
Neljubov, D.N. (1901). Uber die horizontale Nutation der stengel von *Pisum sativum* und eineger anderen Pflanzen. Beit. Bot. Zentralbl. 10, 128–138.
Nitsch, J.P. & Nitsch, C. (1959). Modification du metabolisme des auxines par l'acide gibberellique. Bull. Soc. France Physiol. Veg. 5, 20–23.
Okhuma, K., Lyon, J.L., Addicott, F.T., & Smith, O.E. (1963). Abscission II, an abscission-accelerating substance from young cotton fruit. Science 142, 1592–1593.
Okhuma, K., Addicott, F.T., Smith, O.E., & Thiessen, W.E. (1965). The structure of abscission II. Tetrahedron Lett. 2529–2535.
Osborne, D.J. (1968). Hormonal mechanisms regulating senescence and abscission. Biochemistry and Physiology of Plant Growth Substances (Wightman, F. & Setterfield, G., Eds.), pp. 815–840. The Runge Press, Ottawa.
Osborne, D.J. (1984). Concepts of target cells in plant differentiation. Cell Differentiation 14, 161–169.
Osborne, D.J., Jackson, M.B., & Milborrow, B.V. (1972). Physiological properties of an abscission accelerator from senescent leaves. Nature (New Biol.) 240, 98–101.
Osborne, D.J., Walters, J., Milborrow, B.V., Norville, A., & Stange, L.M.C. (1996). Evidence for a non-ACC ethylene biosynthetic pathway in lower plants. Phytochemistry 42, 51–60.
Overbeek, J. van, Conklin, M.E., & Blakeslee, A.F. (1941). Factors in coconut milk essential for growth and development of *Datura* embryos. Science 94, 350–351.
Paleg, L. (1960). Physiological effects of gibberellic acid. 1. On the carbohydrate metabolism and amylase activity of the barley endosperm. Plant Physiol. 35, 293–299.
Palme, K., Feldwisch, J., Peters, W.S., Schell, J., Zettl, R., Campos, N., & Felle, H. (1992). Auxin binding proteins are located in the ER and in the plasma membrane: Identification by photoaffinity-labelling and characterization. Progress in Plant Growth Regulation (Karssen, C.M., Van Loon, L.C., & Vreugdenhil, D., Eds.), pp. 73–81. Kluwer Academic, The Netherlands.
Parthier, B., Brückner, C., Dathe, W., Hause, B., Herrmann, G., Knöfel, H.-D., Kramell, H.-M., Kramell, R., Lehmann, J., Miersch, O., Reinbothe, S.T., Sembdner, G., Wasternack, C., & Zur Nieden, U. (1992). Jasmonates: Metabolism, biological activities, and modes of action in senescence and stress responses. Progress in Plant Growth Regulation (Karssen, C.M., Van Loon, L.C., & Vreugdenhil, D., Eds.), pp. 276–285. Kluwer Academic, The Netherlands.
Pearce, G., Strydom, D., Johnson, S., & Ryan, C.A. (1991). A polypeptide from tomato leaves activates the expression of proteinase inhibitor genes. Science 253, 895–897.
Phillips, I.D.J. & Wareing, P.F. (1958). Studies in dormancy of sycamore. 1. Seasonal changes in the growth-inhibitors in *Acer pseudoplatanus*. J. Exp. Bot. 9, 350–364.
Phinney, B.O. (1957). Growth response of a single-gene dwarf mutants in maize to gibberellic acid. Proc. Natl. Acad. Sci. USA 42, 185–189.
Rayle, D.L. & Cleland, R.E. (1992). The acid growth theory of auxin-induced cell elongation is alive and well. Plant Physiol. 99, 1271–1274.
Richmond, A. & Lang, A. (1957). Effect of kinetin on protein content and survival of detached *Xanthium* leaves. Science 125, 650–651.
Rick, C.M. & Butler, L. (1956). Phytogenetics of the tomato. Adv. Genet. 8, 267–382.
Roberts, J., Kirk, C., & Venis, M. (1990). Hormone Perception and Signal Transduction in Animals and Plants. Symp. Soc. Deptl. Biol. No. XLIV.
Schaller, G.E. & Bleeker, A.B. (1995). Ethylene binding sites generated in yeast expressing the *Arabidopsis* ETR gene. Science 270, 1809–1811.
Schantz, E.M. & Steward, F.C. (1952). Coconut milk factor: the growth promoter substance in coconut milk. J. Am. Chem. Soc. 74, 6133–6135.
Sembdner, G. & Parthier, B. (1993). The biochemistry and the physiological and molecular actions of jasmonates. Ann. Rev. Plant Physiol. Plant Mol. Biol. 44, 569–589.
Seubert, E. (1925). Über Wachstumsregulatoren in der Koleoptile von *Avena*. Zeitschr. Bot. 17, 49–88.

Sharon, A., Fuchs, Y., & Anderson, J.D. (1993). The elicitation of ethylene biosynthesis by a *Trichoderma* xylanase is not related to the cell wall degradation activity of the enzyme. Plant Physiol. 102, 1325–1329.

Sievers, A.F. & True, R.H. (1912). US Dept. Agr. Bur. Plant. Ind. Bull. 232.

Skoog, F. & Miller, C.O. (1957). Chemical regulation of growth and organ formation in plant tissues cultured *in vitro*. Biological Action of Growth Substances, Symp. Soc. Exptl. Biol. 11, pp. 118–131. Cambridge University Press.

Starling, E.H. (1905). Chemical correlation of the functions of the body. Lancet 2, 339–341.

Strugger, S. (1934). Beiträge zur Physiologie des Wachstums. I. Zur protoplasmaphysiologischen Kausalanalyse des Streckungswachstums. Jarhb. Wiss. Bot. 79, 406–471.

Thompson, D.S. & Osborne, D.J. (1994). A role for the stele in intertissue signalling in the initiation of abscission in bean leaves (*Phaseolus vulgaris L.*). Plant Physiol. 105, 341–347.

Varner, J., Ram Chandra, G., & Chrispeels, M.J. (1965). Gibberellic-acid controlled synthesis of α-amylase in barley endosperm. J. Cell. Comp. Physiol. 65, 55–67.

Venis, M. (1977). Solubilization and partial purification of auxin-binding sites of corn membranes. Nature (London) 266, 268–269.

Ververidis, P. & John, P. (1991). Complete recovery *in vitro* of ethylene-forming enzyme activity. Phytochemistry 30, 725–727.

Vriezen, W.H., van Rijn, C.P.E., Voesenek, L.A.L.J., & Mariani, C. (1997). A homologue of the *Arabidopsis thaliana* ERS gene is actively regulated in *Rumex palustris* upon flooding. The Plant Journal 11, 1265–1271.

Went, F.W. (1928). Wuchsstoff und Wachstum. Rec. Trav. Bot. Néerl. 25, 1–116.

Went, F.W. & Thimann, K.V. (1937). Phytohormones. Macmillan, New York.

Wilkinson, J.Q., Lanahan, M.B., Yen, H-C, Giovannoni, J.J., & Klee, H.J. (1995). An ethylene-inducible component of signal transduction encoded by Never-ripe. Science 270, 1807–1809.

Wright, S.T.C. & Hiron, R.W.P. (1969). (+)-Abscisic acid, the growth inhibitor induced in detached wheat leaves by a period of wilting. Nature (London) 224, 719–720.

Yabuta, T. & Sumiki, Y. (1938). Biochemical studies on "Bakanae" fungus. Crystals with plant growth promoting activity. J. Agric. Chem. Soc., Japan 14, 1526.

Yang, S.F. & Hoffman, N.E. (1984). Ethylene biosynthesis and its regulation in higher plants. Ann. Rev. Plant Physiol. 35, 155–189.

Zeevart, J.A.D. (1958). Flower formation as studied by grafting. Mededelingen Landbouwhogesschool, Wageningen, 58, 1–88.

Chapter 7

PUMPS, CHANNELS, AND CARRIERS:
FROM "ACTIVE PATCHES" TO MEMBRANE TRANSPORT PROTEINS

Richard Boyd

INTRODUCTION

From c. 1900 experiments, many of them with mammalian erythrocytes, indicated cells were selectively permeable to a range of low molecular weight neutral solutes. If molecules with similar structures were added to the suspension medium, entry of the test solute might be reduced.

In 1943, Davson and Danielli introduced, in their seminal book "The Permeability of Natural Membranes," the idea that solute permeability was not a generalized property of the plasma membrane but rather was associated with discrete and

Foundations of Modern Biochemistry,
Volume 4, pages 245–265.

ISBN: 0-7623-0351-4

specific areas of the plasma membrane, the "active patches." During the last 50 years, there has been a huge increase in our understanding of the nature, role and significance of many membrane transport proteins; in retrospect, this progress can be seen to have been based on a series of important discoveries, both experimental and theoretical. The importance of these central investigations extends well beyond "transport" studies influencing, for example Peter Mitchell's[1] considerations about bioenergetics and the mechanisms by which chloroplasts and mitochondria function to synthesize ATP. Neurobiology and the molecular mechanisms responsible for the electrical basis of information processing in the nervous system have been profoundly illuminated by studies of solute penetration as have tumor biology, the regulation of cell function by membrane transport, pharmacology and the specificity of action of therapeutic agents acting on membrane targets. The experimental and theoretical insights leading to present understanding of membrane transport are of general importance to many themes in contemporary cell and molecular biology. It is for this reason that these earlier developments need to be accessible to and understood by, the current researcher.

The field has been fortunate in that in addition to the primary literature scattered in a wide range of biochemical journals there has been a bedrock of important monographs. In addition to Davson and Danielli (1943) Stein has written three monographs at approximately ten year intervals. The first, *The Movement of Molecules across Cell Membranes*, (1967) sought to be in the "spirit of Davson and Danielli" and indeed was published as one of a series of volumes of which Danielli was the editor. It expressed the hope that "no more monographs of this type will need to be written. . . . since the current exciting progress in the isolation of specific carrier molecules from cell membranes must surely lead in the near future to the classification of membrane transport as a branch of enzymology." That this was a hope ahead of its time is indicated by the arrival in 1986 of Stein's monumental *Transport and Diffusion across Cell Membranes* which covered kinetic aspects of solute transport across the cell membrane in particular depth. A highly readable introduction to membrane transport, *Channels Carriers and Pumps*, followed in 1990. Despite its *Scientific American* style, this volume had an unusual and important strength, it emphasized what was not known (e.g. "we do not have in a single instance, a clear understanding of how these molecules—channels, carriers and pumps—function"). The strength of these books is matched by some outstanding text books on channels. The classic is Hille's *Ionic Channels of Excitable Membranes* (1984), a masterful account of experimental and theoretical work; additionally Katz's *Nerve, Muscle and Synapse* (1966) and, more recently, Aidley and Stanfield's *Ion channels, Molecules in Action* (1996) give beautiful accounts of developments respectively leading to the "classical" and "modern" view of ion channel structure and function.

CARRIERS

Early Theories

Davson (see a very readable account in Davson, 1989) as a young post-doctoral student working with Jacobs in the USA in the 1930s, studied the effects of heavy metals and alcohols on red cell permeability to nonelectrolytes such as glycerol (lipid-insoluble). The technique used to measure permeability was indirect (isotopically-labelled substrates were not then available) and depended on the hemolysis technique. In this, the time taken for solute-induced water flux to cause cell lysis and hence increase the hemoglobin concentration in the supernatant, was determined experimentally—the more rapid the hemolysis the greater the permeability of the membrane to the solute under study. Many of the chemicals used simply lysed the cells, but some, such as ethyl alcohol or copper ions, specifically inhibited the permeation of the membrane by glycerol. This inhibition was only found in the red cells of those species such as the rabbit, that intrinsically had a high membrane permeability to the glycerol. The temperature dependence of the glycerol permeability was also studied. The conventional expression for this is Q_{10}, the ratio of the membrane permeability of a given solute when studied at two temperatures differing by 10° Celsius. Q_{10} was found to be different for different solutes and higher for those solutes whose permeation could be inhibited by the inhibitors mentioned above. Furthermore Danielli (see appendix A in *The Permeability of Natural Membranes*) theoretically explored the relationship between permeability (P) and Q_{10} for molecules of different molecular mass (M) and showed that for a homogeneous membrane the product of the relationship $(P.M)^{0.5} \times Q_{10}$ would be constant. Experimentally, for a number of highly permeant molecules, this was not the case, showing that the membrane could not be homogeneous. It was this combination of experimental and theoretical studies that attracted Davson and Danielli to the notion, given the inhomogeneity of the membrane permeability to specific solutes, that there were specific processes in the cell membrane able to facilitate the permeability of a chemically defined group of solutes. "Facilitated diffusion" (see Danielli, 1982) was the term originally used to describe this, the idea being that some membrane property was helping specific solutes to cross the membrane.

Out of this concept grew the cardinal idea of "carrier mediated transport." Necessary for this was the development of a more coherent theoretical analysis built upon the general notion of facilitated diffusion. The major insight here came from Widdas who proposed in 1952 that carrier mediated transport would explain earlier data such as the transport of glucose across the sheep placenta, as well as his own observations on glucose entry into the erythrocyte. There were three assumptions made in developing this quantitative hypothesis:

1. Interaction between the transported substrate and the carrier was chemically specific, e.g., the carrier could interact with glucose but not with fructose;

2. The rate of carrier "reorientation" across the membrane was slower than the attainment of equilibrium with the glucose in solution at each side of the membrane; and
3. The carriers moved across the membrane "by thermal agitation" irrespective of whether or not they were loaded with substrate. Movement of the substrate-loaded carrier in this way and return of the unloaded carrier would therefore result in the movement of solute across the membrane—so-called "net solute transport."

Widdas's quantitative model of the "simple" carrier was able to explain a number of earlier observations and to make predictions about what would be observed in more complex experiments on membrane transport. Thus it was a highly productive scientific insight. One of the earlier, apparently anomalous, results that the theory explained was the dramatic fall of membrane permeability found for solutes which were rapidly transported as solute concentration was increased. For example, in the human red blood cell, Wilbrandt and colleagues had previously measured a permeability "constant" for glucose which was 1000 times higher in dilute solutions of glucose than it was in a concentrated solution. This phenomenon, subsequently called "saturation kinetics," is formally equivalent to the fall, as substrate concentration increases, in the proportion of substrate converted to product by a limited amount of an enzyme.

As originally argued by Fisher and Parsons in 1953 (see Kleinzeller, 1996), this aspect of carrier theory was thus an extension to transport studies of Michaelis'[1] enzyme kinetics. Similarly, carrier theory, as with the theory of enzyme kinetics, also provided a basis for understanding why the movement across the membrane of one substrate, e.g. sorbose entry into the red cell, might be inhibited by another substrate e.g. glucose. The theory also allowed the concentration of glucose which inhibited sorbose flux by half to be interpreted as the concentration of glucose needed to half saturate the carrier system. As Widdas later remarked (1988), the ability of carrier theory to integrate these experimental observations "enhanced the credibility of the underlying concept for the mode of transfer."

But the real power of carrier theory was its ability to predict what would be found experimentally if more complex experiments were carried out. For example, Widdas, in his original report (1952), discussed what would be seen if one of the assumptions of the simplest model was relaxed and the mobility of the carriers in their saturated and unsaturated forms were different. If the loaded carrier translocated more rapidly than did the unloaded form, this necessarily would lead to an "accumulation" of carriers at the "trans" face of the membrane; under certain conditions (see below) this would allow "transfer against the gradient" to be seen. The important point is that this property emerges in a way that is quite different from predictions arising from kinetic analysis based on analogy with the catalytic activity of enzymes. For enzymes, there is no vectorial component and hence only a single active site to which substrate can bind. For carrier-mediated membrane transport there is a minimum of two quite distinct "sites," one "outward-facing" and the other "inward-facing." The proportion of carrier present as each of these two forms at any one time depends on the intrinsic symmetry of the carrier and—the critical point—on the chemical environment on either side of the cell membrane.

For example, if substrate is present at a high concentration on the outside of the membrane but absent from the inside and if the loaded form of the carrier translocates more rapidly than does the unloaded form, then carrier will be pushed towards the "inner" face of the membrane and depleted from the "outer" face. This accounts for some of the very striking functional properties of carrier-mediated transport. It is a fundamental concept which readily distinguishes carriers from enzymes although surprisingly it is often overlooked even in otherwise excellent current textbooks. It is also the basis for understanding why the effects both of competing substrates and of inhibitors depend on whether they are present on the same or the opposite side of the membrane as the substrate. Furthermore, the values of the measured kinetic constants used to characterize carrier mediated transport, the equivalent of enzymologists' Km and Vmax, will depend completely on the conditions under which the constants have been determined. These are equilibrium-exchange with the substrate present at the same concentration inside and outside the cell, and the labelled substrate present only on one side, or zero-trans conditions with the substrate present only on the cis side of the membrane, i.e. the side in which labelled substrate is present. Thus under zero-trans conditions, when free carrier reorientation is rate limiting, return of the carrier will be slow and carrier depletion at the external face will occur, leading to a "high" Km. The depleted free carrier at the external face will readily be saturated by a low concentration of substrate at the same side of the membrane. Under equilibrium-exchange, carrier depletion will not occur at the external face so the observed Km for substrate entry must be higher.

In Widdas's seminal paper, other more generalized properties of carrier kinetics are discussed. He notes the need to understand the energy requirements for each of the four possible transitions (i.e., the inward and outward movements of the loaded and unloaded forms of the carrier). He also, with considerable prescience (see below), remarks on the possibility that, for some carriers, the energetics of these transition steps might be influenced, for example, by the ionic environment at either side of the membrane or by the electrical potential across the membrane. "If the four energy requirements are different from one another, the number of carriers at each interface would no longer tend to remain the same, and this fact, as well as the energy requirements, would influence the rate of transfer."

Following this logic mathematically led to the astonishingly accurate prediction that "a competitive system could effect *against the gradient* transfer across the membrane [when] there was no difference in the free energies of the saturated and unsaturated carriers at the two sides, if a competitor was placed at high concentration on one side of the membrane only." This was subsequently demonstrated (Park et al., 1956) in rabbit red cells using xylose as the test sugar and glucose as the competitor. Xylose was "actively" extruded from pre-loaded cells by an inwardly-directed gradient of glucose.

Further demonstrations of this sort of "counterflow" phenomenon for many different substrates in virtually every type of cell have been used as functional "hallmarks" of carrier-mediated transport. Experimental demonstration of this effect precludes transport being mediated either by simple diffusion or by "fixed pores" in the membrane. In reviewing 20 years of experimental work related to the carrier hypothesis, LeFevre (1975) lists a number of key functional properties of carrier mediated transport, all of which have stood the test of the subsequent 20 years. These include saturation of transport with increased substrate concentration and associated phenomena such as competition between similar substrates, high rates of unidirectional transport, and countertransport. Also covered are flux coupling (including trans effects and cotransport), chemical specificity, inhibition by protein-specific reagents, hormonal regulation, and a steep dependence of the rate of transport on temperature (included only to bemoan its common inclusion in textbooks!).

Experimentally, the introduction of radioactively labelled solutes into biochemistry in the early 1950s led to a serendipitous observation that provided additional support for the concept of carrier-mediated transport. The Vmax for glucose–glucose exchange in red cells was three fold higher than that for net entry (Britton, 1957; see Widdas, 1988). This unexpected result was readily explained by a specific version of the carrier model, with glucose being transported by a carrier with a three times greater mobility in its loaded form than in the unloaded state. Later studies on "obligatory exchange"—for example, of chloride influx associated with bicarbonate efflux as carbon dioxide was taken up by, and then hydrated within, the red cell—would extend this idea by postulating that, for such systems, the return across the membrane of the unloaded carrier was so slow compared to that of the loaded form that only exchange could occur, net transport being precluded.

Although the general properties of carrier mediated transport are qualitatively identical in the two cases, the quantitative difference gives the two processes absolutely distinct properties and, thus, distinct biological functions. The obligatory exchanger is unable to perform "net" transport, for example, it is unable to translocate a substrate into a cell. What it can do is exchange, and hence maintain the relative concentrations of, the intracellular and extracellular pools of that group of molecules which are substrates for that specific obligatory exchange transporter (Frohlich, 1989).

Evidence for Two Distinct Forms of the Carrier: Inward and Outward Facing

Martin (1971) found that the transport of choline across the red cell membrane was inhibited by N–ethylmaleimide (NEM), a sulphydryl reagent which covalently modifies cysteine residues in proteins. The extent of the inhibition depended on whether choline was present during the period during which the reagent was reacted with the inhibitor. Also, the effect was different depending on whether the choline was present outside or inside the cells during incubation. Following this lead, a number of workers used irreversible inhibitors to investigate the effect of substrate localization on the interaction between the carrier, which was clearly a protein given the chemical specificity of the reagents, and the inhibitor.

Using fluorodinitrobenzene, FDNB, which reacts with amino and thiol side chains of certain amino acids to form covalent dinitrophenyl derivatives, to inhibit glucose transport, Edwards (1973; see Stein & Lieb, 1986) showed that the extent of inhibition depended on the way in which the experiment was performed. If cells were preincubated with the substrate (glucose) and the rate of reaction with FDNB subsequently measured by determining the fraction of glucose transport which remained, then the pattern of reactivity was found to depend upon the location of the substrate during the pre-incubation step. Thus, compared to the rate of irreversible inhibition in the absence of glucose—when glucose was present outside but not inside the cell—the rate of inactivation by FDNB of glucose transport was accelerated; with glucose inside but not outside the cell, it was slowed. With glucose present at both faces, the rate of inactivation was the same as in the absence of glucose.

Edwards proposed that these experiments must mean that there were two quite distinct forms of the carrier protein, and that the inhibitor acted preferentially with the carrier in its conformation at the inner face of the membrane. Hence, given that the loaded form of the carrier translocated more rapidly than did the unloaded form, the presence of substrate at the external face recruited more carriers into the inner-facing form which readily reacted with FDNB. When glucose was present inside the cell the reverse would hold, accounting for the protective effect of substrate on the inside of the cell. With glucose on both sides of the membrane, carrier distribution would not alter, hence the rate would not differ from that found for the cell when it was not pre-exposed to substrate. This type of experimental approach, using irreversible inhibitors, was subsequently widely used and was important in bringing together the somewhat abstract, experimentally-inferred existence of carriers with the empirical field of protein chemistry. These experiments thus provided strong evidence of a chemical basis for carrier theory.

As well as attempting to inhibit differentially the inward and outward facing forms of a number of separate carrier proteins, attempts were made in the mid-1970s to investigate the substrate specificity of the inward and outward facing conformations of carrier proteins. Of particular note was the finding by Barnett et al. (1973), using a number of different chemically-modified glucose analogues present either inside or outside the cell, that there were measurable changes in the detailed structure of the sugar binding site associated with the conversion of the glucose carrier from an inward to an outward facing conformation. Of considerable interest was the finding that glucose molecules with a substituent at carbon-6 were highly effective inhibitors at the external but not the internal face whereas carbon-1 substituents were well able to react with the inward, but not with the outward, facing form of the carrier. This indicated that during carrier function (i.e. the translocation step of the loaded form of the carrier species) the conformational interconversion of the two forms occurs with the normal substrate (glucose) entering the binding pocket of the protein at the external face C1 first, and leaving from the internal carrier conformation C6 last. These authors were able to draw a model suggesting how such a conformational change of a carrier might produce translocation, based on asymmetry of the binding sites of the inward and outward conformations and the conformational change produced in the protein during transport. In the absence of direct experimental evidence of the structure of the two forms of the carrier protein, this sort of "conformational change" model of carrier function continues to be valuable as providing some sort of idea as to how such a carrier might function.

CHANNELS

The Pore

The simple pore was originally considered in the context of osmosis as an explanation of how water might move across a biological structure (e.g. an epithelium) in the absence of solute movement. This notion introduced by Brucke in the mid 19th century, (see Hille, 1984) was subsequently extended by Boyle and Conway (1941) to consider the selective ionic permeability of the resting cell membrane. Here the explanation for the high membrane permeability to potassium and to chloride, as compared to sodium, was simple. The hydrated ionic radius of sodium was greater than that of either the hydrated potassium or chloride ion, hence the pores postulated to be present in the membrane would act as a molecular sieve and permit the movement of potassium and of chloride but not of sodium.

Mullins in 1959 (see Hille, 1984) suggested that the problem of how a pore could be permeant to the larger hydrated sodium ion and not to the smaller hydrated potassium ion could be solved if the wall of the particular pore in question was able to solvate the permeant ionic species. He proposed that ions shed all but an innermost layer of water molecules on entering the pore, and that the pore walls fit closely, thereby providing solvation. Ions not fitting closely are not sufficiently solvated and therefore cannot enter the pore. Thus, pores could be designed to account for selectivity favoring one particular ion. This paved the way for a more general acceptance in the 1970s that ionic channels are pores, an acceptance well exemplified by the changing use of the term over the last twenty years. In 1970, there were three papers with the word channel in their title, in 1980 well over 100 and in 1990 over 2000. At present, the figure is probably over 3000 titles per year with some 10,000 papers being published annually that refer to ion channels somewhere in their content (Aidley and Stanfield, 1996).

Excitable Cells

A major landmark in the developments leading up to a modern view of electrical activity in excitable cells, was the analysis of the ionic movements responsible for the action potential, work carried out in the late forties by Hodgkin,[25] Huxley[3] and Katz (see Hodgkin, 1993). The aim of these experiments was to gain an insight into the mechanisms responsible for allowing ionic currents to flow across the membrane. To permit the investigation of these currents, which were controlled by the electrical potential difference across the cell membrane, it was necessary to use a feed-back circuit ("voltage clamping") which had been developed by Cole and Curtis some five years earlier (see Agin, 1972).

In fact, Hodgkin and Huxley's experiments did not give insight into the mechanism of ion flow across the membrane as they had hoped. Hodgkin later described how they had aimed to test the hypothesis that ion permeability changes associated with the action potential might be via a "carrier" type of coupled mechanism allowing sodium influx and potassium efflux to be coupled. In fact, the experimental evidence was decisive in that it excluded a carrier mechanism and thus such coupling. The evidence for this included the time course of the ion permeability changes, the inward sodium current preceding the outward potassium current, and the ready dissection and identification of the individual ionic currents both thermodynamically—by altering the electrochemical gradient independently either for sodium or for potassium—and chemically. Sodium currents were modified by substituting external sodium with glucose, and potassium currents by inhibition with tetraethylammonium (TEA).

These "negative results" (for which a Nobel prize was subsequently awarded) thus forced Hodgkin and Huxley to "settle for the more pedestrian aim of providing a mathematical description of the action potential" based on the voltage and time dependence of the individual sodium and potassium conductances. Since they had found the relationship between ionic conductance and membrane potential to be very steep, it seemed likely that the changes of ion permeability were associated with the movement of some electrically charged particles in the membrane. As they could not detect the movement of the charged particles, whereas the currents produced by the ion movements were readily observed, they deduced that there must be many ions moving across the membrane for each moveable membrane charge. They therefore proposed that the ionic currents were localized at particular sites, "active patches," in the membrane which later became known as "voltage-gated sodium" and "voltage-gated potassium channels."

Clear understanding of the differences between channels and carriers only came two decades later with the realization that the turnover number, the rate of maximal ion permeation, for the fastest carrier was at least two orders of magnitude lower that of a typical channel (see Hille, 1984, and below). In retrospect, it is odd to find that the experiments underpinning the two conceptual pillars (channels and carriers) which nurtured membrane transport biology through the second half of the 20th century were undertaken and published (in the same journal) almost contemporaneously. It would, however, take a very long time for the two fields to acknowledge each other, let alone meet as primary sequences of cloned membrane proteins.

The introduction of the "ion channel" notation was due to the experimental discovery of "the long pore effect" by Hodgkin and Keynes[26] in 1955. Radioactive potassium fluxes across the resting cell membrane in both inward and outward directions were measured at different membrane potentials and at different potassium concentrations outside the cell, (i.e. with different electrochemical gradients for potassium entry or exit). The cells, squid giant axons, were treated with dinitrophenol to block ATP production and hence abolish active ion pumping. What was found was qualitatively what would be expected, so that with an outwardly directed electrochemical gradient potassium ions moved out more rapidly than they moved in i.e. the "flux ratio" was greater than 1, whereas with an inwardly directed electrochemical gradient the reverse was true.

Quantitatively, however, there was a greater effect of the electrochemical gradient on the flux ratio than would be predicted from theory for a monovalent cation like potassium. Theory, later developed by Ussing (see below) suggested that the driving force on the ion, the electrochemical gradient, should be directly correlated with the logarithm of the flux ratio, with a slope of unity. The observed slope had a gradient of 2.5, as if potassium ions in the pore had a valency of 2.5. This analysis led Hodgkin and Keynes to propose that the potassium ions must be interacting in a confined space as they move across the membrane. They used the term *channel* to describe such a space. The observation they described is now called the long pore effect (see Hille, 1984) since the essence of the explanation of the phenomenon is that in order to permeate the channel the ions have to line up as if in a single-file.

At the neuromuscular junction, the electrical changes following the action of the neurotransmitter substance acetylcholine on the muscle cell were investigated by Katz and Fatt in 1951. It was reasonable to suppose that the ionic currents passed through channels that were activated by acetylcholine. Both here and with the nerve action potential only the currents produced by flow through some hundreds or thousands of channels at once could be measured.

The 1960s saw the discovery of a number of specific channel-blocking agents. Tetrodotoxin, for example, from the fugu puffer fish, specifically blocks voltage-gated sodium channels. This provided very convincing confirmation that the sodium and potassium channels of nerve axons really are separate from each other. It also allowed potassium channels in nerves to be studied on their own, permitted estimates of channel densities in the membrane to be made, and ultimately proved crucial in the biochemical isolation of sodium channels.

The 1960s and 1970s also saw increasingly sophisticated approaches to the investigation of ion channels in nerve and muscle. Armstrong used quaternary ammonium ions as blocking agents to probe the nature of potassium channels and Hille (1984) measured the permeability of channels to ions of different sizes, and so was able to estimate the minimum dimensions of the channel pore. These indirect methods gave some idea of channel properties and were highly influential in the models they produced, but it was still not possible to examine the activity of the individual channels directly.

Around 1970, some clues as to individual channel action emerged. Katz and Miledi (1972) discovered that the end-plate membrane potential becomes markedly "noisy" in the presence of acetylcholine, and they interpreted this as a series of "elementary events" produced by the opening and closing of individual channels as they bound and released acetylcholine molecules. This led to a series of studies by Stevens and others using techniques of fluctuation analysis to gain information about the size and duration of these events.

A parallel development came from studies on artificial lipid bilayer membranes. Hladky and Haydon (1984) found that when very small amounts of the antibiotic gramicidin were introduced into such a membrane, its conductance to electrical current flow fluctuated in a stepwise fashion. It looked as though each gramicidin molecule contained an aqueous pore that would permit the flow of monovalent cations through it. Could the ion channels of natural cell membranes act in a similar way? To answer this question, it was first necessary to solve the difficult technical problem of how to record the tiny currents that must pass through single channels.

The breakthrough came in 1976 with the development of the patch clamp technique by Neher[28] and Sakmann.[27] They used a glass microelectrode with a polished tip that could be applied to the surface of a cell so as to isolate a small patch of membrane. The voltage across this patch was held steady ("clamped") by a feedback amplifier so that they could measure the currents flowing through the individual ion channels. This technique, in essence a technique allowing the kinetics of conformational changes in a single protein molecule to be followed in real time, proved to be increasingly productive, especially after further technical improvements. In 1991, Neher and Sakmann were awarded the Nobel Prize for this work.

In a record obtained by the patch clamp technique, the channel is closed for much of the time (i.e. no current flows across the patch of membrane that contains it), but at irregular intervals the channel opens for a short time, producing a pulse of current. Successive current pulses are always of much the same size in any one experiment, suggesting that the channel is either open or closed, and not half open (there are exceptions to this rule). The durations of the pulses, however, and the intervals between them, vary in an apparently random fashion from one pulse to the next. Hence the openings and closings of channels are *stochastic* events. This means that, as with many other molecular processes, we can predict when they will occur only in terms of statistical probabilities. But one of the most useful features of the patch clamp method is that it allows observation of these stochastic changes in single ion channels as they actually happen; individual protein molecules can be observed in action.

One of the most significant events to have occurred in the ion channel field in the last decade has been the discovery of the primary amino acid sequences for a number of families of channels. In 1982, the first ion channel sequence (the acetylcholine receptor, AChR, of the nicotinic subtype) was published (Noda et al., 1982; 1984). This ground-breaking work from Numa's laboratory was based on the high density of such channels in the electric organ of the eel which allowed the purification of very small amounts of channel protein using the specific binding of a snake toxin (bungarotoxin).

Direct protein sequencing of the large proteins which constituted the channel was not possible both because of the very limited amount of material available and because of the not unexpected very hydrophobic nature of these membrane proteins. However short sequences from the amino terminal were obtained and these were sufficient to allow synthesis of degenerate oligonucleotide probes which were then used to probe a library of cDNA synthesized from a small quantity of mRNA extracted from the electric organ of the eel. The cDNA from the positive plasmids identified in this way was then sequenced and the primary amino acid sequence of all the subunits so obtained.

A very similar strategy was applied by the same group to obtain the primary sequence of the voltage gated sodium channel from the electric organ of the eel. For this, the initial minimal protein sequences were obtained using the highly specific toxin from the puffer fish, tetrodotoxin, as a probe to allow biochemical fractionation and purification of the native protein. Again oligonucleotide probing of an eel electric organ cDNA library provided the basis for obtaining the sequence of the whole protein. The channel is composed of a large protein of molecular weight approximately 260,000; the 1820 long amino acid sequence has four homologous domains of very similar sequence. Hydropathy analysis suggests that within each of these four domains there are six segments that probably form transmembrane alpha helices, now called S1-S6.

This highly important work initiated the molecular era of channel studies and, coupled with the simultaneous revolution in functional studies of single channels, has ushered in an era of highly productive studies on sequence analysis which is still in full swing. The absence of 3D-structural information on these channels makes the unambiguous interpretation of present work almost impossible, but the combined thrust of current studies has quite revolutionized understanding of channel function, not least because of the ability to see evolutionary conserved motifs within different channels that previously had not been understood at all. In 40 years, channels have changed from being concepts to amino acid sequences of protein in related families.

PUMPS

Although it has been known from the mid-19th century that cations were not distributed equally between the intra and extra cellular compartments of animal cells—the cytoplasm of which contained much more potassium and less sodium than extracellular fluid—for more than 100 years the possibility that this was a static, innate property of cells which they "inherited" during their biogenesis could not readily be excluded. The introduction of isotopes into biological research after 1945 allowed the high permeability of cells to ions such as K^+ and Cl^- to be demonstrated directly.

A study of major importance in establishing the fundamental processes of cation pumping was performed by Hodgkin and Keynes (1955a). The experiments followed the realization that the discovery of the ionic basis of the action potential in nerve cells had raised an embarrassing and obvious question: If sodium ions enter the cell during the upstroke of an action potential and if potassium ions leave the cell during the repolarization following an action potential, how then is ionic homeostasis of the neurone maintained over the time during which a cell conducts millions of action potentials? Using isotopically labelled sodium injected into the giant axon of the squid, the very cell on which the earlier studies (1952) of Hodgkin and Huxley on the action potential had been performed, although in that case by voltage clamping, the efflux of Na^+ from the cell was studied. Hodgkin and Keynes showed most convincingly that metabolism was required to drive this process since exposure of the cell to metabolic poisons (dinitrophenol, cyanide or azide) strongly inhibited the efflux of labelled sodium into the extracellular fluid. The inhibition was reversible on washing the axon free of the metabolic inhibitor.

This simple experiment was important in that it clearly established the key notion that cellular extrusion of sodium ions by the "sodium pump" was coupled to metabolism. Because in this and subsequent experiments of the same sort the electrochemical gradient for sodium was known precisely, and since the fluxes of sodium (and later potassium) both into and out of the cell could be measured independently, this study also laid the groundwork for a theoretical definition of active transport, a theory worked out independently by Ussing in the "flux ratio" equation for transepithelial active transport of ions (see below).

Sodium extrusion by the sodium pump was unambiguously active, being opposed both by the inwardly directed chemical gradient and the polarity (inside negative) of the cell potential. For potassium influx, more careful analysis was required to establish whether an "active" process was involved since although the ion was being moved into the cell against a chemical gradient, the membrane potential favored cation entry. Analysis of the flux ratio, however, established that K^+ was indeed being moved into the cell actively since, quantitatively, the asymmetry of influx over efflux was greater than the observed electrochemical gradient could account for.

Later studies from the same group established, using microinjection, that ATP within the cell was required for this process, and also showed—with axons from which the cytoplasm had been extruded mechanically and replaced by an artificial intracellular ionic medium—that the sodium pump was a property of the axon membrane since such cells were still able to extrude this cation when supplied with an appropriate source of ATP. In their original paper, Hodgkin and Keynes (1955b) found that in addition to needing metabolism, sodium pumping was dependent on the presence in the external solution of potassium ions. They postulated that the influx of potassium was coupled to the efflux of sodium by the pump, and noted that the number of sodium ions extruded was greater than the number of potassium taken up into the cell, something that was to be of considerable interest 20 years later when the question of stoichiometry and electrogenicity of the pumping process was re-examined.

Following this groundbreaking work on the metabolic basis of active cation transport by an excitable cell, Glynn (1956; 1959) showed that a similar process of coupled sodium extrusion and potassium uptake linked to metabolism was present in a non-excitable cell, the human erythrocyte. Fresh red cells, incubated with glucose to allow ATP synthesis, were compared to cold-stored cells incubated without glucose. The former had an additional route for radioactive potassium uptake. This pathway was saturated at low concentrations (mM) of external potassium and was coupled to sodium efflux. The latter was strongly inhibited following removal of external potassium, directly confirming that the two events were coupled.

This pioneering work showed that the sodium pump was not confined to the plasma membrane of excitable cells (the pump is in fact found in virtually all animal cells); it also paved the way for an avalanche of mechanistic studies examining very many aspects of the function of the sodium pump. These included partial reactions, kinetic affinities for the three substrates at both sides of the membrane and their mutual interdependence, ATP utilization and its stoichiometric relationship to cation fluxes. Such experiments were readily performed on the red cell since these are available in large quantities and the development of simple technical procedures such as red cell "ghost" formation by transient hypotonic shock allowed ready access to, and control of, the intra- as well as extra-cellular face of the pump.

Taken together, the work on the giant axon and the erythrocyte laid the groundwork for all future studies on the sodium pump by demonstrating its cardinal features—the simultaneous requirement for intracellular ATP and sodium together with potassium at the extracellular face of the membrane. It was these functional features that allowed a major, but quite fortuitous, leap to be made in 1957 in understanding the biochemical basis of active cation pumping. This was based on some beautifully simple experiments carried out by Jens Skou[23] (awarded the Nobel prize in 1997), then a young Danish surgeon interested in the mechanism of action of local anesthetics.

Skou had established previously that such anesthetics inhibited certain membrane-bound enzymes found in the membrane fraction isolated from nerve homogenates. He decided to study the endogenous ATPase activity as a possible target for future work, initially thinking that it might be the sodium channel. Luckily, since he lacked access to squid giant axons, he used the easily-obtained mixed nerve from the claw of the crab for this purpose. This proved to be a fortuitous choice since, in contrast to mammalian membrane fragments, these did not spontaneously reseal to form closed vesicles. This allowed substrates added to the incubation medium access both to the inner and outer face of the membrane simultaneously. Skou quickly established that the presence of magnesium ions was essential for ATP hydrolysis. He then investigated the effect of sodium in order to see whether this cation, which he realized was involved in action potential production, would increase hydrolytic activity. The results were variable; on some days a substantial stimulation of hydrolysis was observed, above the control with magnesium alone; on other days it was not. It then emerged that this variability had its source in the presence of potassium in variable amounts depending on whether the substrate used was the barium or the potassium salt of ATP. Once this had been realized, Skou quickly showed that addition of potassium in low concentrations relative to sodium produced a pronounced increase in ATP hydrolysis; higher concentrations of potassium not only reversed this activation but also inhibited the low activity found with sodium alone.

The results suggested that there were different sites for the activating effect of Na^+ and K^+ and that the inhibition by K^+ was due to competition for the Na^+ binding site. Skou (1989) has written that had he then known of the literature on the physiological properties of the pump, he would have realized much sooner that this novel enzyme activity was a candidate for the biochemical expression of the membrane sodium pump. But he did not know the literature and his paper, published as "The Influence of some Cations on an ATPase from Peripheral Nerve," did not catch the attention of others working on cation pumps. Nevertheless, the paper's conclusion is unambiguous: ". . . the crab nerve ATPase fulfills a number of the conditions which must be imposed on an enzyme which is thought to be involved in the active extrusion of sodium ions from the nerve fibre."

The key experiment was one which would not be carried out until Skou learned of a paper published in German four years earlier, which showed that the movement of cations across the red cell membrane was inhibited by cardiac glycosides such as ouabain, plant alkaloids used for some 200 years in the therapy of heart failure. In 1957, Skou was unaware of this important finding, and wrote later that "he had not done the crucial experiment to show Na/K ATPase as the transport system." When done, the experiment was decisive; ouabain inhibited the ATPase activity exactly as it did the cation fluxes. This led to a flurry of activity in many biochemical laboratories and allowed Skou (nine years after his original publication) to write a review in which he concluded that the enzyme fulfilled the requirements for a system responsible for active transport of Na^+ and K^+ across the cell membrane. Thus the Na/K ATPase had the following properties:

1. It was membrane bound.
2. Its affinity on the cytoplasmic side was greater for Na^+ than for K^+.
3. Its affinity on the extracellular side is greater for K^+ than for Na^+.
4. It hydrolyzed ATP.
5. The rate of hydrolysis depended on cytoplasmic Na^+ and extracellular K^+.

6. The enzyme was found in all cells that had coupled active transport of Na^+ and K^+.
7. There was a quantitative correlation between the rate of cation transport in the intact cell and the magnitude of the hydrolytic activity in membrane preparation prepared from such cells.
8. The enzyme was inhibited by cardiac glycosides with a Ki quantitatively similar to that for inhibition of the cation fluxes.

The availability of a specific inhibitor (ouabain) allowed extensive investigation of the role of the sodium pump in cell function. For example Whittam and Ager (see Parsons, 1975) showed that approximately one third of resting oxygen utilization of most tissues *in vitro* was inhibited by addition of ouabain to the incubation medium suggesting that sodium pumping was highly significant for the overall energy flow and ATP utilization by cells.

Secondary Active Transport

With the discovery of the biochemical basis of the sodium pump and the realization that this was a major source of energy utilization by virtually all animal cells, there came a subsequent flowering of many areas of cell physiology. The developments were based on the implicit notion that the energy available in the ion gradients developed by the sodium pump could be used for many purposes in addition to the generation of nerve action potentials. Such experiments became relatively easy to perform given the availability of the cardiac glycosides as specific inhibitors of active sodium pumping. Additionally, the conceptual framework between 1960 and 1980 changed in a fundamental way as the chemiosmotic principles of Mitchell were seen to be broadly applicable to areas well beyond the field of classical bioenergetics (see Nicholls and Ferguson, 1993). The idea of the interconvertability of energy flows by flux coupling was also underpinned thermodynamically by Kedem and Katchalsky (1958).

One example of such a secondary active transport system (in which the flow of sodium ions into the cell down its electrochemical gradient was coupled to the movement of a solute in the opposite direction) was significant in understanding the therapeutic action of the cardiac glycosides themselves. Cardiac glycosides at low concentrations increase the force of cardiac muscle contraction although they are highly toxic at higher concentrations. The force of contraction was known to depend on the intracellular calcium ion concentration, being increased by a rise in Ca_i^{2+} which activated the contractile protein machinery. How could a small decrease in the rate of sodium extrusion from cells cause increased contraction? The answer (see Eisner et al., 1986) lay in flux coupling via an antiporter, the Na/Ca exchanger. It was intellectually highly reassuring to find that interactions between two membrane transport systems (Na/K ATPase and Na/Ca antiport) could provide the key to understanding how an important and widely used drug had its therapeutic effect.

Historically important as an example of flux coupling, and one that was investigated in detail becoming a paradigm for coupled transport, was the sodium coupled glucose transport system of the small intestine and kidney (see below). This was a symport (or co-transport) rather than an antiport, normally carrying glucose into the cell coupled to a flow of sodium ions in the same direction.

Sodium Glucose Cotransport

Although there were reports in the literature in the first half of the 20th century of the inhibitory effects on glucose absorption of sodium removal from the medium in the lumen of the intact small intestine, the experimental demonstration by Csaky and Thale (1960) of this effect on the small intestine *in vitro*, and the demonstration that glucose absorption was inhibited by ouabain, rekindled interest in this phenomenon. Cotransport was originally suggested by Crane and Bihler in an appendix to a symposium report on the specificity of intestinal glucose transport. Remarkably, at the same symposium on Membrane Transport and Metabolism held in Prague in 1960, not only had Mitchell discussed early ideas on chemiosmosis and group translocation, but both Keynes and Skou had reviewed their own contributions to active sodium transport. Out of this milieux, secondary active transport was delivered.

The essence of the original scheme was simple. Glucose was co-transported into the cell and sodium ions extruded at the apical membrane in contact with the intestinal lumen, by two separate transporters which were functionally interconnected. Crane has discussed (see Crane, 1975) how this notion arose during discussion on the beach at Woods Hole, and the importance of sand for drawing and erasing as many models as you wished! In essence, the model was based on the same finding as that of Csaky but Crane and his colleagues had used the technically superior "everted sac" developed in the mid 1950s in Sheffield by T.H. Wilson (Wilson and Wiseman, 1954), to examine the effects of ionic substitution in the mucosal solution on sugar transport into the serosal solution. The normal accumulation of glucose within the serosal compartment to a higher concentration than was present in the mucosal solution (i.e. against a concentration gradient) was abolished by luminal substitution of sodium ions with potassium.

The model nicely explained this experimental finding, but at this early stage was really no more than a working hypothesis, and was only one of a number of plausible possibilities. For example, coupling glucose influx to potassium efflux rather than sodium influx would have explained the data adequately. However, as Crane and his colleagues went on to demonstrate with a series of well designed experiments, the fundamental idea of what became known as the Crane hypothesis remained plausible while other obvious hypotheses were excluded (Crane, 1975).

Acceptance of the hypothesis was slow because there remained a fundamental simplification built into the model which was misconceived. The Crane hypothesis emerged from experiments carried out in a "biochemically oriented" laboratory.

(Crane himself had trained with Cori in St. Louis.) Understanding active transport required underpinning based on adequate theory. As soon as co-transport of glucose with an ion was proposed, it required a theory involving the use of electrochemical rather than chemical gradients for that ion. As late as 1973, Kimmich was able to write a full review entitled "The Inability of the Sodium Gradient to Account for Active Intestinal Sugar Transport." His argument was based on a series of elegant experiments using isolated enterocytes, experiments which showed that it was possible to find some degree of sugar accumulation in such cells under conditions where it could be shown that the chemical gradient of sodium (intracellular to extracellular) had been reversed experimentally. This seemed quite incompatible with the key idea of the Crane model, namely that the movement of sugar "uphill" was driven by the movement of sodium "downhill." What was missing was the strict application of "Ussing theory"—the theory developed to analyze the extent to which the movement of an ion across a biological membrane could or could not be accounted for by prevailing external physicochemical forces. Only if the observed movements of the solute in question were incompatible with the observed driving forces would it be reasonable to invoke "active transport."

Ussing was interested in the transport of ions across the frog skin. The Ussing equation (Ussing, 1980) was derived after Ussing had realized that many of the biological complexities which might complicate a complete analysis would be removed if the bi-directional fluxes across a membrane or an epithelium were directly measured. The ratio of these two experimentally determined fluxes could then be compared to the prevailing electrical and chemical forces, also determined experimentally, i.e. the chemical concentrations of the relevant species on either side of the membrane and the potential difference across the membrane or epithelium. In essence, the electrical driving force had been overlooked in the Crane model and it was this that made, for example, Kimmich's data incompatible with the original model. In fact, in the experiments with the reversed sodium gradient the high internal sodium concentration strongly activated the Na/K ATPase (see above) resulted in the generation of a substantial outward current. Thus, in these unusually hyperpolarized cells, only the inward chemical—but not the electrochemical—gradient for sodium was reversed, accounting for the observed maintained accumulation of glucose.

An experimental advance which clarified the status of the Crane model was the introduction of isolated membranes to the study of membrane transport. These were first employed in intestinal transport of sugars by Isslebacher's laboratory. They used the method, having seen its value in the field of mitochondrial bioenergetics, (see Murer and Kinne, 1980). Membranes from a complex epithelium, for example from the brush border face of the intestinal or renal epithelium, are isolated and purified. Such membranes reseal spontaneously following exposure to high shear forces (e.g. passage through a fine needle) and then provide a very useful experimental system to analyze driving forces controlling transport. With such a system, the experiment can be set up with precisely defined chemical conditions on the inside and the outside of the membrane vesicle and with specified electrical potentials created by ion diffusion potentials (e.g. using ionophores such as the potassium ion specific antibiotic valinomycin). Moreover, since the vesicles lack cytoplasm, there is no ATP to drive primary, ATP-driven, active transport.

From the results of experiments with these biochemical techniques, and incorporating a more analytical view of the relevant driving forces which needed to be controlled, several laboratories rapidly established the nature of intestinal and renal cotransport of not only glucose, but of many other solutes which were transported by secondary active transport in an analogous way. Although flawed in part, the Crane hypothesis was instrumental in driving forward a revolution in the way that trans-epithelial solute transport was viewed. Compare for example Wilson (1960) with Giebisch (1979) to see the extent to which the cotransport of solutes with sodium ions "arrived" during those 20 years.

Sodium glucose cotransport remained an experimental system at the forefront of the next paradigm, the cloning of transporters. In a seminal paper Hediger and co-workers (1987) described the isolation and sequence of a cDNA obtained from a rabbit small intestinal mRNA library which encoded the SGLT1 protein. The basis of this experiment was the chromatographic fractionation of mRNA and the purification and isolation of the relevant product by bioassay of its transport function (i.e. the measurement of sodium dependent glucose transport with the expression system microinjected *Xenopus* oocytes). The membrane protein encoded for by the cDNA was 615 amino acids long and had multiple membrane spanning regions. This protein remains at the forefront of structure function studies on membrane transporters and is on the way to becoming one of the first transport proteins from an animal cell membrane to have its 3-D structure solved.

CONCLUSIONS

The idea that membrane transport could usefully be considered as occurring through the participation of one of three different categories of integral membrane proteins (channels, carriers, pumps) has been extremely important conceptually to the development of thinking in this field over the last 50 years. One of the ironies of very recent work (e.g. Fairman et al., 1995) is that, at least in a number of instances, this classification is likely to be incorrect. Thus, there are cloned proteins which when expressed (e.g. in *Xenopus* oocytes) undoubtedly can be both carriers and channels. An example is EATT4 which indubitably is a high-affinity carrier for anionic amino acids, yet which is also a chloride channel, physiologically gated by glutamate at synapses.

At present, at least 10 different proteins have been shown to behave in this bi-functional way, but how rare this property is for transporters generally is, at present, unclear. The examples reinforce the dictum of Jacques Monod—"Nature is a tinkerer." While it is indeed remarkable that a protein can be engineered to have more than a single transport mode, the discovery of this phenomenon does not diminish the value of the classification anymore than would the nature of the genetic code be discredited by the discovery of overlapping open reading frames in the mitochondrial genome.

What is clear from the nature of the development of the subject so far is the critical importance of experimentalists in one area being alert to, and sufficiently knowledgeable about, advances, both theoretical and empirical, in other areas. When, this has not happened, general understanding has either been delayed or distorted.

REFERENCES

Agin, D.P. (1972). Perspectives in Membrane Biophysics: A Tribute to K.C. Cole. Gordon & Breach, New York.

Aidley, D.J. & Stanfield, P.R. (1996). Ion Channels: Molecules in Action. Cambridge University Press, Cambridge, UK.

Barnett, J.E.G., Holman, G.D., & Munday, K.A. (1973). Structural requirements for binding the sugar-transport system of the human erythrocyte. Biochem. J. 131, 211–221.

Boyle, P.J. & Conway, E.J. (1941). Potassium accumulation in muscle and associated changes. J. Physiol. Lond. 100, 1–63.

Crane, R.K. (1975). The gradient hypothesis and other models of carrier-mediated active transport. Rev. Physiol. Biochem. Pharmacol. 78, 101–149.

Csaky, T.Z. & Thale, M. (1960). Effect of environment on intestinal sugar transport. J. Physiol. Lond. 151, 59–65.

Danielli, J.F. (1982). Experiment, hypothesis and theory in the development of concepts of cell membrane structure 1930–1970 (Martonosi, A.N., Ed.), Vol. 1., pp. 3–14. In: Membranes and Transport. Plenum, New York.

Davson, H. (1989). Biological membranes as selective barriers to diffusion of molecules (Tosteson, D.C., Ed.), pp. 15–50. In: Membrane Transport: People and Ideas. American Physiological Society, Bethesda, MD.

Davson, H. & Danielli, J.F. (1943). Permeability of Natural Membranes. Cambridge University Press, Cambridge, UK.

Eisner, D., Allen, D.G., & Wray, S.C. (1986). Birthday present for digitalis. Nature (London) 316, 674–675.

Fairman, W.A. et al., (1995). A glutamate transporter which also has properties of a gated chloride channel. Nature 375, 599–603.

Frolich, O. (1989). Antiporters. Curr. Opin. Cell Biol. 1, 729–734.

Giebisch, G.H. (1979). Membrane Transport in Biology. Springer.

Glynn, I.M. (1956). Sodium and potassium movements in human red cells. J. Physiol. Lond. 134, 278–310.

Glynn, I.M. (1959). Sodium and potassium movements in nerve, muscle and red cells. Intern. Rev. Cytol. 8, 449–481.

Hediger, M.A., Coady, M.J., Ikeda, T.S., & Wright, E.M. (1987). Expression cloning and cDNA sequencing of the Na^+/glucose cotransporter. Nature (London) 330, 379–381.

Hille, B. (1984). Ionic channels of excitable membranes. Sinauer Assoc., Sunderland, MA.

Hladky, S. & Haydon, D.A. (1984). Ion movements in gramicidin channels. Curr. Topics in Membranes and Transport 21, 327–372.

Hodgkin, A.L. (1993). Recollections and Reminiscences. Cambridge University Press, Cambridge, UK.

Hodgkin, A.L. & Huxley, A.F. (1952). A quantitative description of membrane current and its application to excitation and conduction in nerve. J. Physiol. Lond. 117, 500–544.

Hodgkin, A.L. & Keynes, R.D. (1955a). Active transport of cations in giant axons from *Sapia* and *Loligo*. J. Physiol. Lond. 128, 28–60.

Hodgkin, A.L. & Keynes, R.D. (1955b). The potassium permeability of a giant nerve fibre. J. Physiol. Lond. 128, 61–88.

Katz, B. (1966). Nerve, Muscle and Synapse, 2nd ed. McGraw-Hill, New York.

Katz, B. & Fatt, P. (1951). An analysis of the end-plate potential recorded with an intracellular electrode. J. Physiol. Lond. 115, 320–370.

Katz, B. & Miledi, R. (1972). The statistical nature of the acetylcholine potential and its molecular components. J. Physiol. Lond. 224, 665–700.

Kedem, O. & Katchalsky, A. (1958). Thermodynamic analysis of the permeability of biological membranes to non-electrolytes. Biochim. Biophys. Acta 27, 229–246.

Kepner, G.R. (1979). Benchmark Papers in Human Physiology, Vol. 12. Do.vden, Boston.

Kimmich, G.A. (1973). Coupling between Na^+ and sugar transport in small intestine. Biochim. Biophys. Acta 300, 31–78.

Kleinzeller, A.(1995). A history of Biochemistry: exploring the cell membrane; conceptual developments. Comprehensive Biochemistry (Neuberger, A. & van Deemen, L.L.M., Eds.), Vol. 39. Elsevier, Amsterdam.

Le Fevre, P.G. (1975). The present state of the carrier hypothesis. Curr. Topics in Membranes and Transport. 7, 109–215.

Martin, K.A.C. (1971). Some properties of an SH group essential for choline transport in human erythrocytes. J. Physiol. Lond. 213, 647–664.

Murer, H. & Kinne, R. (1980). The use of isolated membrane vesicles to study epithelial transport processes. J. Memb. Biol. 55, 81–95.

Neher, E. & Sakmann, B., Eds. (1983). Single Channel Recording. Plenum, New York.

Nicholls, D.G. & Ferguson, S.J. (1992). Bioenergetics: an introduction to the chemiosmotic theory. 2nd ed. Academic Press.

Noda, M. et al. (1982). primary structure of the α subunit precursor of *Torpedo* acetylcholine receptor deduced from cDNA sequence. Nature (London) 299, 793–797.

Noda, M. et al. (1984). Primary structure of *Electrophorus* sodium channel deduced from c DNA sequence. Nature (London) 312, 121–127.

Park, C.R., Regan, H., & Morgan, E.H. (1956). Trans-Stimulation of Xylose Uphill by Glucose Gradients in the Red Cell. CIBA Foundation Symp. 9, 240–260.

Parsons, D.S. (1975). Multimembrane systems: membrane function and energetics in intestinal mucosal epithelium (Parsons, D.S., Ed.), pp. 172–192. In: Biological Membranes. Clarendon Press, Oxford, UK.

Skou, J. (1989). The identification of the sodium pump as the membrane bound Na/K ATPase. Biochim. Biophys. Acta 1000, 435–438.

Stein, W.D. (1967). The Movement of Molecules across Membranes. Academic Press, New York.

Stein, W.D. (1990). Channels, Carriers and Pumps: An Introduction to Membrane Transport. Academic Press, New York.

Stein, W.D. & Lieb, W.R. (1986). Transport and Diffusion across Cell Membranes. Academic Press, New York.

Ussing, H.H. (1980). Life with tracers. Annu. Rev. Physiol. 42, 1–16.

Widdas, W.F. (1952). Inability of diffusion to account for placental glucose transport in the sheep and the consideration of the kinetics of a possible carrier transfer. J. Physiol. Lond. 118, 23–39.

Widdas, W.F. (1988). Old and new concepts of the membrane transport for glucose in cells. Biochim. Biophys. Acta 947, 385–404.

Wilson, T.H. (1960). Intestinal Transport. Saunders, Philadelphia.

Wilson, T.H. & Wiseman, G. (1954). The use of sacs of everted small intestine for the study of the transference of substances from the mucosal to the serosal surface. J. Physiol. Lond. 123, 116–125.

Chapter 8

BIOCHEMISTRY THEN AND NOW

Margery G. Ord and Lloyd A. Stocken

INTRODUCTION

A biochemist active in the 1900s or 1950s, contemplating Biochemistry today, would be struck, first of all, by the advent and speed of the molecular biological revolution and by the impact and universal application of molecular genetic techniques (Ord and Stocken, 1996; Witkowski, 1996). For the pioneers, biochemistry was the study of the properties and metabolism of the components of living cells in physico-chemical terms (see Ord and Stocken, 1995). Now, with our increasingly detailed study of the genome, its expressed proteins and the regulation of its transcription, it is becoming possible to understand the behavior of the cell as a whole. We can also interpret how the cell responds to changes in its environment and how the consequential changes in the cytoplasm modulate and regulate events in the nucleus.

Foundations of Modern Biochemistry,
Volume 4, pages 267–280.

ISBN: 0-7623-0351-4

TECHNIQUES

In the 19th century physiologists began what were to prove enormously fruitful investigations using nerve–muscle preparations which are organized systems able to be maintained in a viable state for considerable periods. It was to be many years before individual cells could be equally successfully examined, but from earliest times the range and complexity of natural products challenged and extended the analytical and synthetic skills of organic chemists. Before 1940, biochemistry was primarily concerned with the identification of the components and properties of cytoplasm, with sugars and proteins as the major constituents of interest. Enzymes were the most obvious class of proteins and their general properties were intensively studied. Purification and analysis of proteins, particularly enzymes, were the most common biochemical procedures and were essential elements in training undergraduate and graduate students. The chemical background of many of the pre-1960 biochemists underlined this approach. Quantitative investigations were natural and enzymes an obvious area for study.

The technical scene could hardly have changed more. While purification and structural determinations still feature in major areas of research (see Campbell, 1996; Dwek, 1996), analyses of all sorts, often fully automated, are now performed on μg (or smaller) amounts of material. In many cases the existence of hitherto unsuspected proteins is indicated by DNA sequencing.

There have been other major developments in analytical techniques—and from the application of these methods to more diverse living materials. In animal biochemistry, at least into the 1950s, nutritional observations were made on intact animals, usually rodents. Metabolic studies were performed using tissue slices but mostly disorganized cell preparations. Before the second World War, attempts were being made to fractionate cell preparations; relatively simple procedures did not become available until differential centrifugation was introduced in the 1950s. Although the advantages of cell culture were appreciated, it was not until the 1960s that conditions were established for the maintenance of differentiated somatic cell cultures, especially the need for the appropriate substratum.

Additionally, particular species have been introduced for special purposes. *Drosophila* came onto the scene for classical genetic studies early in this century (Morgan,[29] 1911). *Caenorhabditis elegans*, a flatworm, was introduced by Brenner in the 1970s. It is a relatively simple multicellular organism whose development and behavior proved highly suitable for genetic analysis. The success of this imaginative innovation is now well established. From the 1950s Fischberg and Gurdon in the United Kingdom and Brown and Dawid in the United States explored the use of microinjection into *Xenopus laevis* eggs and oocytes for a variety of investigations into different aspects of nuclear function—including the demonstration of mRNA, rDNA amplification, the properties of Maturation Promoting Factor, and the existence of proteins with Nuclear Locating Signals. A very recent development has been the generation by Wilmut et al. (1997) of a cloned lamb, "Dolly," by transplanting nuclei from the udder of an adult ewe into enucleated, metaphase-arrested, eggs.

The use of cell cultures and *Xenopus* eggs accelerated the development of microanalytical procedures. The (relative) ease with which experienced operators can microinject materials into eggs (and now into many other types of cell) allows studies of single cells using fluorescent or photoluminescent markers for proteins or Ca^{2+}. The primitive microscopes were continually improved during the 19th century, and the introduction of electron microscopy by the 1950s allowed the fine structure of cells to be "seen" for the first time. Time-lapse photography coupled with interference techniques emphasized the dynamic behavior of cells and their organelles. Techniques have now culminated in microscopes with confocal optics (see White et al., 1987) and digitized imaging. Changes over time can be followed automatically, allowing sites of reactions to be localized and, for example, movement of Ca^{2+} through the cell to be tracked. The presence of intracellular fluorescent labels has facilitated the use of Fluorescence Activated Cell Sorting (FACS), so permitting populations of cells with defined properties and/or at defined stages in the cell cycle, to be selected for further experiments. Antibodies with attached magnetic microbeads offer an alternative means of selection, most frequently used for selecting cells in the hemopoietic system (Vartdal et al., 1986).

Initially plant cell culture techniques progressed more slowly than those for animal cells, as did procedures for the isolation of plant cell organelles. Technical difficulties have now largely been overcome and culturing plant embryos has now found important commercial applications.

Many other technical changes and innovations could be mentioned. During the period when enzyme investigations were at their height the importance of pH as a controlling influence of their activity was recognized. Sorensen's technique for estimating pH depended on the colors of certain dyes which changed with hydrogen ion concentration. In pre-war days sets of capillaries containing these dyes (capillators) were available commercially, covering the pH range 1–14. The distinguished inorganic chemist Sidgwick was once heard to remark "Biochemists say their prayers to pH." pH meters and glass electrodes were introduced by about 1940 and have become fully automated.

A major innovation was the introduction of radioactive tracers by Hevesy[1] in 1923. The method could not be fully exploited until after the second World War when isotopes such as ^{32}P, ^{35}S, ^{14}C and ^{3}H became generally available and procedures for their detection and estimation had been developed. Between 1945 and 1955, details of all the major metabolic pathways were confirmed and many new pathways were discovered. By the 1960s, the need for more precise localization of reactions required greater amounts of radioisotopes, especially ^{3}H thymidine and ^{32}P, to be employed in each experiment, potentially threatening both the health of the operator and of the preparation. Through the 1980s alternative labelling methods of great sensitivity have been developed, such as the use of biotinylated reagents whose presence can subsequently be detected using fluorescently labelled avidin-antibody techniques.

SURPRISES

The Cytoplasm

Edouard Buchner's[1] experiments with yeast extracts (1894–1897) heralded the start of systematic analyses of metabolic pathways. It was not inherently obvious that reactions in disorganized tissue preparations would reflect actual intracellular events. Disrupting cell membranes and destroying cell ultrastructure (Peters,[30] 1962; see Peters, 1963) with the consequential release of hydrolases could have so interfered with particular steps in the pathways that their resolution would have been precluded. Physiologists in particular were sceptical of the relevance of results obtained with cell homogenates, querying their relation to events in cells *in vivo*. In fact, disorganizing the cell structure did remove some normal constraints. Mechanisms by which many cytoplasmic reactions are regulated were disturbed, and did not become apparent until gentler procedures were devised to isolate and study organelles, particularly mitochondria, chloroplasts and lysosomes. Later, Earl Sutherland,[3] Cuatracacas, and others identified the cell membrane as the site of action of many hormones, and showed that its disruption significantly affected rates of intracellular reactions (see Irvine, 1997).

Much uncertainty reigned over the nature of proteins, the best known of which were hemoglobin, the digestive enzymes, and later, insulin. Properties of individual amino acids and the peptide bond were studied early in this century, but it was not until urease was crystallized by Sumner[1] in 1926, followed by the isolation of other "pure" enzymes, that it was finally accepted in the 1930s that enzymes were proteins and that their catalytic properties were not the function of some adsorbed low molecular weight entity. Somewhat later, towards the end of the 1930s, coenzymes were isolated and their roles established.

Studies on the mechanisms of reactions in organic chemistry in the 1930s and 1940s were extended to the problems of enzyme catalysis. As protein structures emerged in the 1950s, attention was focused on the sites and functions of the different amino acid residues. It is interesting to note that in 1956 E. Fischer wrote, "I hope that in speaking of what is known today of the active sites of enzymes I have succeeded in creating in your minds the utmost confusion. If so, I have given you a truthful account of our present knowledge of the structure of enzyme loci." Demonstrating the participation of amino acid residues in partial covalent reactions in the course of enzyme activity greatly clarified our understanding of biological catalysis.

The discovery since the 1970s of large numbers of proteins that are neither primarily cytoplasmic nor extracellular structural proteins nor enzymes, but regulatory factors present individually in small amounts in cells, and exerting their effects at the genome, has revolutionized our views on the function of proteins. Many, numerically, have no catalytic function, but make weak specific, often multiple, interactions with other proteins and/or cell components.

The demonstration of the importance of inorganic phosphate in carbohydrate breakdown, the discovery of ATP (1928–1929) and the successful elucidation of the mechanisms by which P_i was incorporated into glycolytic intermediates profoundly influenced interpretation of mitochondrial and photosynthetic phosphorylation (see Ord and Stocken, 1995; Whatley, 1997). The importance of the proton motive force was aggressively disputed for several years by "traditional" biochemists (see Ferguson, 1997).

Another field in which obscurity reigned until the early 1950s was the pathway for carbon dioxide fixation in higher plants. This was not resolved until the seminal experiments of Calvin[32] (1955) with $^{14}CO_2$. One of us can still remember her school chemistry instructor in the early 1940s discussing the possible polymerization of formaldehyde, CH_2O, to $C_6H_{12}O_6$ (Whatley, 1997).

A dramatic and unexpected consequence of the introduction of isotopic labelling was the demonstration by Schoenheimer[1] and Rittenberg[1] that most body constituents were in continuous turnover (Schoenheimer, 1942). While limb regeneration in *Amphibia* had been known for many years and the capacity for compensatory hyperplasia by, for example, rodent liver, the finding that most tissue constituents were constantly being broken down and resynthesized greatly stimulated the study of synthetic pathways, and the processes by which intracellular structures, and organs, are formed. Labelling studies established the origins of the different molecules. Much more work was needed (see Siekevitz,[2] 1996) to show how proteins were programmed and assembled.

The Genome

The relative stability of DNA compared to RNA and the demonstration by Gurdon (1958; see Gurdon, 1991) that nuclei from some differentiated tissues could, after serial transfer, promote the development of enucleated eggs to adult frogs, emphasized the stability of the genome and thus the suitability of DNA as the carrier of genetic information. The extreme conservation of proteins such as histones was shown by the 1960s and has been confirmed now by the homologies in histone genes through the plant and animal kingdom. Since the 1920s, ultraviolet and ionizing radiation have been known to induce mutations. That the genome could be more radically changed by introducing substantial pieces of DNA and even intact genes, was indicated from the experiments of Lederberg and Tatum (1946) on conjugation in *E. coli*, and the transfer of the sex factor from F^+ to F^- cells. Further experiments identified Hfr strains of *E. coli* as those in which genetic material had been incorporated into the genome of the recipient.

Further examples of plasmids (autonomously replicating extranuclear DNA) emerged through the 1950s and 1960s (see Day and Poulton, 1996); those for drug resistance factors soon achieved notoriety. Also unexpected was the discovery of DNA in mitochondria and chloroplasts, its probable prokaryotic origin, and the possibility of gene exchange between the organelles and the nucleus (see Day and Poulton, 1996; Palmer, 1997).

In contrast to the apparent stability of the genome, was the demonstration in the 1960s that amounts of rDNA in Amphibian nuclei were amplified during oogenesis (see Ord and Stocken, 1996); later, the excess DNA is lost. Conversely, during lymphocyte development, DNA sequences are eliminated and the gaps rejoined (see Steiner, 1998, this volume). Other surprises include the existence of multiple copies of genes for products such as histones which must be synthesized during brief periods of the cell cycle, and repetitive, non-coding, DNA apparently idiosyncratically distributed in different species. Also unexpected was the finding in eukaryotes of "split" genes where the transcribed RNA requires intron excision and splicing before the correct mRNA is produced. Even less expected was the discovery of RNA-based catalysis and the emergence of the "RNA world" (Cech,[7] 1986; Csermely, 1997; Zhang and Cech, 1997; Schuster, 1998).

The ability since the 1980s to manipulate and sequence DNA (see Witkowski, 1996) and large RNAs has consequences which will continue to emerge well into the next century. Genomic sequences for *Hemophilus influenzae* (1.7 Mbp), *Escherichia coli* (4.7 Mbp), the archaebacterium *Methanococcus jannaschii* (1.8 Mbp), *Saccharomyces cerevisiae* (12 Mbp) and *Bacillus subtilis* (4.2 Mbp) are now available, that for *Caenorhabditis elegans* is expected and those for mouse and man are progressing rapidly. It is therefore possible to identify very large numbers of genes whose protein products are already known, or can be predicted from the gene sequence. Additionally, in *S. cerevisiae* there may be up to 2000 "orphan genes" with no homologies with genes of presently known function (Clayton et al., 1997). Site-directed mutagenesis and other procedures allow specific and localized gene manipulations. Besides the obvious potential applications for human therapeutics and agriculture, there have been unexpected "spin-offs" for evolutionary theory and molecular taxonomy (see Maynard Smith, 1993). As well as the peculiarities of their environments, rRNA sequences of halophilic and thermoacidophilic bacteria show that these are members of a new kingdom—the archaebacteria—distinguishable both from other better known *prokaryotes*, the *eubacteria*, and from *eukaryotes*.

Some remaining problems are obvious. Plausible explanations can be offered for the codon degeneracy of the more common amino acids and for the emergence in, for example *mitochondria*, of exceptions to the universality of the genetic code. When polycistronic messages and gene clusters were first observed in *prokaryotes* it was easy to explain these as economical regulatory mechanisms. Overlapping genes in ϕX174 (Smith et al., 1977) were remarkably convenient for a small virus. Gene order in *eukaryotes*, their chromosomal allocation, the variability of chromosome number between closely related species, and the modulation of chromosome structure to allow gene expression, are not understood. More fundamentally, our increasing knowledge of the machinery underlying the central dogma in bacteria, archaea and eukarya is raising questions about the Universal Ancestor/Cenancestor (Edgell and Doolittle, 1997; Olsen and Woese, 1997; Schuster, 1998) to complement those raised earlier by Oparin (e.g., 1957) and Bernal (1967).

Integrated Systems

The diversity of protein structures has already been mentioned. It is a property of proteins to bind other molecules—water, ions, neutral solutes such as sugars, lipid soluble molecules, other proteins and nucleic acids. Binding sites may be on the surface of the molecule, in a crevice or, with membrane bound proteins, projecting into the extracellular fluid or cytoplasm. Over time, many binding sites have become highly specific, with binding constants $\geq 10^6$ or indeed $> 10^9$. The structure of many such sites are now known (Campbell, 1996).

It was originally believed that protein shape only depended on its amino acid sequence. As the mechanism for protein synthesis was clarified in the 1950s it was suggested that the shape was assumed as the nascent peptide rolled off the ribosome. But there were already difficulties with this notion. Even before World War I physiologists had observed the sigmoid shape of the oxygen saturation curve for hemoglobin and had suggested "cooperativity" in the binding—binding of the first O_2 molecule promoted the binding of the subsequent molecules. Koshland (1958) proposed that substrate binding to an enzyme often produced conformational changes in the protein ("Induced fit"). The discovery of allosterism (Monod,[3] Changeux and Jacob,[3] 1963) showed that ligands other than substrates could also affect the dynamic structure of proteins. Very extensive studies have now been made of protein folding (see for example, Jaenicke, 1987; Creighton, 1990). It was not until the late 1970s that it became apparent that many proteins require other proteins as assistants (chaperonins) with which the nascent peptide chain becomes associated, subsequently folding to achieve its three-dimensional form (Laskey et al., 1978; Ellis, 1987). Perhaps the most dramatic demonstration of the importance of "shape" is provided by the conversion of the natural prion protein PrP^c from its soluble α helical form to an insoluble β-pleated sheet, PrP^{sc}, with the consequent production of infectivity in the spongiform encephalopathies (see Horwich and Weissman, 1997).

Contemporary analytical methods (see Campbell, 1996) permit the ligand binding sites and the conformational changes they produce on the native protein to be identified. Even with proteins which have not been isolated but with amino acid sequences predicted from DNA studies, searching data banks allows comparisons with proteins whose structure and function are already known.

In the 1960s the dynamic aspects of biological macromolecules also became apparent from studies on cell membranes. The demonstrable movements of both lipid and protein constituents led to the Fluid Mosaic theory of membrane structure (see Singer, 1976) and radically changed our views on cell membranes. These were no longer seen as relatively passive barriers but contained constituents actively influencing intracellular events, whose participation leads both to major effects on cytoplasmic/nuclear interactions and also to alterations in the microstructure of the membrane and associated components such as actin.

Until the advent of isotopes the cell membrane was thought to be a relatively impermeable barrier, only permitting the entry of some low molecular weight solutes. Insulin was known to facilitate the uptake of glucose and thyroxine to affect tadpole metamorphosis; how these hormones worked was quite unknown. It is now accepted that receptor molecules on the outer surface of the cell bind hormones and growth factors (see Irvine, 1997) which in general produce their effects without entering the cell. In other cases ligand binding causes channels to open in the protein allowing the passage of the solute (Boyd, 1998, this volume). Details of the traverse are frequently still unclear. Proteins also interact with intracellular receptors, those destined for export for example, docking through their signal peptide sequence onto ribonucleoprotein particles which link ribosomes to the (rough) endoplasmic reticulum (see Siekovitz, 1996). In some instances proteins are modified post-translationally, for example by prenylation or with myristic acid, and the introduced group then docks onto a receptor surface. Some intracellular proteins destined for destruction by proteolysis are ubiquitinized, a reaction first described in the 1970s as a modification to some classes of histone. Proteins which function intranuclearly may have nuclear location sequences (NLS) which interact with components in nuclear pores, permitting the proteins to enter (Laskey et al., 1997).

Parallels between everyday mechanisms and cellular processes extend also to myosin head groups racheting to actin subunits to produce the sliding filaments of skeletal muscle (see Perry, 1997), rotary motors for bacterial flagella (see Armitage, 1997) or movements along microtubules, powered by the proton motive force or ATP (see also Block, 1997). The existence of a mitochondrial rotary motor has recently been established. Boyer's proposal that the γ subunit of the mitochondrial ATP synthase rotated within the other subunits has been vindicated by X-ray crystallography (Walker, 1997) and direct demonstration of the rotation.

For different functions to coexist within a cell, the reactions must be highly regulated. Probably the first protein activation processes to be recognized were those for the extracellular digestive enzymes—pepsinogen conversion to pepsin, trypsinogen to trypsin etc.—by cleavage of masking peptides. Improvements in the techniques for protein isolation and analysis allowed the blood clotting mechanism to be elucidated in the 1950s. This again involved proteolysis, activated at least in part by Ca^{2+}. The blood clotting cascade was probably the first to be recognized, each step causing the further activation of an enzyme, with consequent amplification (Macfarlane, 1964). Components in the complement cascade were identified through the 1960s and 1970s.

The 1960s saw the emergence of the secondary messenger hypothesis (Rall et al., 1957). Binding the messenger to a regulatory protein could cause its dissociation and thus the unmasking of the active kinase (cAMP, protein kinase A), or on the other hand, the binding was necessary for the activity of the holoenzyme (δ subunit phosphorylase kinase). More and more protein kinases were discovered, some activated by cAMP, cGMP, Ca^{2+}/calmodulin, etc. (Irvine, 1997; Randle, 1997). Tyrosine phosphorylation is of central importance in cell signaling but was more difficult to demonstrate than that on serine or threonine, in part because of the limited availability of tyrosine residues in proteins. Histidine N-phosphorylation proved to be of particular significance in some bacterial systems (see Armitage, 1997). Recent studies on the cell cycle have led to the identification of a large number of cell cycle dependent kinases (cdc kinases) activated *inter alia* by specific cyclins—proteins which often have short half-lives and which are destroyed by intracellular proteases (Nurse, 1990; Murray, 1995; Wuarin and Nurse, 1996).

Thus, from the late 1950s it became evident that the activity of many intracellular proteins was regulated by phosphorylation (see Randle, 1997). There are other protein modifications like glycosylation (Dwek, 1996), and acetylation which, on histones, since the 1960s has been known to be associated with transcriptionally active regions of the genome. Control of this acetylation is still uncertain (see Roth and Allis, 1996; Pazin and Kadonaga, 1997) but some transcriptional activators appear to have histone acetylase activity and some co-repressors are histone deacetylases (see Wolffe, 1997). Histone phosphorylation has been described by Ord and Stocken (1968) but its role, if any, is unknown. Another protein modification was first described by Rapkine in fertilized sea urchin eggs in the 1930s and confirmed by Mazia[3] in the 1950s. These were cyclical changes in the ratio of protein thiol to disulfide as the cells went through mitosis. These effects and their possible relation to changes in NAD(P)H/NAD(P)$^+$ ratios in the cells, and the other, now better known, changes through the cycle, have not been re-examined recently.

SOCIOLOGY

Paralleling the changes in the fields in which biochemists are now working, are the changes in the manner and places in which research is carried out. Into the 1950s, numbers of biochemists worldwide were relatively small. Biochemistry was a major degree subject only in a few universities. Until the 1950s, physiological chemistry was usually a small, though obligatory, part of the syllabus for medical students. In some universities "biochemistry" might be available as an ancillary subject taken with chemistry. Research groups were also usually small—a senior person who often continued working at the bench, with one or two graduate students, and perhaps a visiting post-doc. Large teams like that working on stress with Hans Selye or that of Du Vigneaud in New York in the 1950s, were the exception.

Now, however, university "Biochemistry" departments have expanded; established biochemists having groups which may be upward of 20 people. Many of these will be post-docs, including some with first degrees in subjects such as computing or physics. The post-docs will often be funded by "soft" money associated with project or program grant grants. Commercial employment of graduate biochemists has increased dramatically, particularly in pharmaceutical firms, and in entrepreneurial specialist enterprises exploiting genetic and/or technical developments.

As now, communication between biochemists was good, but different. Because numbers were modest most biochemists in the United Kingdom knew, or had at least met, most other biochemists. (Membership of the British Biochemical Society in 1946 was 1017, in 1973, 5877, and in 1997, over 9000). Meetings were smaller with multiple sessions uncommon. It wasn't too difficult to be aware of developments outside one's own field. There were fewer journals, and many fewer specialized ones. The first International Congress of Biochemistry was in Cambridge, UK, in August, 1949. Most biochemists attended the 5th International in Moscow in 1961, and many the 7th in Tokyo in 1967. By then, there had to be parallel sessions.

Informal, limited circulation "papers" appeared in the 1960s, initially to ensure rapid dissemination of topics in molecular biology. These disappeared with the growth of specialized journals and general pressure for rapid publication. It remains to be seen for how long papers will continue to be printed, and what will be the fate of information stored in this way. Will people continue to consult anything before the 1980s data bases? And will the present "Web" be capable of accommodating within acceptable access times, even the data expected in the next ten years?

One of the most significant changes has been in equipment costs. Up to c. 1940, a well-founded laboratory was provided with a balance ("method of swings"), glass-blowing facilities (glassware was commonly made by the researcher), a centrifuge, (though hand-centrifuges were still in use), a "comparator kit" with standard indicators to estimate pH, perhaps a Dubosq spectrophotometer or an early "Hilger" and a gas supply for the Bunsen burner(s). The expansion of biochemistry in the 1950s was accompanied by an increase in commercially available apparatus. In 1963, our new laboratory was equipped with an automatic balance, a pH meter, a recording UV spectrophotometer, a fraction-collector, a large-scale centrifuge (1.5 L capacity), a preparative ultracentrifuge capable of spinning c. 0.25 L at 10^5 g and an automatic scintillation counter, for £10–15,000. Now, basic equipment may require c. £50,000. Highly specialized equipment like a state-of-the art NMR (750 megaherz) or DNA sequencing facilities may cost the best part of £1m ($1.6 m).

Much of the increased cost is associated with the need for greatly increased sensitivity (that for protein sequencing has risen c. 1000 fold in 30 years), automation, speed and convenience of use, and facilities for digitization of the results and their computer analysis. For centrifuges, electrophoretic apparatus, radioisotopes and cell culturing much more stringent safety regulations are a further factor. Even 45 years ago enzymes and reagents such as ATP had to be made by the researcher if they were needed for an assay. Reagents, enzymes and finally "kits" became commercially available from the 1960s.

Increased costs, frequently met in part by support from commercial firms for special projects (or other forms of collaboration) increased pressure for quick returns, and thus, rapid publication. Conflicts may potentially arise, however, over publication, with delays imposed to protect commercial rewards and patenting (see for example Marshall, 1997; Vogel, 1997). It is difficult to judge if competition has really increased. The "Double Helix" (Watson,[2] 1968) revealed to the general reader the intensity of competition in science, though one has only to read the concerns of Darwin and his friends over Wallace's work to realize fears of this sort did not start in the twentieth century.

A further consequence of increased costs is evident, at least in Europe. It has become increasingly difficult for universities to fund research. The "well-founded base" evident 20 years ago may no longer hold. Facilities and staff available for training future biochemists may also be diminished and the institution already has difficulty in attracting finance for exciting and potentially important projects. Exploration of new ideas and support for relatively unknown, young, post-docs are jeopardized. On the other hand the enormous costs of, for example, the Human Genome Project, has strengthened international collaboration and planning.

APPLIED BIOCHEMISTRY

As has already been mentioned, as much, or perhaps more, biochemical research is now performed outside rather than within universities or research institutes. New discoveries are rapidly assessed for their commercial application, particularly to medicine or agriculture. Following the analysis by Fildes and Woods in the 1940s of the basis of action of sulfonamides, and by Park and Strominger, (1965; see Waxman and Strominger, 1983) of the effects of penicillin on cell-wall synthesis in gram-positive bacteria, it became apparent that many "drugs" affected cells by competing effectively with normal metabolic reactions (see Work & Work, 1948; Ord & Stocken, 1995). As our understanding of ligand/protein interactions has grown, so have the possibilities of utilizing computer-aided design and combinatorial syntheses to create inhibitors active even at specific isoform levels. Solutions to the problem of delivering compounds to selected cells have been proposed, including localized administration or the use of specifically targeted liposomes. The efficiency of the latter still has to be improved.

Of even greater impact may be attempts to correct diseases of genetic origin, most particularly those showing simple Mendelian inheritance. Even though various techniques for specific gene inactivation, replacement or enhancement are now available (see Witkowski, 1996; Verma and Somia, 1997) and constantly being improved and simplified, problems of delivery are still formidable. Currently, cell removal from the host, correction of the genetic defect *in vitro*, selection of repaired cells, their propagation and reintroduction, possibly after depletion of the original host population, seems the favored approach. Presently only certain populations can be approached in this way, with results which are promising rather than established.

A further consequence of our increased understanding of the genome and its regulation is our greater knowledge of the cell cycle and of apoptosis (programmed cell death) and potentially of the switch from cell proliferation to differentiation. It may even become possible to promote the capacity to repair or replace damaged tissue. Conditions favoring correct axonal restoration have been reported as has the ability of fetal cells to recolonize locally damaged areas in the brain.

As our knowledge of the genome increases, so has it become clear that many conditions involve both multiple genes and environmental factors. Identification of individuals whose lifestyle or genetic makeup puts them at risk for particular diseases, raises ethical and social problems which are unresolved and have brought molecular and cell biology firmly into the social arena (see Epstein, 1997).

As compared with the early part of this century, "biochemists" pay less attention to the study of nutrition. This is now a discipline in its own right and is extended to include not only humans but also all animals and plants in commercial production. Some topics are still under investigation like the roles of trace elements such as zinc or selenium, and the importance in the diet of antioxidants like vitamin E to combat deleterious, possibly carcinogenic, effects of oxidizing free radicals. Noninvasive examination of living material by NMR and the insights which can be thereby obtained into the metabolic status of tissues, has led to the presence in hospitals of magnetic resonance imaging (MRI) facilities and their expanding importance in diagnosis.

We hope this review and the series in general have drawn attention to the almost incredible advances in biochemical knowledge during the past fifty years. But it is now goodbye to Bio-chemistry and welcome to Molecular and Cellular Biochemistry.

ACKNOWLEDGMENTS

We are grateful to Professors Newell and Radda for continuing to extend to us the hospitality of the department, and to Professors Armitage and Whatley and doctors Gary Brown and Bruce Henning for helpful suggestions.

REFERENCES

Armitage, J. (1997). Three hundred years of bacterial motility. Foundations of Modern Biochemistry, Vol. 3, pp. 107–164. JAI Press, Greenwich, CT.

Bernal, J.D. (1967). The Origin of Life. Weidenfeld & Nicholson, U.K.

Block, S.M. (1997). Real engines of creation. Nature (London) 386, 217–219.

Calvin, M. (1955). The photosynthetic carbon cycle. In: Proceedings of 3rd International Congress (1956) (Liebecq, C. Ed.) pp. 211–227. Vaillant-Carmanne, Liege.

Campbell, I.D. (1996). Structural Biology. Foundations of Modern Biochemistry, pp. 133–151. JAI Press, Greenwich, CT.

Cech, T. (1986). RNA as an enzyme. Sci. Am. 255, 64–75.
Clayton, R.A., White, O., Ketchum, K.A., & Venter, J.C. (1997). The first genome for the third domain of life. Nature (London) 387, 459–462.
Creighton, T.E. (1990). Protein folding. Biochem. J. 270, 1–16.
Csermely, P. (1997). Proteins, RNAs and chaperones in enzyme evolution: a folding perspective. Trends in Biochem. Sci. 22, 147–149.
Day, A. & Poulton, J. (1996). Extranuclear DNA. Foundations of Modern Biochemistry, Vol. 2, pp. 59–97. JAI Press, Greenwich, CT.
Dwek, R.A. (1996). Glycobiology. Foundations of Modern Biochemistry, Vol. 2, pp. 153–202. JAI Press, Greenwich, CT.
Edgell, D.R. & Doolittle, W.F. (1997). Archaea and the origin(s) of DNA replication proteins. Cell 89, 995–998.
Ellis, R.J. (1987). Proteins as molecular chaperones. Nature (London) 328, 378–379.
Epstein, C.J. (1997). ASHG presidential address: towards the 21st century. Am. J. Hum. Genet. 60, 1–9.
Ferguson, S. (1997). Bioenergetics from 1960. Foundations of Modern Biochemistry, Vol. 3, pp. 3–22. JAI Press, Greenwich, CT.
Fischer, E. (1958). In: Outlines of Enzyme Chemistry, 2nd ed. (Neilands, J.B. & Stumpf, P.K., Eds.), p. 194. Wiley, New York.
Gurdon, J. (1991). Nuclear transplantation in *Xenopus*. Methods in Cell Biol. 36, 299–309.
Horwich, A.L. & Weissman, J.S. (1997). Deadly conformations—Protein misfolding in Prion Disease. Cell 89, 499–510.
Irvine, R.F. (1997). Talking to cells—cell membrane receptors and their modes of action. Foundations of Modern Biochemistry, Vol. 3, pp. 173–201. JAI Press, Greenwich, CT.
Jaenicke, R. (1987). Folding and association of proteins. Prog. Biophys. Mol. Biol. 49, 117–237.
Koshland, D.E. (1958). Application of a theory of enzyme specificity to protein synthesis. Proc. Natl. Acad. Sci. USA 44, 98–104.
Laskey, R.A., Gorlich, D., Madina, M.A., Makkerh, J.P.S., & Romanowski, P. (1997). Regulatory roles of the nuclear membrane. Exptl. Cell Res. 229, 204–211.
Laskey, R.A., Honda, B.M., Mills, A.D., & Finch, J.T. (1978). Nucleosomes are assembled by an acidic protein which binds histones and transfers them to DNA. Nature (London) 275, 416–420.
Lederberg, J. & Tatum, E.L. (1946). Gene recombination in *E. coli*. Nature (London) 158, 558.
Macfarlane, R.G. (1964). An enzyme cascade in the blood clotting mechanism and its function as a biochemical amplifier. Nature (London) 202, 498–499.
Marshall, E. (1997). Publishing sensitive data. Science 276, 523–526.
Mazia, D. (1961). Mitosis and the physiology of cell division. In: The Cell (Brachet, J. & Mirsky, A.E., Eds.), Vol. 3, pp. 77–412. Academic Press, New York.
Maynard Smith, J. (1993). The Theory of Evolution (Canto, Ed.). Cambridge University Press, Cambridge.
Monod, J., Changeux, J.P., & Jacob, F. (1963). Allosteric proteins and cellular control processes. J. Mol. Biol. 6, 306–329.
Morgan, T.H. (1911). Random segregation versus coupling in Mendelian inheritance. Science 34, 384.
Murray, A. (1995). Cyclin ubiquitinization: the destructive end of mitosis. Cell 81, 149–152.
Nurse, P. (1990). A universal control mechanism in regulation: the onset of M-phase. Nature (London) 344, 503–508.
Olsen, G.J. & Woese, C.R. (1997). Archaeal genomics: an overview. Cell 89, 991–994.
Oparin, A.I. (1957). The origin of life on earth, 3rd ed. Oliver & Boyd, Edinburgh & London.
Ord, M.G. & Stocken, L.A. (1968). Changes in the phosphorylation of histones during liver regeneration. Biochem. J. 103, 5–6 p.
Ord & Stocken, L.A. (1995). Early adventures in biochemistry. Foundations of Modern Biochemistry, Vol. 1. JAI Press, Greenwich, CT.

Ord, M.G. & Stocken, L.A. (1996). The coding properties of DNA and the central dogma. Foundations of Modern Biochemistry, Vol. 2, pp. 3–26. JAI Press, Greenwich, CT.
Palmer, J.D. (1997). The mitochondria that time forgot. Nature (London) 387, 454–455.
Pazin, M.J. & Kadonaga, J.T. (1997). What's up and down with histone deactylation and transcription? Cell 89, 325–328.
Perry, S.V. (1997). Muscle contraction and relaxation. Foundations of Modern Biochemistry, Vol. 3, pp. 67–105. JAI Press, Greenwich, CT.
Peters, R.A. (1963). Biochemical Lesions and Lethal Synthesis. Pergamon Press, Oxford, London, New York, Paris.
Rall, T.W., Sutherland, E.W., & Berthet, J. (1957). The relationship of glucagon and epinephrine to liver phosphorylase. J. Biol. Chem. 224, 463–475.
Randle, P. (1997). Mechanisms in regulation: protein phosphorylation. Foundations of Modern Biochemistry, Vol. 3, pp. 203–237. JAI Press, Greenwich, CT.
Rapkine, L. (1931). Sur les processus chimique au cours de la division cellulaire. Ann. Physiol. Physico-Chim. Biol. 7, 382.
Roth, S.Y. & Allis, C.D. (1996). Histone acetylation. Trends in Cell Biol. 6, 371.
Schoenheimer, R. (1942/1964). The dynamic state of body constituents. Hafner Publishing, New York.
Schuster, P. (1998). In Foundations of Modern Biochemistry, Vol. 4, pp. 159–198. JAI Press, Greenwich, CT.
Siekevitz, P. (1996). Protein synthesis and the ribosome. Foundations of Modern Biochemistry, Vol. 2, pp. 109–131. JAI Press, Greenwich, CT.
Singer, S.J. (1976). Molecular organization of membranes. Annu. Rev. Biochem. 43, 805–833.
Smith, M. et al., (1977). DNA sequence at the C termini of the overlapping genes a and B in bacteriophage ΦX 174. Nature (London) 265, 702–705.
Vartdal, V. et al. (1986). Using cells positively selected from blood by immunomagnetic isolation—a fast and reliable technique. Tissue Antigens 28, 301–312.
Verma, I.M. & Somia, N. (1997). Gene therapy-promises, problems and prospects. Nature (London) 389, 239–242.
Vogel, G. (1997). Publishing sensitive data. Science 276, 523–526.
Walker, J. (1997). See Surridge, C. Biologists' work on protein energy converters. Nature (London) 389, 771.
Watson, J.D. (1968). The double helix. Weidenfeld & Nicolson, London.
Whatley, F.R. (1997). Changing views of photosynthesis. Foundations of Modern Biochemistry, Vol. 3, pp. 23–65. JAI Press, Greenwich, CT.
White, J.G., Amos, W.B., & Fordham, M. (1987). An evaluation of confocal versus conventional imaging of biological structure by fluorescent light microscopy. J. Cell Biol. 105, 41–48.
Wilmut, I., Schnieke, A.E., McWhir, J., Kind, A.J., & Campbell, K.H.S. (1997). Viable offspring derived from fetal and adult mammalian cells. Nature (London) 385, 810–813.
Witkowski, J.A. (1996). Manipulating DNA: from cloning to knockouts. Foundations of Modern Biochemistry, Vol. 2, pp. 27–57. JAI Press, Greenwich, CT.
Wolffe, A.P. (1997). Sinful repression. Nature (London) 387, 16–17.
Work, T.S. & Work, E. (1948). The Basis of Chemotherapy. Wiley, New York.
Wuarin, J. & Nurse, P. (1996). Regulating S phase: cdks, licensing and proteolysis. Cell 85, 785–787.
Zhang, B. & Cech, T. (1997). Peptide bond formation by selected ribozymes in vitro. Nature (London) 390, 96–100.

AUTHOR INDEX

SUBJECT INDEX

www.ingramcontent.com/pod-product-compliance
Ingram Content Group UK Ltd.
Pitfield, Milton Keynes, MK11 3LW, UK
UKHW020242290926
15470UKWH00002B/679

9 780762 303519